LONDON MATHEMATICAL SOCIETY LECTURE NOTE S

Managing Editor: Professor J.W.S. Cassels, Department of Pure Mathem
Statistics, University of Cambridge, 16 Mill Lane, Cambridge CB2 1SB,

The books in the series listed below are available from booksellers, or, in case of difficulty, from Cambridge University Press.

34 Representation theory of Lie groups, M.F. ATIYAH *et al*
36 Homological group theory, C.T.C. WALL (ed)
39 Affine sets and affine groups, D.G. NORTHCOTT
46 p-adic analysis: a short course on recent work, N. KOBLITZ
49 Finite geometries and designs, P. CAMERON, J.W.P. HIRSCHFELD & D.R. HUGHES (eds)
50 Commutator calculus and groups of homotopy classes, H.J. BAUES
57 Techniques of geometric topology, R.A. FENN
59 Applicable differential geometry, M. CRAMPIN & F.A.E. PIRANI
66 Several complex variables and complex manifolds II, M.J. FIELD
69 Representation theory, I.M. GELFAND *et al*
74 Symmetric designs: an algebraic approach, E.S. LANDER
76 Spectral theory of linear differential operators and comparison algebras, H.O. CORDES
77 Isolated singular points on complete intersections, E.J.N. LOOIJENGA
79 Probability, statistics and analysis, J.F.C. KINGMAN & G.E.H. REUTER (eds)
80 Introduction to the representation theory of compact and locally compact groups, A. ROBERT
81 Skew fields, P.K. DRAXL
82 Surveys in combinatorics, E.K. LLOYD (ed)
83 Homogeneous structures on Riemannian manifolds, F. TRICERRI & L. VANHECKE
86 Topological topics, I.M. JAMES (ed)
87 Surveys in set theory, A.R.D. MATHIAS (ed)
88 FPF ring theory, C. FAITH & S. PAGE
89 An F-space sampler, N.J. KALTON, N.T. PECK & J.W. ROBERTS
90 Polytopes and symmetry, S.A. ROBERTSON
91 Classgroups of group rings, M.J. TAYLOR
92 Representation of rings over skew fields, A.H. SCHOFIELD
93 Aspects of topology, I.M. JAMES & E.H. KRONHEIMER (eds)
94 Representations of general linear groups, G.D. JAMES
95 Low-dimensional topology 1982, R.A. FENN (ed)
96 Diophantine equations over function fields, R.C. MASON
97 Varieties of constructive mathematics, D.S. BRIDGES & F. RICHMAN
98 Localization in Noetherian rings, A.V. JATEGAONKAR
99 Methods of differential geometry in algebraic topology, M. KAROUBI & C. LERUSTE
100 Stopping time techniques for analysts and probabilists, L. EGGHE
101 Groups and geometry, ROGER C. LYNDON
103 Surveys in combinatorics 1985, I. ANDERSON (ed)
104 Elliptic structures on 3-manifolds, C.B. THOMAS
105 A local spectral theory for closed operators, I. ERDELYI & WANG SHENGWANG
106 Syzygies, E.G. EVANS & P. GRIFFITH
107 Compactification of Siegel moduli schemes, C-L. CHAI
108 Some topics in graph theory, H.P. YAP
109 Diophantine analysis, J. LOXTON & A. VAN DER POORTEN (eds)
110 An introduction to surreal numbers, H. GONSHOR
111 Analytical and geometric aspects of hyperbolic space, D.B.A. EPSTEIN (ed)
113 Lectures on the asymptotic theory of ideals, D. REES
114 Lectures on Bochner-Riesz means, K.M. DAVIS & Y-C. CHANG
115 An introduction to independence for analysts, H.G. DALES & W.H. WOODIN
116 Representations of algebras, P.J. WEBB (ed)
117 Homotopy theory, E. REES & J.D.S. JONES (eds)
118 Skew linear groups, M. SHIRVANI & B. WEHRFRITZ
119 Triangulated categories in the representation theory of finite-dimensional algebras, D. HAPPEL
121 Proceedings of *Groups - St Andrews 1985*, E. ROBERTSON & C. CAMPBELL (eds)
122 Non-classical continuum mechanics, R.J. KNOPS & A.A. LACEY (eds)
124 Lie groupoids and Lie algebroids in differential geometry, K. MACKENZIE
125 Commutator theory for congruence modular varieties, R. FREESE & R. MCKENZIE
126 Van der Corput's method of exponential sums, S.W. GRAHAM & G. KOLESNIK
127 New directions in dynamical systems, T.J. BEDFORD & J.W. SWIFT (eds)

London Mathematical Society Lecture Note Series. 171

Squares

A. R. Rajwade
Panjab University, Chandigarh

Published by the Press Syndicate of the University of Cambridge
The Pitt Building, Trumpington Street, Cambridge CB2 1RP
40 West 20th Street, New York, NY 10011-4211, USA
10 Stamford Road, Oakleigh, Melbourne 3166, Australia

First published 1993

Printed in Great Britain at the University Press, Cambridge

Library of Congress cataloguing in publication data available

British Library cataloguing in publication data available

ISBN 0 521 42668 5

Contents

Preface

The aim of this book is to introduce the reader to this fascinating topic of squares, which, to quote Olga Taussky [T1], "has one of the longest history and begins with Pythagoras' theorem and the Pythagorian triangles". Our text, in particular, includes the most elegant and astonishing recent results of Albrecht Pfister about the 2^n identities, the Stufe of fields (the *Stufe* $s(K)$ of a field K is the least positive integer s for which the equation $-1 = a_1^2 + \ldots + a_s^2$ $(a_j \in k)$ is solvable. In case of non-solvability, we put $s(K) = \infty$) and the sums of squares in function fields over the real numbers (or real closed fields).

Sums of squares can be regarded as a special case of the diagonal (quadratic) forms (in particular of the beautiful Pfister forms) which are themselves a special case of general quadratic forms. Some results, which look beautiful when viewed from the point of view of squares, appear quite artificial as results on quadratic forms; e.g. to say that in a certain field K,

$$-1 \text{ can be written as a sum of two squares} \tag{1}$$

is the same thing as saying that for given a, b, c, $d \in K$, with $ad - bc \neq 0$, -1 can be written as

$$-1 = (a^2 + c^2)x^2 + (b^2 + d^2)y^2 + 2(ab + cd)xy. \tag{2}$$

Whereas (2) is artificial, (1) is a striking result.

In the present exposition, we have made it a point to stick to squares as far as possible and to appeal to the theory of general quadratic forms only when necessary. And indeed it is impossible totally to dispense with the general theory; the use of its basic ideas, at least in parts, becomes absolutely essential.

Historically quadratic forms were regarded as a topic in number theory. However, E.Witt in his classical paper "Theorie der quadratischen Formen in beliebigen Körpern", which appeared in Crelle's Journal in 1937 [W2], opened up a new chapter in the theory of quadratic forms: that of combining the number theoretic aspect with the algebraic development, by the creation of the famous Witt ring.

Then, triggered off by Cassels' paper "On the representation of rational functions as sums of squares" which appeared in Acta Arithmetica in 1964 [C2], A.Pfister, about 1966, suddenly came up with his celebrated structure theorems, giving birth to a purely algebraic theory of quadratic forms. Special cases of the arithmetical aspect of Pfister's theory are his beautiful results about sums of squares and Pfister forms.

After Cassels' and Pfister's discoveries the subject picked up fast and excellent books and survey articles have been written on the subject by many great authors like T.Y.Lam [L2], W.Scharlau [S1], Olga Taussky [T1], Daniel Shapiro [S6], Emil Grosswald [G3] to name a few.

So, what exactly is new in the present treatment? Simply that most of the exposition is extremely elementary, only requiring familiarity with, fields, polynomials and matrices. What we find most surprising is that such outstanding questions as

(I) What numbers can (and do) occur as Stufe of fields?

(II) For what values of n can there be identities like

$$(x_1^2+\ldots+x_n^2)(y_1^2+\ldots+y_n^2)=z_1^2+\ldots+z_n^2$$

with $z_j \in K(x_1,\ldots,x_n,y_1,\ldots,y_n)$?

require practically no further prerequisites than those mentioned above and yet all these years these simple elegant results, had eluded great mathemticians.

Chapters 1 through 16 and the appendices bear out the above remarks; they are all very elementary and include much of the arithmetical aspects of Pfister's work and yet can be easily understood by undergraduate level students. Chapters 17 and 18 make essential use of the Hasse-Minkowski theorem, but otherwise keep to the spirit of the rest of the book.

Most of the treatment is for fields, except for Chapters 6, 7, 8 and 9 which look at positive semi-definite forms and sums of squares in the ring $\mathbb{R}[X_1,X_2,\ldots,X_n]$. This topic is of great historical importance as Hilbert's pioneering work began with it and already in 1888, he proved [H3] that the set $\mathcal{P}_{n,m}$ of all positive semi-definite forms of degree m (m even necessarily) in $\mathbb{R}[X_1,\ldots,X_n]$ is the same as the set $\Sigma_{n,m}$ of all sums of squares of forms of degree $m/2$ in $\mathbb{R}[X_1,\ldots,X_n]$ if and only if $n=2$, for all (even) $m\geq 2$; for all $n\geq 2$, $m=2$ and $n=3$, $m=4$.

Another reason for including this topic here is that it fits admirably into the spirit of the book and it is of current research interest, a large number of comparitively easy results being still to be discovered and some strikingly beautiful theorems have been proved by Choi, Lam, Reznick and Robinson, to name a few. We therefore give a large number of references concerned with this topic to bring the reader up to date with it.

A survey of elementary ideas about quadratic forms is almost indispensible in any work on squares and we have added one chapter (Chapter 11) to cover this. Before Chapter 11, we hardly use this theory.

The only place where squares feature systematically is the Artin-Schreier theory of formally real fields, which, in a way is a very helpful background to the subject matter of squares, and so to that of this book. So although it is not needed directly for the understanding of the rest of the book, except perhaps Chapter 16, we thought we really ought to include it in a treatment on squares. This is done in Chapter 15, also treated in an elementary way.

My sincere and grateful thanks are due to Professor J.W.S. Cassels for suggestions which have gone a long way in improving the text. I thank him especially for many of the proofs too, for example Hilbert's proof (1888) of Theorem 6.1, which I would never have been able to sort out on my own.

My grateful thanks are also due to Professor A.Pfister and Professor D.B.Shapiro for giving many helpful suggestions and proofs of, for example, Theorems 13.6 and 3.5, and to Professor T.Y.Lam and Professor K.Y.Lam, for sending me some reprints and other useful material.

Finally I thank my colleague Dr. J.C. Parnami who helped me extensively with the manuscript.

I wish you all a happy reading on squares, but I warn you: don't become one yourself.

A.R.Rajwade

Notation

1. The set of all

(i) natural numbers is $\mathbf{N}$,
(ii) integers is $\mathbf{Z}$,
(iii) rational numbers is $\mathbf{Q}$,
(iv) real numbers is $\mathbf{R}$,
(v) complex numbers is $\mathbf{C}$,
(vi) quaternions is $\mathbf{H}$,
(vii) octonions is Ω.

2. $\mathbf{F}_q$ is the finite field of $q = p^\alpha$ elements.

3. For a field K denote by
(i) $s(K)$, the Stufe of K (see Definition 2.1),
(ii) $P(K)$ the pythagoras number of K,
(iii) $K((t))$ the field of formal power series in t over K

$$= \{a_N t^N + a_{N+1} t^{N+1} + \ldots \mid N \in \mathbf{Z}, a_j \in K\}$$

4. For a quadratic form f defined over a field K, denote by
(i) $M_f(K)$ the set of all similarity factors of f over K

$$= \{c \in K^* \mid cf \sim f\}$$

(ii) $V_f(K)$ the set of all non-zero values taken by f over K

$$V_f(K) = \{c \in K^* \mid c = f(v), v \in K^n\}.$$

In the special case when $f = X_1^2 + \ldots + X_n^2$, we write $V_f(K)$ as $G_n(K)$.

5.
(i) $K[X_1,\ldots,X_n]$ is the ring of polynomials in $X_1,\ldots,X_n$ with coefficients in K.
(ii) $K(X_1,\ldots,X_n)$ is the field of rational functions in $X_1,\ldots,X_n$ with coefficients in K.

6. SOS stands for sum of squares. PSD stands for positive semi-definite.

7.
(i) $\Sigma_{n,m}$ = set of all forms of degree m in n variables which are SOS in $\mathbb{R}[X_1,\ldots,X_n]$.
(ii) $\mathcal{P}_{n,m}$ = set of all PSD forms in $\mathbb{R}[X_1,\ldots,X_n]$ of degree m.
(iii) $\mathcal{E}(S)$ = set of all extremal forms of any convex set S.

8. $\Phi_m(X_1,\ldots,X_{2^m})$ = m-fold Pfister form. The 2-fold Pfister form $\Phi_2 = X^2+aY^2+bZ^2+abW^2$ is denoted by $[a,b]$.

9. For $n \in \mathbb{N}$, if $n = 2^m u$ (u odd), the Radon function $\rho(n)$ is defined by

$$\rho(n) = \begin{cases} 2m+1 \\ 2m \\ 2m \\ 2m+2 \end{cases} \text{according as } m \equiv \begin{cases} 0 \\ 1 \\ 2 \\ 3 \end{cases} \pmod 4.$$

Equivalently if $m = 4a+b$ $(0 \le b \le 3)$, then $\rho(n) = 8a + 2^b$.

10. For a form f, we denote by $\mathfrak{S}(f)$ the zero set of f in the relevant projective space.

1

The theorem of Hurwitz (1898) on the 2,4,8-identities

The curious identity

$$(X_1^2 + X_2^2)(Y_1^2 + Y_2^2) = (X_1Y_1 - X_2Y_2)^2 + (X_1Y_2 + X_2Y_1)^2 \qquad (1.1)$$

tells us that a product of two sums of two squares is itself a sum of two squares. Known to the Greeks, (1.1) appears at high school level as one learns about complex numbers and their norms and proves that the norm of the product of two complex numbers Z_1, Z_2 is the product of their norms:

$$|Z_1Z_2|^2 = |Z_1|^2|Z_2|^2 \qquad (1.1)'$$

Now writing $Z_1 = X_1 + iX_2, Z_2 = Y_1 + iY_2$, so that we have

$$Z_1Z_2 = (X_1Y_1 - X_2Y_2) + i(X_1Y_2 + X_2Y_1)$$

one sees that (1.1) and (1.1)′ are the same.

This identity enables one to prove the following curious result: Let K be a field and let

$$G_2(K) = \{a \in K^* | a = x^2 + y^2, x, y \in K\}.$$

Then $G_2(K)$ is a group under multiplication. Indeed, the closure property *is* the identity (1.1) while if $a = x^2 + y^2 \in G_2(K)$, then

$$1/a = a/a^2 = (x^2 + y^2)/a^2 = (x/a)^2 + (y/a)^2 \in G_2(K)$$

as required.

The following striking identity was already known to Euler in 1770 [E2] and he used it to prove Lagrange's theorem that every positive integer is a sum of four squares:

$$(X_1^2 + X_2^2 + X_3^2 + X_4^2)(Y_1^2 + Y_2^2 + Y_3^2 + Y_4^2) = Z_1^2 + Z_2^2 + Z_3^2 + Z_4^2 \quad (1.2)$$

where

$$\begin{aligned} Z_1 &= X_1Y_1 - X_2Y_2 - X_3Y_3 - X_4Y_4 \\ Z_2 &= X_1Y_2 + X_2Y_1 + X_3Y_4 - X_4Y_3 \\ Z_3 &= X_1Y_3 + X_3Y_1 - X_2Y_4 + X_4Y_2 \\ Z_4 &= X_1Y_4 + X_4Y_1 + X_2Y_3 - X_3Y_2 \end{aligned}$$

The discovery of quaternions by William Hamilton in 1843 [H2] brought out the real significance of the identity (1.2) in as much as (1.2) is simply the fact that the norm of a product of two quaternions is equal to the product of their norms.

Almost immediately after Hamilton's discovery of the quaternions, Arthur Cayley [C4] in 1845, discovered the octonions (the Cayley numbers) which give rise to the incredible looking identity

$$(X_1^2 + \ldots + X_8^2)(Y_1^2 + \ldots + Y_8^2) = Z_1^2 + \ldots + Z_8^2 \tag{1.3}$$

where

$$\begin{aligned} Z_1 &= X_1Y_1 - X_2Y_2 - X_3Y_3 - X_4Y_4 - X_5Y_5 - X_6Y_6 - X_7Y_7 - X_8Y_8, \\ Z_2 &= X_1Y_2 + X_2Y_1 + X_3Y_4 - X_4Y_3 + X_5Y_6 - X_6Y_5 - X_7Y_8 + X_8Y_7, \\ Z_3 &= X_1Y_3 + X_3Y_1 - X_2Y_4 + X_4Y_2 + X_5Y_7 - X_7Y_5 + X_6Y_8 - X_8Y_6, \\ Z_4 &= X_1Y_4 + X_4Y_1 + X_2Y_3 - X_3Y_2 + X_5Y_8 - X_8Y_5 - X_6Y_7 + X_7Y_6, \\ Z_5 &= X_1Y_5 + X_5Y_1 - X_2Y_6 + X_6Y_2 - X_3Y_7 + X_7Y_3 - X_4Y_8 + X_8Y_4, \\ Z_6 &= X_1Y_6 + X_6Y_1 + X_2Y_5 - X_5Y_2 - X_3Y_8 + X_8Y_3 + X_4Y_7 - X_7Y_4, \\ Z_7 &= X_1Y_7 + X_7Y_1 + X_2Y_8 - X_8Y_2 + X_3Y_5 - X_5Y_3 - X_4Y_6 + X_6Y_4, \\ Z_8 &= X_1Y_8 + X_8Y_1 - X_2Y_7 + X_7Y_2 + X_3Y_6 - X_6Y_3 + X_4Y_5 - X_5Y_4. \end{aligned}$$

Although the identity emerges most naturally from Cayley numbers, it was discovered nearly a quarter of a century earlier by C.F. Degan (1822) with minor sign differences (see [D2] p.164).

Degan stated (erroneously of course) that there is a like formula for 2^n squares. For the case of 16 squares, he gave the literal parts of the 16 bilinear functions $Z_1, Z_2, \ldots Z_{16}$ but left most of the signs undetermined, saying that the only difficulty is the prolixity of the ambiguities of signs.

Degan was also aware of the 2– and 4– variable Pfister forms whose detailed study we shall take up in Chapter 12.

As before, if we define

$$G_4 = \{a \in K^* | a = x_1^2 + \ldots + x_4^2, x_j \in K\}$$

and G_8 similarly, then it follows from (1.2) and (1.3) respectively that G_4 and G_8 are groups under multiplication, so that we have the chain of inclusions

$$K^{*^2} \subseteq G_2 \subseteq G_4 \subseteq G_8 \subseteq K^*.$$

A great many unsuccessful attempts followed Degan's discovery of (1.3), to extend formulae (1.1),(1.2) and (1.3) to a similar 16 term identity, and many workers, realizing the impossibility of such an extension, tried giving convincing arguments to prove the impossibility. Hamilton's and Cayley's discoveries had reduced the problem to the determination of the so-called normed algebras over the real numbers $\mathbf{R}$; the four known ones being $\mathbf{R}$ (of dimension 1), the complex numbers $\mathbf{C}$ (of dimension 2), the quaternions $\mathbf{H}$ (of dimension 4) and the octonions $\mathbf{O}$ (of dimension 8). It is an astonishing observation how the axioms of the ordered field $\mathbf{R}$ gradually drop off as we move up these higher dimensional hypercomplex systems: $\mathbf{C}$ is, no doubt a field, commutative and associative (under multiplication) and a division ring, but the order property is lost. $\mathbf{H}$ is only an associative division ring; thus commutativity, and order are both lost. Finally $\mathbf{O}$ is not even associative – it is merely a division ring; thus commutativity, associativity and order are all lost.

The half century following the discovery of these quaternions and octonions saw many attempts to find a 16-dimensional hypercomplex system over the reals and several erroneous affirmations were given. Finally in 1898, Hurwitz [H7] gave a decisive solution to the problem about the dimensionality of all possible normed algebras over $\mathbf{R}$ and so also about the possible values of n for which there is an identity of the type (1.3) with n terms. More precisely we have the following.

Theorem 1.1 (Hurwitz-1898). *Let K be a field with* $\mathrm{char}K \neq 2$. *The only values of n for which there is an identity of the type*

$$(X_1^2 + \ldots + X_n^2)(Y_1^2 + \ldots + Y_n^2) = Z_1^2 + \ldots + Z_n^2 \tag{1.4}$$

where the Z_k are bilinear functions of the X_i and the Y_j, coefficients in K are $n = 1, 2, 4, 8$.

Actually Hurwitz proved this only over $\mathbf{C}$, but his proof generalizes to any field K with $\mathrm{char}K \neq 2$. We give here a proof given by Dickson in his beautiful expository paper [D2] of 1919. A proof using normed algebras can be found in A.A. Albert's *Studies in Modern Algebra* [A5].

The idea is to convert (1.4) into a system of matrix equations. The bilinearity condition on the Z_k can be written as

$$\begin{pmatrix} Z_1 \\ Z_2 \\ \vdots \\ Z_n \end{pmatrix} = \begin{pmatrix} a_{11} & a_{12} & \ldots & a_{1n} \\ a_{21} & a_{22} & \ldots & a_{2n} \\ \ldots & \ldots & \ldots & \ldots \\ a_{n1} & a_{n2} & \ldots & a_{nn} \end{pmatrix} \begin{pmatrix} Y_1 \\ Y_2 \\ \vdots \\ Y_n \end{pmatrix} = A\mathbf{Y},$$

where the a_{ij} are linear functions of $X_1, X_2, \ldots, X_n$. Then (1.4) becomes

$$(X_1^2 + \ldots + X_n^2)(Y_1, \ldots, Y_n)\begin{pmatrix} Y_1 \\ Y_2 \\ \vdots \\ Y_n \end{pmatrix} I_n = (Z_1, \ldots, Z_n)\begin{pmatrix} Z_1 \\ \vdots \\ Z_n \end{pmatrix}$$
$$= \mathbf{Y}'A'A\mathbf{Y},$$

i.e.

$$(Y_1, Y_2, \ldots, Y_n)[(X_1^2 + X_2^2 + \ldots + X_n^2)I_n - A'A]\begin{pmatrix} Y_1 \\ Y_2 \\ \vdots \\ Y_n \end{pmatrix} = 0,$$

and since this is true for all $Y_1, Y_2, \ldots, Y_n$, it follows that

$$A'A = (X_1^2 + \ldots + X_n^2)I_n \tag{1.5}$$

Now

$$A = \begin{pmatrix} a_{11} & a_{12} & \cdots \\ a_{21} & a_{22} & \cdots \\ \multicolumn{3}{c}{\cdots\cdots\cdots\cdots} \end{pmatrix}$$
$$= \begin{pmatrix} b_{11}^{(11)}X_1 + b_{11}^{(12)}X_2 + \ldots, b_{12}^{(11)}X_1 + b_{12}^{(12)}X_2 + \ldots \\ b_{21}^{(11)}X_1 + b_{21}^{(12)}X_2 + \ldots, b_{22}^{(11)}X_1 + b_{22}^{(12)}X_2 + \ldots \\ \cdots\cdots\cdots\cdots\cdots\cdots\cdots\cdots\cdots\cdots\cdots\cdots \end{pmatrix}$$
$$= A_1X_1 + A_2X_2 + \ldots + A_nX_n \quad \text{say}$$

By (1.5), $(A_1'X_1 + A_1'X_2 + \ldots + A_n'X_n)(A_1X_1 + A_2X_2 + \ldots + A_nX_n)$
$$= (X_1^2 + X_2^2 + \ldots + X_n^2)I_n.$$

Since this is true for all X_j, we have

(1) $A_j'A_j = I_n (j = 1, 2 \ldots, n)$, hence also $A_jA_j' = I_n$.
(2) $A_j'A_k + A_k'A_j = 0, 1 \le j, k \le n, j \ne k$.

Conversely, the existence of such a system implies that (1.4) holds with Z_k bilinear in the X_i and the Y_j. Note also that if $n = 1$, (2) is vacuous, and (1) can be trivially satisfied so we may suppose $n > 1$.

Now let $B_i = A_n'A_i (i = 1, 2, \ldots, n-1)$. The B's are easily seen to satisfy

(1) $B_i'B_i = I_n$
(2) $B_i' + B_i = 0 \qquad (i, j = 1, 2, \ldots, n-1)$
(3) $B_i'B_j + B_j'B_i = 0 \quad (i \ne j)$

Hence we have

$$\left.\begin{array}{lll}\text{(i)} & B_i' = -B_i \quad (i = 1, 2, \ldots, n-1) \text{ i.e. the } B_i \\ & \quad\text{are skew-symmetric matrices} \\ \text{(ii)} & B_i^2 = -I \quad (i = 1, 2, \ldots, n-1) \\ \text{(iii)} & B_i B_j = -B_j B_i \quad i, j = 1, 2, \ldots, n-1 \quad i \neq j. \end{array}\right\} \qquad (1.6)$$

It follows that $|B_i| = |B_i'| = |-B_i| = (-1)^n |B_i|$, and since $|B_i| \neq 0$ we must have n even. Hence

Proposition 1. *There is no identity of the type* (1.4) *if* $n(>1)$ *is odd.*

In future, therefore, we suppose n to be even. Now consider the following set $\mathcal{G}$ of $n \times n$ matrices:

$$\{I, B_{i_1}, B_{i_1}B_{i_2}, B_{i_1}B_{i_2}B_{i_3}, \ldots, B_{i_1}B_{i_2}\ldots B_{i_{n-2}}$$
$$\text{and } B_1 B_2 \ldots B_{n-1} \ (i_1 < n, i_1 < i_2 < n, \ldots)\}.$$

Here B_{i_1} takes $n-1$ values viz. $B_1, B_2 \ldots B_{n-1}$, while $B_{i_1}B_{i_2}$ takes $\binom{n-1}{2}$ values viz. $B_1B_2, B_1B_3, \ldots$ etc. So altogether there are $1 + \binom{n-1}{1} + \ldots + \binom{n-1}{n-1} = 2^{n-1}$ elements in the set $\mathcal{G}$. Let $G = B_{i_1}B_{i_2}\ldots B_{i_r} \in \mathcal{G}$. Then we have

Lemma 1. *G is symmetric if $r \equiv 0$ or 3 modulo 4, and skew-symmetric if $r \equiv 1$ or 2 modulo 4.*

Proof.

$$G' = B_{i_r}' \ldots B_{i_1}' = (-1)^r B_{i_r} \ldots B_{i_1}$$
$$= (-1)^r(-1)^{r-1} B_{i_1}(B_{i_r} \ldots B_{i_2})$$

by (iii) of (1.6) to commute B_{i_1} successively with $B_{i_2}, \ldots, B_{i_r}$,

$$= (-1)^r(-1)^{r-1}(-1)^{r-2} B_{i_1}B_{i_2}(B_{i_r} \ldots B_{i_3})$$

and so on

$$= (-1)^r(-1)^{r-1} \ldots (-1)^2(-1) B_{i_1}B_{i_2} \ldots B_{i_r}$$
$$= (-1)^{1+2+\ldots+r} G$$
$$= (-1)^{r(r+1)/2} G$$
$$= \begin{cases} G \text{ if } r \equiv 0, 3, (4) \\ -G \text{ if } r \equiv 1, 2, (4) \end{cases}$$

□

Lemma 2. *Let $M \in \mathcal{G}$. Then the set $M\mathcal{G} = \{MG | G \in \mathcal{G}\}$ is simply a permutation of $\mathcal{G}$ with each term prefixed with either* $+1$ *or* -1.

Proof. The result is clear if the multiplier M is B_1, since then the product will contain or lack B_1 according as the multiplicand of $\mathcal{G}$ lacks or contains B_1 (use again (1.6)).

If the multiplier is B_2, we first replace $B_1B_2\ldots$, wherever it appears, by $B_2B_1\ldots$ and see that the former argument applies.

After thus proving our statement when the multiplier is any B_i, we see that it holds when the multiplier is any product of the B's. □

An Example: $n = 4$.

$$\mathcal{G} = \{I, B_1, B_2, B_3, B_1B_2, B_2B_3, B_1B_3, B_1B_2B_3\}.$$

Then

$$\begin{aligned} B_3\mathcal{G} &= \{B_3, B_3B_1, B_3B_2, B_3^2, B_3B_1B_2, B_3B_2B_3, B_3B_1B_3, B_3B_1B_2B_3\} \\ &= \{B_3, -B_1B_3, -B_2B_3, -I, B_1B_2B_3, B_2, B_1, -B_1B_2\} \\ &= \{-I, B_1, B_2, B_3, -B_1B_2, -B_2B_3, -B_1B_3, B_1B_2B_3\}. \end{aligned}$$

Our aim is now the following.

Proposition 2. *At least half of the elements of $\mathcal{G}$ are linearly independent.*

With this in view, we look for any linear relations that can exist amongst the elements of $\mathcal{G}$.

Definition. A relation $\lambda_1G_1 + \lambda_2G_2 + \ldots + \lambda_sG_s = 0$, $G_j \in \mathcal{G}$, $\lambda_j \in \mathbf{R}$, or $R = 0$ for short, is called irreducible if it is not possible to express R as $R_1 + R_2$, where $R_1 = 0$, $R_2 = 0$ represent two linear relations that hold between the subsets R_1 and R_2 of R with $R_1 \cap R_2 = \emptyset$, i.e. there are no matrices common to R_1 and R_2.

We have the following.

Lemma 3. *An irreducible relation $R = 0$ cannot involve both symmetric and skew-symmetric matrices.*

Proof. Let M_1 be the subset of all symmetric matrices in R and M_2 the set of all skew-symmetric matrices in R. Then $M_1 + M_2 = 0$, i.e. $M_1 = -M_2$. Hence $M_1 = M_1' = -M_2' = M_2$, i.e. $M_1 = M_2$. It follows that $M_1 = 0$, $M_2 = 0$ which contradicts the irreducibility of $R = 0$. □

Now let $R = 0$ be any irreducible relation between the matrices of $\mathcal{G}$. By multiplying R by a suitable λG ($\lambda \in \mathbf{R}$, $G \in \mathcal{G}$) we get a new relation $T = 0$, one term of which is I and all the remaining terms are products of matrices of $\mathcal{G}$ by real constants. For suppose μG ($\mu \in \mathbf{R}$, $G \in \mathcal{G}$) is a term in R which we wish should become I in the relation $T = 0$. One just multiplies $R = 0$ by $\pm\mu^{-1}G^{-1}$ and notes that one of $\pm G^{-1} \in \mathcal{G}$.

For example if $4B_2B_3$ is one term of R, then on multiplying $R = 0$ by

$$-\frac{1}{4}(B_2B_3)^{-1} = -\frac{1}{4}B_3^{-1}B_2^{-1} = -\frac{1}{4}(-B_3)(-B_2)$$
$$= -\frac{1}{4}B_3B_2 = \frac{1}{4}B_2B_3,$$

we get what is required.

This new relation $T = 0$ is also irreducible, for if $T = 0$ were to split as $T_1 = 0$, $T_2 = 0$, then since $T = \lambda GR$ we have $\lambda^{-1}G^{-1}T = R$ and so $R = 0$ splits as $\lambda^{-1}G^{-1}T_1 = 0, \lambda^{-1}G^{-1}T_2 = 0$, which gives a contradiction.

Hence we may suppose that $T = 0$ looks like

$$I = \sum c_{i_1i_2i_3}B_{i_1}B_{i_2}B_{i_3} + \sum d_{i_1i_2i_3i_4}B_{i_1}B_{i_2}B_{i_3}B_{i_4} + \dots \qquad (*)$$

where by Lemma 3, each of the matrices $B_{i_1}B_{i_2}B_{i_3}, B_{i_1}B_{i_2}B_{i_3}B_{i_4}$, etc. is symmetric since I is symmetric. That is why no singleton B_i nor any of the products B_iB_j of two B's can be involved in $(*)$ since B_i and B_iB_j are skew-symmetric by Lemma 1.

Now multiply $(*)$ throughout on the right by B_i to obtain an irreducible relation which then involves only skew-symmetric matrices since one term (on the left side) is the skew-symmetric matrix B_i. But by Lemma 1, $B_{i_1}B_{i_2}B_{i_3}B_{i_4}$ is symmetric. So all the c_j are 0 if only i is distinct from $i_1i_2i_3$. Since i may have any value $\leq n-1$, we see that each c is 0 unless $n - 1 = 3$ for then i cannot be chosen different from i_1, i_2, i_3.

Next we show that all the d's are 0 too; for multiply $(*)$ by B_{i_4} and it becomes

$$B_{i_4} = \sum d_{i_1i_2i_3i_4}B_{i_4}B_{i_1}B_{i_2}B_{i_3}B_{i_4} + \dots.$$

But $B_{i_4}B_{i_1}B_{i_2}B_{i_3}B_{i_4} = (-1)^3B_{i_1}B_{i_2}B_{i_3}B_{i_4}^2 = B_{i_1}B_{i_2}B_{i_3}$. So $(*)$ becomes

$$B_{i_4} = \sum d_{i_1i_2i_3i_4}B_{i_1}B_{i_2}B_{i_3} + \dots.$$

Here $B_{i_1}B_{i_2}B_{i_3}$ is symmetric, while B_{i_4} is skew-symmetric (by Lemma 1). It follows that all the d's are 0 too.

The method used in proving $c = 0$ applies when the number r of factors in $B_{i_1}B_{i_2}\dots B_{i_r}$ is $\equiv 3(4)$ and $r < n-1$. Similarly the method used in proving $d = 0$ also applies when $r \equiv 0(4)$.

Hence if our relation exists, it has the form

$$I = kB_1B_2\dots B_{n-1}$$

the right hand term being the only survivor. Now I is symmetric so $B_1B_2\dots B_{n-1}$ is symmetric i.e. $n-1 \equiv 0$ or $3(4)$, but n is even so $n - 1 \equiv 3(4)$ i.e. $n \equiv 0(4)$.

We have thus proved the following.

$$\left.\begin{array}{l}\textit{If an irreducible relation between the elements}\\ \textit{of } \mathcal{G} \textit{ does exist, then } n \equiv 0(4).\end{array}\right\} \qquad (1.7)$$

Now square this relation to get

$$\begin{aligned} I &= k^2 B_1 B_2 \dots B_{n-1} B_1 B_2 \dots B_{n-1} \\ &= k^2(-1)^{n-2} B_2 \dots B_{n-1} B_2 \dots B_{n-1} \\ &= \dots\dots\dots\dots\dots \\ &= k^2(-1)^{\frac{1}{2}(n-1)n} I. \end{aligned}$$

Since $n \equiv 0(4)$, we see that $k^2 = 1$ i.e. $k = \pm 1$. Hence we have the following.

Lemma 4. *If $n \equiv 2(4)$ then the 2^{n-1} matrices of $\mathcal{G}$ are linearly independent, while for $n \equiv 0(4)$, they are either linearly independent or are connected by the relations which arise from the relation $I = \pm B_1 B_2 \dots B_{n-1}$ through multiplication by the various elements of $\mathcal{G}$, but are connected by no further irreducible linear relations.*

Example. Let $n = 4$. Then

$$\mathcal{G} = \{I, B_1, B_2, B_3, B_1B_2, B_2B_3, B_1B_3, B_1B_2B_3\}$$

and these eight matrices are either linearly independent or are connected by the following four irreducible linear relations and no others:

$$I = \pm B_1B_2B_3, B_1 = \mp B_2B_3, B_2 = \pm B_1B_3, B_3 = \mp B_1B_2.$$

These express $B_1B_2B_3$, B_2B_3, B_1B_3, B_1B_2 linearly in terms of I, B_1, B_2, B_3; so that these latter matrices are, in any case, linearly independent.

Now consider all the irreducible linear relations that exist between the element of $\mathcal{G}$. As we have seen, they are all of the type

$$G \cdot I = \pm G \cdot B_1 B_2 \dots B_{n-1} (G \in \mathcal{G})$$

and no others. Now reduce the right side of this using (1.6). Then one of G or the reduced right side obviously contains fewer than half of the B's while the other contains more than half of the B's.

Thus these irreducible linear relations merely serve to express the products containing more than half of the B's in terms of those with less than half of the B's.

So in every case (i.e. irrespective of whether $n \equiv 0$ or 2 (mod 4)) the 2^{n-2} matrices of $\mathcal{G}$, which are products of less than $\frac{n-1}{2}$ B's, are linearly independent. Hence for all values of n (necessarily even) if there is to be an identity of the type (1.4), the 2^{n-2} matrices of $\mathcal{G}$ consisting of the product of at most $\frac{n-2}{2}$ B's are linearly independent.

This completes the proof of Proposition 2. □

We can now give a proof of our main result.

The elements of $\mathcal{G}$ are all $n \times n$ matrices and the maximum number of linearly independent $n \times n$ matrices is n^2 since they form, over the reals, a vector space of dimension n^2. Hence by the proposition we get

$$2^{n-2} \leq n^2.$$

This is satisfied if $n \leq 8$ but fails if $n = 10$. Now if it fails for $n = m$, then it fails for $n = m + 1$ for we have

$$\begin{aligned} 2^{m+1-2} &= 2 \cdot 2^{m-2} > 2 \cdot m^2 \quad \text{(since the relation fails for } m) \\ &> (m+1)^2 \text{ if } m \geq 3. \end{aligned}$$

It follows that if an identity of the type (1.4) exists, then $n \leq 8$ (and n is even). For $n = 2$, 4, 8 we already have the required type of identities. It remains to dispose off the case $n = 6$.

Suppose an identity exists for $n = 6$. Then since $6 \equiv 2(4)$, we see that

the 2^5 *matrices of* $\mathcal{G}$ *are linearly independent.* (i)

Of these 32 matrices, 16 are skew-symmetric by Lemma 1, viz. the ones that are products of 1, 2 or 5 B's. But

between any 16 *skew-symmetric* 6×6 *matrices there exists a linear relation.* (ii)

This is because the 15 matrices

$$\begin{pmatrix} 0 & 1 & 0 & \dots \\ -1 & 0 & 0 & \dots \\ 0 & 0 & 0 & \dots \\ \dots & \dots & \dots & \dots \end{pmatrix}, \begin{pmatrix} 0 & 0 & 1 & \dots \\ 0 & 0 & 0 & \dots \\ -1 & 0 & 0 & \dots \\ \dots & \dots & \dots & \dots \end{pmatrix}, \dots$$

with a 1 in the one place above the main diagonal, -1 in the corresponding place below, and 0's elsewhere, form a basis for the subspace of all 6×6 skew-symmetric matrices and so this subspace has dimension 15. This proves (ii).

(i) and (ii) above are contradictory. Hence no identity of type (1.4) can exist for $n = 6$.

That at last completes the proof of Hurwitz's theorem. □

Remark 1. The proof works for any field K of characteristic $\neq 2$.

Remark 2. There are three obvious ways of generalizing the identity (1.4); they are

(a) Allow the Z_i to be rational functions of the X_j and the Y_k: $Z_i \in K(X_1, \dots, X_n, Y_1, \dots, Y_n)$, rather than be just bilinear functions and then find the possibilities for n.

(b) Consider the (r, s, n)-identity

$$(X_1^2 + \dots + X_r^2)(Y_1^2 + \dots + Y_s^2) = Z_1^2 + \dots + Z_n^2, \tag{1.8}$$

where the Z_i are bilinear in the X_j and the Y_k (or more generally rational functions) and determine, for given r, s the least value of n for which (1.8) holds; or alternatively determine the maximum value of r (given s and n) for which (1.8) holds.

(c) Instead of a 'product formula' (1.4) for the form $X_1^2 + \ldots + X_n^2$, look for such a formula for more general quadratic forms, i.e.

$$q(X_1, \ldots, X_n) \cdot q(Y_1, \ldots, Y_n) = q(Z_1, \ldots, Z_n)$$

where $Z_i \in K(X_1, \ldots, X_n, Y_1, \ldots, Y_n)$.

Pfister solved (a) and (c) completely, whereas for the special case $s = n$, Hurwitz and Radon solved (b). In the next chapter, we shall describe Pfister's solution of (a) and its singularly beautiful consequences. In later chapters, (b) and (c) will also be covered. Actually little is known about (b) when r, s, n are all different.

Exercises

1. Show from first principles that an identity of the type

$$(X_1^2 + X_2^2 + X_3^2)(Y_1^2 + Y_2^2 + Y_3^2) = Z_1^2 + Z_2^2 + Z_3^2$$

where Z_1, Z_2, Z_3 are bilinear functions of $X_1, X_2, X_3, Y_1, Y_2, Y_3$, with coefficients in the field $\mathbf{Q}$ of rational numbers, cannot hold. [Hint: $3 \cdot 5 = 15$.]

2. Determine the most general bilinear functions Z_1, Z_2 in X_1, X_2, Y_1, Y_2 for which (1.1) holds.

3. Show that for any field K with $\mathrm{char} K \neq 2$, the existence of the more general identity

$$(X_1^2 + \ldots + X_r^2)(Y_1^2 + \ldots + Y_s^2) = Z_1^2 + \ldots + Z_n^2$$

where the Z_k are bilinear functions of the X_i's and the Y_j's with coefficients in K is equivalent to the existance of $n \times s$ matrices $A_1, A_2, \ldots A_r$ satisfying

$$A_i' A_i = I_s \ (1 \leq i \leq r)$$

$$A_i' A_j + A_j' A_i = 0 \ (i \neq j, 1 \leq i, j \leq r).$$

Definition. We then say that the triple (r, s, n) is admissible over K.

4. Show that $(3, 5, 7)$ is admissible over any field K.

Hint:

$$(X_1^2 + X_2^2 + X_3^2)(Y_1^2 + \ldots + Y_5^2) =$$
$$(X_1^2 + \ldots + X_3^2 + 0^2)(Y_1^2 + \ldots + Y_4^2) + (X_1^2 + X_2^2 + X_3^2)Y_5^2.$$

Now use (1.2) to write this as a sum of seven squares.

5. Prove that if the triple (r, n, n) is admissible over K then so is $(r + 1, 2n, 2n)$.

6. For the real number field $\mathbf{R}$, show that in an identity of the form

$$(X_1^2 + \cdots + X_r^2)(Y_1^2 + \cdots + Y_s^2) = Z_1^2 + \cdots + Z_n^2$$

with $Z_j \in \mathbf{R}(X_1, \ldots, X_r, Y_1, \ldots, Y_s)$, the Z_j are necessarily bilinear in the X's and the Y's.

2

The 2^n-identities and the Stufe of fields: theorems of Pfister and Cassels

Although the impossibility of the identity (1.4) for $n \neq 1, 2, 4, 8$ has been proved, it was under the stringent restriction that the Z_k are bilinear polynomials in the X_i and the Y_j. One could look into the possibility of the existence of other values of n for which (1.4) holds, if we allow the Z_k to be more general polynomials in the X_i and the Y_j. However, in 1966, Frank Adams [A1] showed that when n is not 1, 2, 4, 8, there are no identities of the type (1.4) even if the Z_k are allowed to be any bi-skew, continuous functions of the X_i and Y_j (see the Exercises in Chapter 13, Definition (iv)p. 187). It was thus totally unexpected when in 1967, Albrecht Pfister [P5] proved the following remarkable

Theorem 2.1. *Let K be a field and let $n = 2^m$ be a power of 2. Then there are identities*

$$(X_1^2 + \ldots + X_n^2)(Y_1^2 + \ldots + Y_n^2) = Z_1^2 + \ldots + Z_n^2 \qquad (2.1)$$

where the Z_k are linear functions of the Y_j with coefficients in $K(X_1, \ldots, X_n)$:

$$Z_k = \sum_{j=1}^{n} T_{kj} Y_j \quad \textit{with} \quad T_{kj} \in K(X_1, \ldots, X_n).$$

Conversely suppose n is not a power of 2. Then there is a field K such that there is no identity (2.1) with the $Z_k \in K(X_1, \ldots, X_n, Y_1, \ldots, Y_n)$. Here the Z_k are not even demanded to be linear in the Y_j.

Intimately connected with this result is the notion of the Stufe $s = s(K)$ of a field K:

Definition 2.1. The smallest positive integer s for which the equation

$$-1 = a_1^2 + a_2^2 + \ldots + a_s^2 \ (a_j \in K)$$

is solvable is called* the *Stufe* $s(K)$ of K. If the equation has no solution, we put $s = \infty$ and call K formally real.

The following beautiful result is again due to Pfister [P1]

Theorem 2.2. *For any field K, $s(K)$, if finite, is always a power of 2. Conversely every power of 2 is the Stufe of some field K.*

In this chapter, we shall give a proof of both these theorems. In the process we shall get other results which are interesting in their own right.

The proof of the first part of Theorem 2.1 requires no elaborate algebraic machinery and is indeed remarkably simple. We dispose of it first.

Proof of the first part of Theorem 2.1. We use induction on m. We know that (2.1) holds for $m = 1, 2, 3$ (see (1.1), (1.2), (1.3) respectively). Suppose that it holds for m. Write $T = (T_{ij})$ so that

$$\begin{pmatrix} Z_1 \\ Z_2 \\ \vdots \\ Z_n \end{pmatrix} = T \begin{pmatrix} Y_1 \\ Y_2 \\ \vdots \\ Y_n \end{pmatrix} \tag{2.2}$$

Then (2.1) can be written as

$$(X_1^2 + \ldots + X_n^2)(Y_1^2 + \ldots + Y_n^2) = (X_1^2 + \ldots + X_n^2)(Y_1, \ldots, Y_n) \begin{pmatrix} Y_1 \\ Y_2 \\ \vdots \\ Y_n \end{pmatrix}$$

$$= (Z_1^n + \ldots + Z_n^2) = (Z_1, Z_2, \ldots, Z_n) \begin{pmatrix} Z_1 \\ Z_2 \\ \vdots \\ Z_n \end{pmatrix}$$

$$= (Y_1, \ldots, Y_n) T'T \begin{pmatrix} Y_1 \\ Y_2 \\ \vdots \\ Y_n \end{pmatrix}$$

by (2.2), i.e.

$$(X_1^2 + \ldots + X_n^2)(Y_1, \ldots, Y_n) I_n \begin{pmatrix} Y_1 \\ \vdots \\ Y_n \end{pmatrix} - (Y_1, \ldots, Y_n) T'T \begin{pmatrix} Y_1 \\ \vdots \\ Y_n \end{pmatrix} = 0$$

---

* Often referred to as the *level.*

or

$$(Y_1, \ldots, Y_n)\{(X_1^2 + \ldots + X_n^2)I_n - T'T\}\begin{pmatrix} Y_1 \\ \vdots \\ Y_n \end{pmatrix} = 0.$$

Since this is true for all $Y_1, \ldots, Y_n$, we must have

$$TT' = (X_1^2 + \ldots + X_n^2)I_n$$

and so also $T'T = (X_1^2 + \ldots + X_n^2)I_n$ T being orthogonal.

We now prove (2.1) for $2^{m+1} = 2n$. Write

$$(X_1, \ldots, X_{2n}) = (\mathbf{X}^{(1)}, \mathbf{X}^{(2)})$$

where $\mathbf{X}^{(1)} = (X_1, \ldots, X_n)$ and $\mathbf{X}^{(2)} = (X_{n+1}, \ldots, X_{2n})$. By the induction hypothesis there exist two matrices $T^{(1)}$, $T^{(2)}$ say, corresponding to $\mathbf{X}^{(1)}$, $\mathbf{X}^{(2)}$ respectively such that

$$\left.\begin{array}{l} (X_1^2 + \ldots + X_n^2)I_n = \mathbf{X}^{(1)}\mathbf{X}^{(1)'}I_n = T^{(1)}T^{(1)'} = T^{(1)'}T^{(1)} \\ \text{and} \\ (X_{n+1}^2 + \ldots + X_{2n}^2)I_n = \mathbf{X}^{(2)}\mathbf{X}^{(2)'}I_n = T^{(2)}T^{(2)'} = T^{(2)'}T^{(2)} \end{array}\right\} \quad (2.3)$$

and we wish to show that there exists a matrix T say, such that

$$T'T = (X_1^2 + \ldots + X_n^2 + X_{n+1}^2 + \ldots + X_{2n}^2)I_{2n}. \quad (2.4)$$

Try $T = \begin{pmatrix} T^{(1)} & T^{(2)} \\ T^{(2)} & X \end{pmatrix}$ - a partitional matrix, where X will be determined by (2.4). We have

$$T'T = \begin{pmatrix} T^{(1)'} & T^{(2)'} \\ T^{(2)'} & X' \end{pmatrix}\begin{pmatrix} T^{(1)} & T^{(2)} \\ T^{(2)} & X \end{pmatrix},$$

and using block multiplication of matrices this equals

$$\begin{pmatrix} T^{(1)'}T^{(1)} + T^{(2)'}T^{(2)} & T^{(1)'}T^{(2)} + T^{(2)'}X \\ T^{(2)'}T^{(1)} + X'T^{(2)} & T^{(2)'}T^{(2)} + X'X \end{pmatrix}$$
$$= \begin{pmatrix} (X_1^2 + \ldots + X_n^2 + X_{n+1}^2 + \ldots + X_{2n}^2)I_n & A \\ B & C \end{pmatrix}$$

say; we want to choose X so that $A = B = 0$ and

$$C = (X_1^2 + \ldots + X_n^2 + X_{n+1}^2 + \ldots + X_{2n}^2)I_n.$$

To make $A = 0$ we have to have $X = -T^{(2)'^{-1}}T^{(1)'}T^{(2)}$. This automatically makes $B = 0$ (just check). Now it seems too much to expect C to be what we want. But we have

$$\begin{aligned} C &= T^{(2)'}T^{(2)} + T^{(2)'}T^{(1)}T^{(2)'^{-1'}}T^{(2)'^{-1}}T^{(1)'}T^{(2)} \\ &= (X_{n+1}^2 + \ldots + X_{2n}^2)I_n + (X_{n+1}^2 + \ldots + X_{2n}^2)^{-1}T^{(2)'}T^{(1)}T^{(1)'}T^{(2)} \\ &= (X_{n+1}^2 + \ldots + X_{2n}^2)I_n + \\ &\qquad (X_{n+1}^2 + \ldots + X_{2n}^2)^{-1}(X_1^2 + \ldots + X_n^2)T^{(2)'}T^{(2)} \end{aligned}$$

$$= (X_{n+1}^2 + \ldots + X_{2n}^2)I_n + (X_1^2 + \ldots + X_n^2)I_n$$
$$= (X_1^2 + \ldots + X_n^2 + X_{n+1}^2 + \ldots + X_{2n}^2)I_n.$$

This completes the proof of the first part of Theorem 2.1.

We now come to a result of Cassels [C2], which, in a way, was the starting point of this whole business and which is an indispensible tool in our further developments.

Cassels' Lemma (1964). *Let $f(X) \in K(X)$ be a polynomial with coefficients in K. If $f(X)$ is a sum of n squares of elements of the field $K(X)$, then it is a sum of n squares of elements of the ring $K[X]$.*

Note: What is new in this enunciation is that the same number n of squares suffice in $K[X]$; without this condition, the result had been proved by Artin [A6].

Proof. There are three trivial cases of the lemma which we dispose of first.
(i) $n = 1$. Then $f(X) = (p(X)/q(X))^2$, so $q(X)|p(X)$.
(ii) char $K = 2$. Then $a^2 + b^2 = (a+b)^2$ and so if

$$f(X) = \gamma_1^2(X) + \ldots + \gamma_n^2(X)$$

then combining two squares at a time into one, $f(X)$ reduces to a single square, i.e. we land up in case (i).
(iii) -1 is a sum of $n-1$ squares of elements in K.

Say $-1 = b_1^2 + \ldots + b_{n-1}^2$. Then for any $f(X)$, we have

$$f(X) = \left(\frac{f+1}{2}\right) - \left(\frac{f-1}{2}\right)^2$$
$$= \left(\frac{f+1}{2}\right)^2 + \left(b_1\frac{(f-1)}{2}\right)^2 + \ldots + \left(b_{n-1}\frac{(f-1)}{2}\right)^2$$

a sum of n squares of elements of $K[X]$.

So now let us suppose none of these three cases holds and let

$$f(X) = (p_1(X)/q_1(X))^2 + \ldots + (p_n(X)/q_n(X))^2.$$

Dropping the X from now on and clearing the denominators, this gives

$$fZ^2 = Y_1^2 + \ldots + Y_n^2, \quad Z, Y_1, \ldots, Y_n \in K[X], Z \neq 0.$$

Thus the equation

$$fZ^2 = Y_1^2 + \ldots + \ldots + Y_n^2 \tag{1}$$

has a solution $(Z, Y_1, \ldots, Y_n)$ with $Z \neq 0$ and we have to show that there exists a solution of (1) with $Z \in K$ ($Z \neq 0$), i.e. with degree of Z (in X) $= 0$. Now since (1) has a solution with $Z \neq 0$, so there is a solution, call it $(\zeta, \eta_1, \ldots, \eta_n)$, with $\zeta \neq 0$ for which $\deg \zeta$ is as small as possible:

$$f\zeta^2 = \eta_1^2 + \ldots + \eta_n^2. \tag{2}$$

We shall show that this degree is 0 i.e. that $\zeta \in K$, by showing that if not, then there exists a solution, say $(\zeta^*, \eta_1^*, \ldots, \eta_n^*)$ with $\zeta^* \neq 0$ and $\deg \zeta^* < \deg \zeta$.

So suppose $\deg \zeta > 0$. By the division algorithm in $K[X]$, we can write, for $j = 1, 2, \ldots, n$,

$$\eta_j = \lambda_j \zeta + \gamma_j$$

where either $\gamma_j = 0$ or $\deg \gamma_j < \deg \zeta$. i.e.

$$\eta_j / \zeta = \lambda_j + \gamma_j / \zeta = \lambda_j + \Lambda_j, \tag{3}$$

say. Note that not all the γ_j can be zero, otherwise ζ divides all the η_j so that (2) becomes $f = \lambda_1^2 + \ldots + \lambda_n^2$ - a contradiction, since the degree of ζ was least possible.

Now let

$$\zeta^* = \zeta \left\{ \sum_i \lambda_i^2 - f \right\} - 2 \left\{ \sum_i \lambda_i \eta_i - f\zeta \right\}$$

and

$$\eta_j^* = \eta_j \left\{ \sum_i \lambda_i^2 - f \right\} - 2\lambda_j \left\{ \sum \lambda_i \eta_i - f\zeta \right\}.$$

Then visibly, all of $\zeta^*, \eta^*, \ldots, \eta_n^* \in K[X]$. We now claim

(a) that $(\zeta^*, \eta_1^*, \ldots, \eta_n^*)$ is a solution of (1)
(b) $\zeta^* \neq 0$ and
(c) $\deg \zeta^* < \deg \zeta$

This would then contradict the definition of ζ and so would prove the lemma.

We prove (a) by brute force: we must show that $\sum_j \eta_j^{*^2} - f\zeta^{*^2} = 0$ i.e. that

$$\sum_j \left[\eta_j^2 \left\{ \sum_i \lambda_i^2 - f \right\}^2 + 4\lambda_j^2 \left\{ \sum_i \lambda_i \eta_i - f\zeta \right\}^2 \right.$$
$$\left. - 4\lambda_j \eta_j \left\{ \sum_i \lambda_i^2 - f \right\} \left\{ \sum_i \lambda_i \eta_i - f\zeta \right\} \right]$$
$$= f \left[\zeta^2 \left\{ \sum_i \lambda_i^2 - f \right\}^2 + 4 \left\{ \sum_i \lambda_i \eta_i - f\zeta \right\}^2 \right.$$
$$\left. - 4\zeta \left\{ \sum_i \lambda_i^2 - f \right\} \left\{ \sum_i \lambda_i \eta_i - f\zeta \right\} \right]$$

Here the first terms from both sides cancel since $\sum \eta_j^2 = f\zeta^2$ and it remains to prove that

$$4\left\{\sum_i \lambda_i\eta_i - f\zeta\right\}\left[\left(\sum \lambda_i\eta_i - f\zeta\right)\sum \lambda_j^2 - \left(\sum_i \lambda_i^2 - f\right)\sum_j \lambda_j\eta_j\right.$$
$$\left. - \left(\sum_i \lambda_i\eta_i - f\zeta\right)f + \left(\sum \lambda_i^2 - f\right) + f\zeta\right] = 0$$

Here the expression in square brackets just cancels out.

To prove (b) and (c) we substitute for λ_j from (3) in ζ^*. Then

$$\zeta^* = \zeta\left(\sum_i\left(\frac{\eta_i^2}{\zeta^2} + \Lambda_i^2 - \frac{2\eta_i\Lambda_i}{\zeta}\right) - f\right) - 2\left(\sum_i\left(\frac{\eta_i}{\zeta} - \Lambda_i\right)\eta_i - \eta\zeta\right)$$
$$= \zeta(\Lambda_i^2 + \ldots + \Lambda_n^2) \text{ (using } f\zeta^2 = \eta_i^2 + \ldots + \eta_n^2)$$
$$= \zeta\sum_i \gamma_i^2/\zeta^2 = 1/\zeta\sum_i \gamma_i^2.$$

Here not all the γ_i are zero (as already noted) and so $\sum\gamma_i^2$ is non-zero since otherwise by equating the coefficient of the highest power in X to 0, we find that 0 is a sum of at most n squares of elements of K, which is the third trivial case of the lemma. Thus $\zeta^* \neq 0$, which proves (b).

Finally $\zeta^* = 1/\zeta\sum_i\gamma_i^2$ giving $\zeta\zeta^* = \sum_i\gamma_i^2$. Equating degrees, we get $\deg\zeta +\deg\zeta^* = 2\max_i(\deg\gamma_i) < 2\deg\zeta$ since $\deg\gamma_i < \deg\zeta$ (for all i). Thus $\deg\zeta^* < \deg\zeta$, which proves (c).

This completes the proof of Cassels' lemma. □

Remark. The solution $(\zeta^*, \eta_1^*, \ldots, \eta_n^*)$ does not just come out of the blue. It is the second point of intersection Q of the quadric (1) with the line joining the points $P = (\zeta, \eta_1, \ldots, \eta_n)$ (on the quadric) and $P' = (1, \lambda_1, \ldots, \lambda_n)$ (in space) in the n-dimensional projective space over the field $K(X)$. The simplest way to get this point Q is as follows: a general point of the line PP' is

$$(\theta\zeta + \varphi, \theta\eta_1 + \varphi\lambda_1, \ldots, \theta\eta_n + \varphi\lambda_n)$$

θ/φ being a parameter for various points, $\varphi = 0$ giving the point P. To get Q we substitute this general point in the quadric (1):

$$f(\theta^2\zeta^2 + \varphi^2 + 2\theta\varphi\zeta) = \sum_{j=1}^n(\theta^2\eta_j^2 + \varphi^2\lambda_j^2 + 2\theta\varphi\eta_j\lambda_j)$$

i.e. $\theta^2(\zeta^2 f - \sum\eta_j^2) + 2\theta\varphi(f\zeta - \sum\lambda_j\eta_j) + \varphi^2(f - \sum\lambda_j^2) = 0$. But $\zeta^2 f = \sum\eta_j^2$ so this becomes $2\theta\varphi(f\zeta - \sum\lambda_j\eta_j) + \varphi^2(f - \sum\lambda_j^2) = 0$ This has a root $\varphi = 0$ as expected giving the point P. The other root is

$\theta/\varphi = -(\sum \lambda_j^2 - f)/2(\sum \lambda_j \eta_j - \zeta f)$, and substituting this in the general point and multiplying by a suitable factor (allowed in a projective space) we get our point Q as required.

We now deduce a few corollaries from this lemma.

Corollary 1. *Let* char$K \neq 2$ *and let* $f(X_1, \ldots, X_m) \in K(X_1, \ldots, X_m)$ *be a sum of* n *squares of elements of* $K(X_1, \ldots, X_m)$. *Let* $a_1, a_2, \ldots, a_m \in K$ *be such that* $f(a_1, \ldots, a_m)$ *is defined (i.e. the dominator is not* 0*). Then* $f(a_1, \ldots, a_m)$ *is a sum of* n *squares in* K.

Remark. The point is that although $f(X_1, \ldots, X_m)$ is defined at $(a_1 \ldots, a_m)$, it may well happen that the summands $f_j^2(X_1, \ldots, X_m)$ of the right hand side of $f(X_1, \ldots, X_m) = f_1^2 + \ldots + f_n^2$ may not be defined at $(a_1, \ldots, a_m)$, but still acccording to the corollary, $f(a_1, \ldots, a_m)$ is a sum of n squares in K.

Proof. We use induction on m. For $m = 1$, we have

$$f(X) = g(X)/h(X) = \gamma_1^2(X) + \ldots + \gamma_n^2(X).$$

Then $gh = (\gamma_1 h)^2 + \ldots + (\gamma_n h)^2$. Thus gh, which is in $K[X]$, is a sum of n squares in $K(X)$ and so by Cassels' lemma, it is a sum of n squares in $K[X]$:

$$gh = f_i^2 + \ldots + f_n^2 \quad (f_j \in K[X]).$$

Hence $g(X)/h(X) = \left(\frac{f_1(X)}{h(X)}\right)^2 + \ldots + \left(\frac{f_n(X)}{h(X)}\right)^2$. Now by hypothesis, $f(a) = g(a)/h(a)$ is defined; i.e. $h(a) \neq 0$, so each $f_j(a)/h(a)$ is defined. □

Let now $m > 1$. Let $L = K(X_1, \ldots, X_{m-1})$. Assume the result for $m-1$ variables and let $g(X_1, \ldots, X_m)/h(X_1, \ldots, X_m)$ be a rational function which is a sum of n squares in $K(X_1, \ldots, X_m)$. Regard g/h as a rational function of X_m belonging to $L(X_m)$. So by the case $m = 1$, we see that $g(X_1, \ldots, X_{m-1}, a_m)/h(X_1, \ldots, X_{m-1}, a_m)$ is a sum of n squares in $L = K(X_1, \ldots, X_{m-1})$. So by the induction hypothesis $g(a_1, \ldots, a_m)/h(a_1, \ldots, a_m)$ is a sum of n squares in K. This completes the proof of the corollary.

□

Corollary 2. *Suppose* $n = 2^m$. *Let* G_n *be the set of all non-zero elements of* K *which are sums of* n *squares in* K. *Then* G_n *is a group under multiplication.*

Proof. Let $\alpha\beta \in G_n$ say, $\alpha = \alpha_1^2 + \ldots + \alpha_n^2, \beta = \beta_1^2 + \ldots + \beta_n^2$. Then

$\alpha^{-1} = \alpha/\alpha^2 = (\alpha_1/\alpha)^2 + \ldots + (\alpha_n/\alpha)^2 \in G_n$ and it remains to prove that $\alpha\beta \in G_n$. Consider the identity

$$(X_1^2 + \ldots + X_n^2)(Y_1^2 + \ldots + Y_n^2) = Z_1^2 + \ldots + Z_n^2$$

which exists since $n = 2^m$. In this let $X_1 \to \alpha_1, \ldots, X_n \to \alpha_n, Y_1 \to \beta_1, \ldots, Y_n \to \beta_n$. Then the left side is well defined and equal to $\alpha\beta$ and so by Corollary 1, the right side is a sum of n squares of elements of K, i.e. $\alpha\beta \in G_n$ as required. □

We see that it is the identity (2.1) that does the trick.

We can now prove the first part of Theorem 2.2: *$s(K)$ is always a power of 2.*

Proof of the first part of Theorem 2.2. Let

$$n = 2^m \leq s(K) < 2^{m+1} \tag{2.5}$$

Then $a_1^2 + \ldots + a_n^2 + a_{n+1}^2 + \ldots + a_s^2 + 1 = 0 \;\; (a_j \in K)$. Let $A = a_1^2 + \ldots + a_n^2, B = a_{n+1}^2 + \ldots + a_s^2 + 1$. Here A, B are both non-zero, otherwise $s(K) < s$. Also A, B both $\in G_n$ (by adding a suitable number of 0^2's to B if necessary). Then $A + B = 0$ so $A = -B$ i.e. $-1 = B/A \in G_n$ since G_n is a group i.e. $-1 = c_1^2 + \ldots + c_n^2$ giving $s(K) \leq n$. Comparing with (2.5) we get $s(K) = n = 2^m$. □

To prove the remaining parts of Theorems 2.1 and 2.2 we need to deduce some more corollaries from Cassels' lemma; see [C2].

Corollary 3. *Let* char$K \neq 2$. *A necessary and sufficient condition for $X^2 + d \in K[X]$ to be a sum of n squares in $K(X)$ (and so in $K[X]$ by Cassels' lemma) is that either*

(i) -1 is a sum of $n-1$ squares in K or
(ii) d is a sum of $n-1$ squares in K.

Proof. If $-1 = b_1^2 + \ldots + b_{n-1}^2$, then for any polynomial $f(X) \in K[X]$, we have

$$f = \left(\frac{f+1}{2}\right)^2 - \left(\frac{f-1}{2}\right)^2 = \left(\frac{f+1}{2}\right)^2 + \left(\frac{b_1(f-1)}{2}\right)^2 + \ldots + \left(\frac{b_{n-1}(f-1)}{2}\right)^2.$$

In particular $X^2 + d$ is a sum of n squares.

If d is a sum of $n-1$ squares then visibly $X^2 + d$ is a sum of n squares in $K[X]$.

For the converse, suppose $X^2 + d$ is a sum of n squares in $K[X]$. If (i) holds, well and good; otherwise let $X^2 + d = p_1^2(X) + \ldots + p_n^2(X)$ say. Here we may suppose the $p_j(X)$ to be linear poynomials in X for if not, then equating to 0 the coefficient of the highest power of X gives (i). Then

$$X^2 + d = (a_1X + b_1)^2 + \ldots + (a_nX + b_n)^2 \tag{2.6}$$

Now one of the equations $C = \pm(a_nC + b_n)$ is always solvable in K. For if $a_n \neq 1$ then $C = +(a_nC + b_n)$ is solvable, while if $a_n = 1$ then $C = -(a_nC + b_n)$ is solvable since char $K \neq 2$. Now put $X = C$ in (2.6):

$$C^2 + d = (a_1C + b_1)^2 + \ldots + (a_{n-1}C + b_{n-1})^2 + (a_nC + b_n)^2.$$

Cancelling C^2 with $(a_nC + b_n)^2$ we see that d is a sum of $n - 1$ squares in K. This completes the proof. □

Corollary 4. *Let* $\mathbf{R}$ *be the field of real numbers. Then* $X_1^2 + \ldots + X_n^2$ *is not a sum of* $n - 1$ *squares of elements in* $\mathbf{R}(X_1, \ldots, X_n)$.

Proof. We use induction on n. For $n = 1$, the result is trivial. So suppose the result is true for $n - 1$. Let $K = \mathbf{R}(X_1, \ldots, X_{n-1}), X_n = X$ and $d = X_1^2 + \ldots + X_{n-1}^2$. If $X_1^2 + \ldots + X_n^2$ is a sum of $n - 1$ squares in $K(X) = \mathbf{R}(X_1, \ldots, X_n)$, then by Corollary 3, $d = X_1^2 + \ldots + X_{n-1}^2$ is a sum of $n - 2$ squares in K, since -1 is clearly not a sum of $n - 2$ squares in K - indeed not a sum of squares at all in K, which is formally real. This contradicts the induction hypothesis and completes the proof of Corollary 4. □

We are now in a position to complete the proofs of the remaining parts of Theorems 2.1 and 2.2.

Every power of 2 *is the Stufe of some field* K.

Proof. Let $n = 2^m$ and let $K = \mathbf{R}(X_1, \ldots, X_{n+1}, Y)$ where $X_1, \ldots, X_{n+1}$ are independent transcendentals over $\mathbf{R}$ and Y satisfies the equation

$$Y^2 + X_1^2 + \ldots + X_{n+1}^2 = 0 \tag{2.7}$$

We claim that $s(K) = n = 2^m$. In any case by (2.7) $s(K) \leq n + 1$ and so is at most n since $n + 1$ cannot be a power of 2 whereas $s(K)$ *is* (except in the trivial case $n = 1$ i.e. $m = 0$).

If $s(K) < n$ then there exist $t_1, \ldots, t_n \in K$, not all zero such that

$$t_1^2 + \ldots + t_n^2 = 0 \tag{2.8}$$

Let $L = \mathbf{R}(X_1, \ldots, X_{n+1})$ so that $K = L(Y)$. By (2.7) Y is algebraic over L of degree 2 and so each element of K is a linear polynomial in Y with coefficients from L. Write $t_j = a_j + Yb_j, a_j, b_j \in L$. Then by (2.8) we see that

$$\sum a_j^2 + Y^2 \sum b_j^2 = 0$$

and

$$\sum a_j b_j = 0.$$

Here not all the a_j are zero, otherwise $\sum b_j^2 = 0$ and so each $b_j = 0$ since the $b_j \in L = \mathbf{R}(X_1, \ldots, X_{n+1})$ is formally real. Then each t_j would be zero which is not true. Similarly not all the b_j are zero. Hence

$$\begin{aligned} -Y^2 &= \sum_{j=1}^{n} a_j^2 / \sum_{j=1}^{n} b_j^2 \in G_n \text{ (by the group property of } G_n) \\ &= c_1^2 + \ldots + c_n^2, \text{ say } c_j \in L, \end{aligned}$$

i.e. $X_1^2 + \ldots + X_{n+1}^2$ is a sum of n squares in L which contradicts Corollary 4. Thus $s(K)$ is *not* less than n and so equals n. This completes the proof.

□

Remark. The proof also works for $\mathbf{Q}(X_1, \ldots, X_{n+1}, Y)$.

Finally we prove the remaining part of Theorem 2.1.

Suppose n is not a power of 2. Then there is a field K such that there is no identity

$$(X_1^2 + \ldots + X_n^2)(Y_1^2 + \ldots + Y_n^2) = Z_1^2 + \ldots + Z_n^2$$

with $Z_j \in K(X_1, \ldots, X_n, Y_1, \ldots, Y_n)$.

Proof. Let $2^{m-1} < n < 2^m$. Let K be a field having Stufe $2^m = \nu$, say. Then $a_1^2 + \ldots + a_n^2 + a_{n+1}^2 + \ldots + a_\nu^2 + 1 = 0$. Let $A = a_1^2 + \ldots + a_n^2$, $B = a_{n+1}^2 + \ldots + a_\nu^2 + 1$; hence $A, B \in G_n$ and if an identity of the above type exists, then G_n is a group (see the proof of Corollary 2). So $-1 = B/A \in G_n$, i.e. $-1 = C_1^2 + \ldots + C_n^2$ $(C_j \in K)$ hence $s(K) \leq n < \nu$. But K was chosen to have Stufe ν. This gives a contradiction and so completes the proof.

□

Remark 1. In our examples of fields with high Stufe both the fields $\mathbf{R}(X_1, \ldots, X_{n+1}, Y)$ and $\mathbf{Q}(X_1, \ldots, X_{n+1}, Y)$ are of high transcendence degree over $\mathbf{R}$ or $\mathbf{Q}$ as the case may be. We have the following

Problem: *Does high Stufe always imply high degree of transcendence? (over $\mathbf{R}$ or $\mathbf{Q}$).*

Remark 2. A result more general than Cassels' lemma is the following:

let K be a field with char$K \neq 2$ and let $f(x) \in K[X]$ be a polynomial such that

$$f(X) = \alpha_1 \gamma_1^2(X) + \ldots + \alpha_n \gamma_n^2(X)$$

where $\alpha_j \in K$ and $\gamma_j(X) \in K(X)$. Then

$$f(X) = \alpha_1 f_1^2(X) + \ldots + \alpha_n f_n^2(X)$$

where $f_j(X) \in K[X]$.

For a proof see [P5].

Remark 3. In the proof of the Theorem 2.1, instead of taking

$$T = \begin{pmatrix} T^{(1)} & T^{(2)} \\ T^{(2)} & X \end{pmatrix}$$

and wondering why the second row began with $T^{(2)}$, we could take

$$T = \begin{pmatrix} T^{(1)} & T^{(2)} \\ Y & X \end{pmatrix}.$$

Then our requirement (2.4) leads to the equation $T^{(1)\prime}T^{(1)} + Y'Y = (X_1^2 + \ldots + X_n^2 + \ldots + X_{2n}^2)I_n$ i.e. $Y'Y = (X_{n+1}^2 + \ldots + X_{2n}^2)I_n$. Now one solution of this is $Y = T^{(2)}$ by (2.3). So it is natural to take Y to be this solution and not look for any other. The rest now goes through as before.

Remark 4. Corollary 2 about the group property of $G_n(K)$ is so important that we would like a direct proof of it. This we now give.

In Theorem 2.1, we proved we could take the Z_k to be linear forms in the Y_j with coefficients in $K(X_1, \ldots, X_n)$. What is the nature of these coefficients? How simple can we take them? Can we make at least some of them linear forms even in the X_i's? The answers to these questions will be given in Chapter 13. For the time being we prove here a result about the form of the first term Z_1. This approach will lead to an independent proof of the important group property of $G_n(K)$ ($n = 2^m$). We have the following result.

Theorem 2.3. *Let $n = 2^m$ and let $X_1, \ldots, X_n; Y_1, \ldots, Y_n \in K$. Then $(X_1^2 + \ldots + X_n^2)(Y_1^2 + \ldots + Y_n^2) = (X_1Y_1 + \ldots + X_nY_n)^2 + Z_2^2 + \ldots + Z_n^2$ for some $Z_2, \ldots, Z_n \in K$.*

We first need the following easy

Lemma 1. *Let $n = 2^m$ and let $c = c_1^2 + \ldots + c_n^2$ $(c_j \in K)$. Then there exists an $n \times n$ matrix S with first row $= (c_1, \ldots, c_n)$ such that $SS' = S'S = cI_n$.*

Proof. First let $c = 0$. If all the $c_j = 0$, we take S to be the zero matrix.

So suppose, say, $c_1 \neq 0$. Let R be the row vector $(c_1, \ldots, c_n)$ and take $S = c_1^{-1} R'R$, which has first row R as required. Further

$$\begin{aligned} SS' &= c_1^{-1} R'Rc_1^{-1} R'R \\ &= c_1^{-2} R'(RR')R \\ &= 0, \end{aligned}$$

since $RR' = c_1^2 + \ldots + c_n^2 = c = 0$. Similarly $S'S = 0$ and the proof is complete. So we may now suppose that $c \neq 0$ and we proceed by induction on m.

Write

$$\begin{aligned} \underline{R} = (c_1, \ldots, c_{2^m}) &= (c_1, \ldots, c_{2^{m-1}}, c_{2^{m-1}+1}, \ldots, c_{2^m}) \\ &= (\underline{R}_1, \underline{R}_2). \end{aligned}$$

Let $a = c_1^2 + \ldots + c_{2^{m-1}}^2, b = c_{2^{m-1}+1}^2 + \ldots + c_{2^m}^2$, so that $c = a + b$. Here since $c \neq 0$, so a, b, cannot be both zero; say, without loss of generality, that $a \neq 0$. By the induction hypothesis, there exist square matrices S_1, S_2 of size 2^{m-1} such that

$$\begin{aligned} S_1 S_1' = S_1' S_1 = aI_{2^{m-1}} \\ S_2 S_2' = S_2' S_2 = bI_{2^{m-1}}. \end{aligned}$$

Furthermore the first row of S_1 is $(c_1, \ldots, c_{2^{m-1}})$ and that of S_2 is $(c_{2^{m-1}+1}, \ldots, c_2^m)$. Now let

$$S = \begin{pmatrix} S_1 & S_2 \\ -a^{-1} S_1' S_2' S_1 & S_1' \end{pmatrix}.$$

This has first row equal to R as required and an easy matrix computation gives $SS' = S'S = cI_n$, e.g.

$$\begin{aligned} SS' &= \begin{pmatrix} S_1 & S_2 \\ -a^{-1} S_1' S_2' S_1 & S_1' \end{pmatrix} \begin{pmatrix} S_1' & -a^{-1} S_1' S_2 S_1 \\ S_2' & S_1 \end{pmatrix} \\ &= \begin{pmatrix} aI_{2^{m-1}} + bI_{2^{m-1}} & -a^{-1} aS_2 S_1 + S_2 S_1 \\ -a^{-1} S_1' S_2' aI_{2^{m-1}} + S_1' S_2' & a^{-2} S_1' S_2' S_1 S_1' S_2 S_1 + S_1' S_1 \end{pmatrix}. \\ &= \begin{pmatrix} cI_{2^{m-1}} & 0 \\ 0 & bI_{2^{m-1}} + aI_{2^{m-1}} \end{pmatrix} = cI_{2^m} \end{aligned}$$

□

Proof of Theorem 2.3. Write

$$\begin{aligned} X = X_1^2 + \ldots + X_n^2 \\ Y = Y_1^2 + \ldots + Y_n^2. \end{aligned}$$

Then there exist $n \times n$ matrices U, V such that $UU' = U'U = XI_n$. $VV' = V'V = YI_n$ and

$$\begin{aligned} U \text{ has 1st row } &= (X_1, \ldots, X_n), \\ V \text{ has 1st row } &= (Y_1, \ldots, Y_n). \end{aligned}$$

Then

$$XYI_n = XVV' = V(U'U)V' = (VU')(VU')' = V(U'U)V' = WW',$$

where $W = VU'$. This equation says that if $(Z_1, \ldots, Z_n)$ is the first row of W then $XY = Z_1^2, \ldots + Z_n^2$. But since $W = VU'$, we have $Z_1 = X_1Y_1 + \ldots + X_nY_n$. □

Mathematics is rarely humorous enough to provoke laughter. However, on taking K to be the field of real numbers, and writing α_j for X_j, β_j for Y_j and $\zeta_2, \ldots, \zeta_n$ for $Z_2, \ldots, Z_n$, the above gives

$$\sum_{j=1}^{n} \alpha_j^2 \sum_{j=1}^{n} \beta_j^2 = \left(\sum_{j=1}^{n} \alpha_j \beta_j \right)^2 + \zeta_2^2 + \ldots + \zeta_n^2$$

$$\geq \left(\sum_{j=1}^{n} \alpha_j \beta_j \right)^2$$

and as T.Y. Lam remarks, this gives us a hilarious proof of the Cauchy–Schwartz inequality.

The group property of $G_n(K)$ $(n = 2^m)$ now follows, since Theorem 2.3 implies closure.

The simplicity of Z_1 can be used to much advantage as we shall see in the next chapter.

While we are at it we shall give another easy approach to the proof of the group property of $G_n(K)$.

Let $f(X_1, \ldots, X_n) = f(\mathbf{X})$ and $g(\mathbf{X})$ be quadratic forms over K. We call them equivalent over K and write $f \overset{K}{\sim} g$, if there exists a non-singular linear transformation $\mathbf{Y} = T\mathbf{X}$ i.e.

$$Y_i = \sum_{j=1}^{n} t_{ij} X_j \quad (1 \leq i \leq n), t_{ij} \in K$$

such that $g(\mathbf{X}) = f(T\mathbf{X})$ identically.

Definition 2.2. Denoting

$$M_f(K) = M_f = \{c \in K^* | cf \sim f\},$$

we call c a similarity factor of f over K.

Lemma 2.

(i) *M_f is a group under multiplication.*

(ii) $M_f(K) \supset {K^*}^2$

(iii) *if $f \sim g$ then $M_f = M_g$.*

Proof. Let $c_1, c_2 \in M_f$ so that

$$f(T_1\mathbf{X}) = c_1 f(\mathbf{X}),$$
$$f(T_2\mathbf{X}) = c_2 f(\mathbf{X}).$$

Then $f(T_1T_2\mathbf{X}) = c_1 f(T_2\mathbf{X}) = c_1c_2 f(\mathbf{X})$ and $f(T_1^{-1}\mathbf{X}) = c_1^{-1} f(\mathbf{X})$. This proves (i).

Next if $d \in K^*$, then $f(d\mathbf{X}) = d^2 f(\mathbf{X})$ i.e. $d^2 f \sim f$ or $d^2 \in M_f$ giving (ii).

Finally if $f \sim g$ then $cf \sim cg$ $(c \in K)$. Hence $f \sim cf$ iff $g \sim cg$. This proves (iii). □

Definition 2.3. Let $V_f(K) = V_f = \{c \in K^* | c = f(\mathbf{V}), \mathbf{V} \in K^n\}$, i.e. the set of non-zero values taken by f over K.

Lemma 3. *Let $b \in V_f, c \in M_f$. Then $bc \in V_f$.*

Proof. Let $b = f(\mathbf{V}), \mathbf{V} = (V_1, \dots, , V_n)$ $(V_j \in K)$ and $f(T\mathbf{X}) = cf(\mathbf{X})$. Then $bc = f(T\mathbf{V})$ □

Now let us return from general quadratic forms to sums of squares and write $\varphi_n(\mathbf{X}) = X_1^2 + \dots + X_n^2$. Then

Theorem 2.4 (Pfister). *Let $n = 2^m$ be a power of 2; then $M_{\varphi_n} = V_{\varphi_n}$.*

The key to Theorem 2.4 is the following.

Lemma 4. *Let $d \in M_{\varphi_n}$ and suppose $d + 1 \neq 0$ then $1 + d \in M_{\varphi_{2n}}$.*

Proof.

$$\begin{aligned} \varphi_{2n}(\mathbf{X}) &= X_1^2 + \dots + X_n^2 + X_{n+1}^2 \dots + X_{2n}^2 \\ &= \varphi_n(\mathbf{X}') + \varphi_n(\mathbf{X}'') \end{aligned} \tag{2.9}$$

where $\mathbf{X} = (X_1, \dots, X_{2n}), \mathbf{X}' = (X_1, \dots, X_n), \mathbf{X}'' = (X_{n+1}, \dots, X_{2n})$. Here $\varphi_n(\mathbf{X}'') \sim d\varphi_n(\mathbf{X}'')$ since $d \in M\varphi_n$. Hence $\varphi_{2n}(\mathbf{X}) = \varphi_n(\mathbf{X}') + \varphi_n(\mathbf{X}'') \sim \varphi_n(\mathbf{X}') + d\varphi_n(\mathbf{X}'') = \psi$, say. Thus to show $1 + d \in M\varphi_{2n}$, it is enough to show that $1 + d \in M\psi$. Now, putting

$$\begin{cases} Y_1 = X_1 - dX_{n+1} \\ Y_2 = X_2 - dX_{n+2} \\ \dots\dots\dots\dots\dots \\ Y_n = X_n - dX_{2n} \end{cases} \quad \text{and} \quad \begin{cases} Y_{n+1} = X_1 + X_{n+1} \\ Y_{n+2} = X_2 + X_{n+2} \\ \dots\dots\dots\dots\dots \\ Y_{2n} = X_n + X_{2n}, \end{cases}$$

we get

$$\begin{aligned}(1+d)\psi(\mathbf{X}) &= (1+d)(X_1^2+\ldots+X_n^2+dX_{n+1}^2+\ldots+dX_{2n}^2)\\ &= Y_1^2+\ldots+Y_n^2+dY_{n+1}^2+\ldots+dY_{2n}^2\\ &= \psi(Y_1,Y_2,\ldots,Y_n,Y_{n+1},\ldots,Y_{2n}),\\ &= \psi(\mathbf{Y})\end{aligned}$$

as required. □

We also enunciate the trivial

Lemma 5. $M\varphi_n \subset M\varphi_{2n}$.

We now give a

Proof of Theorem 2.4. Here $n = 2^m$. Use induction on m. So suppose $V\varphi_n = M\varphi_n$ and we have to show that $V\varphi_{2n} = M\varphi_{2n}$. Now trivially $1 \in V\varphi_{2n}$ (put $X_1 = 1, X_2 = \ldots = X_{2n} = 0$) so by Lemma 3, $M\varphi_{2n} \subset V\varphi_{2n}$. On the other hand, by (2.9) it follows that any C in $V\varphi_{2n}$ has one of the forms

(i) $C = C_1 \in V\varphi_n$
(ii) $C = C_1 + C_2$ $(C_1, C_2 \in V\varphi_n)$.

In the first case, from the induction hypothesis,

$$C_1 \in V\varphi_n = M\varphi_n \subset M\varphi_{2n}$$

by Lemma 5. In the second case, $C_1, C_2 \in V\varphi_n = M\varphi_n$ by the induction hypothesis and so

$$C = C_1 + C_2 = C_1(1 + C_1^{-1}C_2) = C_1(1+d)$$

say, where $d = C_1^{-1}C_2 \in M\varphi_n$ since $M\varphi_n$ is a group by Lemma 2. Here $C_1 \in M\varphi_{2n}$ (by the first case) and $1 + d \in M\varphi_{2n}$ (by Lemma 4). Hence $C = C_1(1+d) \in M\varphi_{2n}$ as $M\varphi_{2n}$ is a group. So $V\varphi_{2n} \subset M\varphi_{2n}$. □

Now by Lemma 2, $M\varphi_n$ is a group and $V\varphi_n$ is just an alias for G_n; so G_n is a group. □

Exercise. For a quadratic form φ defined over K show:
(i) if $a, d \in K^*$ then $d \in V\varphi(K)$ iff $a^2d \in V\varphi(K)$;
(ii) $V\varphi(K)$ is a union of cosets of K^* modulo K^{*^2}.

We call K^*/K^{*^2} the group of square classes of K. Then, by abuse of notation, $V\varphi(K)$ can be regarded as a subset of K^*/K^{*^2}. We shall have more to do with these square classes in Chapter 16.

Exercises

1. Give examples of quadratic fields of Stufe 1, 2, 4.

2. Prove directly that 3 can not occur as a Stufe. This is the simplest case of van der Waerden's problem (1932) enquiring which numbers can occur as a Stufe. (If $-1 = x^2 + y^2 + z^2$ i.e. $0 = 1 + x^2 + y^2 + z^2$ then multiplying by $1 + x^2$ gives

$$\begin{aligned} 0 &= (1 + x^2)^2 + (1 + x^2)(y^2 + z^2) \\ &= (1 + x^2)^2 + (y + xz)^2 + (z - xy)^2, \end{aligned}$$

by (1.1). Now $1 + x^2 \neq 0$, for otherwise the Stufe equals 1 so the above gives $-1 = \left(\frac{y+xz}{1+x^2}\right)^2 + \left(\frac{z-xy}{1+x^2}\right)^2$ i.e. the Stufe ≤ 2. □

3. Let $\mathsf{F}q$ be the field of $q = p^\alpha$ elements. Prove that

$$s(\mathsf{F}q) = \begin{cases} 1 \text{ if either } p = 2 \text{ or } p \equiv 1(4) \text{ or } p \equiv 3(4), 2|\alpha, \\ 2 \text{ otherwise i.e. } p \equiv 3(4), 2 \not|\alpha. \end{cases}$$

4. Prove that the field $\mathsf{R}(X)$ is formally real where R is the real number field.

5. Let K be a field of Stufe s. Prove that each element of K is a sum of $s + 1$ squares in K.

6. For a natural number n and any field K, let

$$G_n(K) = \{a \in K^* | a = a_1^2 + \ldots, a_n^2, a_j \in K\}$$

so that $K^{*^2} \subseteq G_1(K) \subseteq G_2(K) \subseteq \ldots \subseteq G_n(K) \subseteq \ldots \subseteq K^*$. Prove that if $s(K) = n$ then $G_{n+1}(K) = K^*$ (cf. Exercise 5.)

7. Prove that if $s(K) \geq n$ then, using Corollary 4

$$G_n K(X_1, \ldots, X_n)) \subsetneqq G_{n+1}(K(X_1, \ldots, X_n)).$$

8. Prove that if (r, s, n) is admissible over K, then for any field $F \supseteq K$,

$$G_r(F) \cdot G_s(F) \subseteq G_n(F).$$

9. We know from Exercise 4, Chapter 1, that $(3, 5, 7)$ is admissible over any field K so that $G_3(K) \cdot G_5(K) \subseteq G_7(K)$. Prove that equality holds.

If $a = a_1^2 + \ldots + a_7^2 \in G_7(K)$ then without loss of generality $a_1^2 + a_2^2 + a_3^2 \neq 0$ so

$$\begin{aligned} a &= (a_1^2 + a_2^2 + a_3^2)(1 + (a_4^2 + a_5^2 + a_6^2 + a_7^2)/(a_1^2 + a_2^2 + a_3^2)) \\ &= (a_1^2 + a_2^2 + a_3^2) \times \{\text{a sum of five squares}\}, \end{aligned}$$

since the quotient is an element of the group $G_4(K) \in G_3(K) \cdot G_5(K)$. □

10. Use Exercises 7, 8 and 9 to prove that there exist fields K over which $(3,5,6)$ is not admissible. (Hint; the admissibility of $(3,5,6)$ over K implies $G_3(F) \cdot G_5(F) \subseteq G_6(F)$ for all $F \supseteq K$ (by Exercise 8), so by Exercise 9, $G_6(F) = G_7(F)$ for such F. But if $s(K) \geq 6$, then by Exercise 7, there exists a field $F \supseteq K$ such that $G_6(F) \subsetneqq G_7(F)$. Hence if $s(K) \geq 6$ then $(3,5,6)$ is not admissible over K.)

3

Examples of the Stufe of fields and related topics

Before we take up specific examples, we would like to say something about how representations of an integer as a sum of squares (SOS) in $\mathbb{Q}$ is related to that in $\mathbb{Z}$, and indeed more generally about how the representation of an element a of an integral domain A as an SOS in A is related to the representation of a as an SOS in the field of quotients F of A. We paraphrase our questions more explicitly as follows:

Question 1. If $a \in A$ is an SOS of n elements of F then is a an SOS of elements of A?

Question 2. If the answer to Question 1 is 'yes', then is a an SOS of the same number n of elements of A?

The lemma of Cassels proved in the last chapter is an excellent example of a problem of this nature, where the answer to both the above questions is given in the affirmative (A being the ring $K[t], K$ a field). Note that the answer to the first question was proved to be in the affirmative by Artin already in 1927; but the second was not answered then.

Another very instructive example is provided in the case $A = \mathbb{Z}$, so that $F = \mathbb{Q}$; the answers to both the questions being in the affirmative and this result is often called the Davenport-Cassels lemma. However, the proof of the result as well as the method goes back to Aubry (1912) (see [A7]) and so more appropriately it ought to be called the Aubry Lemma.

Theorem 3.1. *Let n be a positive integer which is a sum of three squares of rational numbers; then n is a sum of three squares of integers.*

Proof. Suppose $n = \lambda_1^2 + \lambda_2^2 + \lambda_3^2$ $(\lambda_j \in \mathbf{Q})$. Clearing denominators we may write this as

$$t^2 n = \mu_1^2 + \mu_2^2 + \mu_3^2 \quad (\mu_j \in \mathbf{Z}) \tag{3.1}$$

with $t \in \mathbf{Z}$ minimal. We shall show that $t = 1$.

Write $\mu_j/t = y_j + z_j$ with $y_j \in \mathbf{Z}, |z_j| \leq 1/2$ $(j = 1, 2, 3)$. If z_1, z_2, z_3 are all 0, then $\mu_j/t \in \mathbf{Z}$ and so by the minimality of $t, t = 1$. So suppose not all z_j are 0 and let

$$a = y_1^2 + y_2^2 + y_3^2 - n \tag{3.2}$$

$$b = 2[nt - (\mu_1 y_1 + \mu_2 y_2 + \mu_3 y_3)] \tag{3.3}$$

$$t' = at + b$$

$$\mu_j' = a\mu_j + by_j \tag{3.4}$$

Then $a, b, t' \in \mathbf{Z}$ and using (3.4) we get

$$\begin{aligned}
\mu_1'^2 + \mu_2'^2 + \mu_3'^2 &= a^2 \sum \mu_j^2 + b^2 \sum y_j^2 + 2ab \sum \mu_j y_j \\
&= a^2(t^2 n) + b^2(a + n) + ab(2nt - b) \\
&\qquad \text{(using (3.1) – (3.3))} \\
&= n(at + b)^2 \\
&= nt'^2
\end{aligned} \tag{3.5}$$

Thus (3.5) gives a representation similar to (3.1) where we shall show that $t' < t$ giving a contradiction to the minimality of t. Substituting for a, b from (3.2), (3.3), and then for nt^2 from (3.1), we have

$$\begin{aligned}
tt' &= at^2 + bt \\
&= t^2(y_1^2 + y_2^2 + y_3^2 - n) \\
&\quad + t(2nt - 2(\mu_1 y_1 + \mu_2 y_2 + \mu_3 y_3)) \\
&= t^2 \sum y_j^2 - 2t \sum \mu_j y_j + \sum \mu_j^2 \\
&= \sum (ty_j - \mu_j)^2 \\
&= t^2(z_1^2 + z_2^2 + z_3^2).
\end{aligned}$$

Thus

$$\begin{aligned}
t' &= t(z_1^2 + z_2^2 + z_3^2) \\
&\leq t((1/2)^2 + (1/2)^2 + (1/2)^2) \\
&= \frac{3}{4} t
\end{aligned}$$

as required. □

Remark 1. There is a striking similarity (in addition to the method of descent used in both cases) between the method of proof of

(i) Cassels' lemma of Chapter 2
(ii) Aubry's lemma proved above.

Remark 2. As may be expected, the answers to both the questions at the start of this chapter depend on the number n and on the nature of the domain A. The following are some examples exhibiting a variety of answers:

(1) $A = K[t]$ (K any field); then the answers to both the questions are 'yes'.

(2) $A = \mathbf{Z}[i]$. Then $i = \left(\frac{1+i}{2}\right)^2 + \left(\frac{1+i}{2}\right)^2$ is a sum of two squares in F, the field of quotients $\mathbf{Q}(i)$ of A, but i is obviously not an SOS of any number of elements of A.

It may be suspected that the example in (2) works that way because 2 is not invertible in A. That this is not really the case is shown by the following example (where 2 is invertible in A).

(3) $A = \mathbf{R}[X_1, X_2, \ldots, X_d]$, $d \geq 2$. According to Hilbert [H3], there exist polynomials $f \in A$, which are positive semi-definite (i.e. $f(a_1, \ldots, a_d) \geq 0$ for all $a_j \in \mathbf{R}$) with the property that f is not an SOS in A. By Artin's solution to Hilbert's 17th problem [A6], f is always an SOS in the quotient field $\mathbf{R}(X_1, \ldots, X_d)$ of A; indeed by a theorem of Pfister's, 2^d squares suffice.

All three statements will be proved in later chapters. As a concrete example we have Robinson's polynomial (form), see [R8],

$$f(X,Y,Z) = X^4Y^2 + Y^4Z^2 + Z^4X^2 - 3X^2Y^2Z^2,$$

which is non-negative by the Arithmetic Geometric Mean Inequality applied to the numbers X^4Y^2, Y^4Z^2, Z^4X^2. But f is not an SOS in $\mathbf{R}[X,Y,Z]$; a proof of this will come later. However, in $\mathbf{R}(X,Y,Z)$ we have explicitly

$$f = \frac{X^4Y^2(X^2+Y^2-2Z^2)^2 + (X^2-Y^2)^2(X^2Y^2+X^2Z^2+Y^4)Z^2}{(X^2+Y^2)^2},$$

which is a sum of four squares.

For further information on this topic see the Choi, Lam, Reznick, Rosenberg paper [C15].

Remark 3. There is a proof of Theorem 3.1 given by Serre in [S2]. He calls it the Davenport-Cassels lemma.

As our first example, we shall calculate the Stufe of quadratic fields. We have the following.

Theorem 3.2. *Let $D > 0$ be a square-free integer; then the Stufe $s(K)$ of*

$K = \mathbb{Q}(\sqrt{-D})$ is

$$\begin{cases} 1 & \text{if } D = 1, \\ 2 & \text{if } D \neq 8b+7, \\ 4 & \text{if } D = 8b+7. \end{cases}$$

If $D < 0$, then K is formally real of course.

Proof. (Rajwade [R3]). Writing $D = a^2 + b^2 + c^2 + d^2$, when $a, b, c, d \in \mathbb{Z}$ we see that $0 = (\sqrt{-D})^2 + a^2 + b^2 + c^2 + d^2$, and it follows that $s(K) \leq 4$. Now $s(K) = 1$ if and only if $\sqrt{-1} \in K$ and this happens only in the case $D = 1$. If $D \not\equiv 7 \pmod 8$, then D is a sum of three squares and so $0 = (\sqrt{-D})^2 + a^2 + b^2 + c^2$ so $s(K) \leq 3$. But now $-1 = \alpha^2 + \beta^2 + \gamma^2$ implies

$$-1 = \left(\frac{\alpha\gamma + \beta}{\alpha^2 + \beta^2}\right)^2 + \left(\frac{\beta\gamma - \alpha}{\alpha^2 + \beta^2}\right)^2 \tag{3.6}$$

so $s(K) = 2$ since $D \neq 1$. Note also that $\alpha^2 + \beta^2 \neq 0$, since otherwise $s(K) = 1$. Finally let $D \equiv 7 \pmod 8$. If $s(K)$ were less than 4, then it would be equal to 2, i.e.

$$-1 = (a_1 + b_1\sqrt{-D})^2 + (a_2 + b_2\sqrt{-D})^2, \quad a_1, b_1, a_2, b_2 \in \mathbb{Q}.$$

Here without loss of generality, we may suppose that $b_1 \neq 0$. Equating reals and imaginaries we get the following two equations:

$$a_1^2 + a_2^2 - D(b_1^2 + b_2^2) = -1$$

$$a_1 b_1 + a_2 b_2 = 0.$$

These imply $D = \left(\frac{a_2}{b_1}\right)^2 + \left(\frac{b_1}{b_1^2+b_2^2}\right)^2 + \left(\frac{b_2}{b_1^2+b_2^2}\right)^2$. Thus D is a sum of three rational squares, which is a contradiction since $D \equiv 7 \pmod 8$. Thus $s(K) \not< 4$. This completes the proof. □

In the proof we have used the 3-square theorem, viz. if $D \not\equiv 7(8)$, then D is a sum of three squares in $\mathbb{Z}$. Actually Theorem 3.2 and the 3-square theorem are equivalent. Indeed we have the following (see [S8]).

Theorem 3.3. *Let $m > 1$ be a square-free integer. Then m is a sum of three integral squares (or equivalently rational squares by the Davenport-Cassels Lemma [S2]) if and only if $s(\mathbb{Q}(\sqrt{-m}) = 2$.*

Proof. First let $m = a^2 + b^2 + c^2$ in $\mathbb{Z}$, and so in $\mathbb{Q}$ and hence in $\mathbb{Q}(\sqrt{-m})$. Then $a^2 + b^2 + c^2 + (\sqrt{-m})^2 = 0$ in $\mathbb{Q}(\sqrt{-m})$ i.e. -1 is a sum of three squares in $\mathbb{Q}(\sqrt{-m})$ and so by (3.6), $s(\mathbb{Q}(\sqrt{-m})) = 2$ (since $s = 1$ gives $m = 1$ and the whole thing is then trivial).

Conversely let -1 be a sum of two squares in $\mathbb{Q}(\sqrt{-m})$. Then 0 is a sum of three squares and so a sum of four squares (trivially by adding a 0^2) in $Q(\sqrt{-m})$:

$$0 = \sum_{j=1}^{4} (a_j + b_j\sqrt{-m})^2 \quad (a_j, b_j \in Q)$$

Then $\sum a_j^2 - m \sum b_j^2 = 0$ and $\sum a_j b_j = 0$; hence

$$\begin{aligned} m(\sum b_j^2)^2 &= m(\sum b_j^2)(\sum b_j^2) = (\sum a_j^2)(\sum b_j^2) \\ &= (\sum a_j b_j)^2 + \text{ the rest of the Euler identity} \\ &= 0 + \text{ a sum of three squares.} \end{aligned}$$

Here $\sum b_j^2 \neq 0$, otherwise $b_j = 0$ for all j; giving $\sum a_j^2 = 0$ i.e. $a_j = 0$ for all j. Thus m is a sum of three squares in Q and so in Z by Theorem 3.1 (the Aubry Lemma).

This completes the proof (see also [R6] in this context).

Our next result deals with finite fields.

Theorem 3.4. *Let F_q be the finite field of $q = p^\alpha$ elements then*

$$s(\mathsf{F}_q) = \begin{cases} 1 & \text{if either } p = 2 \text{ or } p \equiv 1(4), \text{ or } p \equiv 3(4), 2|\alpha \\ 2 & \text{otherwise i.e. if } p \equiv 3(4), 2\not|\alpha \end{cases}.$$

Proof. First let $p = 2$. Then $-1 = 1 = 1^2$ giving $s(\mathsf{F}_{2^\alpha}) = 1$. Next if $p \equiv 1(4)$ then $(-1/p) = 1$ i.e. $1 = x^2$ is solvable in $\mathsf{F}_p \subset \mathsf{F}_{p^\alpha}$ (for all α). So $s(\mathsf{F}_{p^\alpha}) = 1$.

Let now $p \equiv 3(4)$. Let $A = \{-1 - X^2 | X = 1, 2, \ldots, \frac{p-1}{2}, 0\}$ and $B = \{Y^2 | Y = 1, 2, \ldots, \frac{p-1}{2}, 0\}$, both in F_p.

Then $|A| = |B| = (p+1)/2$. Hence by the pigeon-hole principle there exist $X_0, Y_0 \in \mathsf{F}_p$ such that $-1 - X_0^2 = Y_0^2$ i.e. $-1 = X_0^2 + Y_0^2$ in F_p; but -1 is not a square in F_p since $p \equiv 3(4)$. It follows that $s(\mathsf{F}_p) = 2$ if $p \equiv 3(4)$.

Now $\mathsf{F}_{p^2} = \mathsf{F}_p(\sqrt{-1})$ and here $-1 = (\sqrt{-1})^2$, so $s(\mathsf{F}_{p^2}) = 1$. But $\mathsf{F}_{p^\alpha} \supset \mathsf{F}_{p^2}$ if $2|\alpha$ so F_{p^α} has Stufe 1 if $2|\alpha$. If $s(\mathsf{F}_{p^\alpha}) = 1$ even for $2\not|\alpha$ then $-1 = X^2$ is solvable in F_{p^α} $(2\not|\alpha)$, so $\mathsf{F}_{p^\alpha} \supset \mathsf{F}_p(X) = \mathsf{F}_p(\sqrt{-1}) \cong \mathsf{F}_{p^2}$ which is false since $2\not|\alpha$. Hence $s(\mathsf{F}_{p^\alpha}) = 2$ if $2\not|\alpha$.

This completes the proof. □

Our next aim is the cyclotomic fields. Actually it is well known how to compute the Stufe $s(K)$ of any algebraic number field K by using the Hasse-Minkowski theorem, which in the first place tells us that $s(K)$, whenever finite (i.e. whenever K is totally complex), is at most 4 (a special case of Siegel's theorem). Indeed we have the following result:

Let K be a totally complex algebraic number field. Then $s(K) \leq 4$. Further $s(K) = 4$ if and only if for all primes $\mathfrak{p}|2$, the local degree $[K_\mathfrak{p} : \mathbb{Q}_2]$ is odd.

As an immediate corollary we can deduce the following two results:

(1) *Let $d > 0$ be a rational and let $K = \mathbb{Q}(\sqrt{d})$, then $s(K) = 4$ iff $d \neq$ a sum of three squares in $\mathbb{Q}$.*

(2) *Let $K = \mathbb{Q}(\zeta), \zeta$ a primitive p^{th} root of 1, then $s(K) = 4$ iff the order of $2 \bmod p$ is odd.*

These results use the powerful Hasse-Minkowski theorem. We are interested in elementary proofs of (1) and (2) (we have already proved (1)) allowing the use of Galois theory, the four square theorem in $\mathbb{Q}$, but not beyond $\mathbb{Q}$, etc. Parts of the complete theorem can be done very simply and elegantly and we shall not miss giving any such elegant proofs in their proper place. Our aim is the following.

Theorem 3.5 (Shapiro and Leep). *Let p be an odd prime, ζ a primitive p^{th} root of unity and $K = \mathbb{Q}(\zeta)$. Let d be the order of $2 \bmod p$ i.e. the least positive integer such that $2^d \equiv 1 \pmod{p}$. Then $s(K) = 4$ iff d is odd (i.e. $s(K) = 2$ iff d is even)*

First of all it is not obvious that -1 is a sum of four squares in $\mathbb{Q}(\zeta) = K$ (i.e. it is not obvious without using the Hasse-Minkowski theorem) and this is the first thing that we must prove.

For $p \equiv 3(4)$, this can be proved very elegantly as follows. Let τ be the Gauss sum so that $\tau^2 = (-1/p) \cdot p = -p$ since $p \equiv 3(4)$. If $p \equiv 3(8)$ then $p = a_1^2 + a_2^2 + a_3^2$ i.e. $a_1^2 + a_2^2 + a_3^2 + \tau^2 = 0$ or -1 is a sum of three squares in K since $\tau \in K$ and so a sum of two squares in K, indeed $a_1^2 + a_2^2 + a_3^2 + (\sqrt{-p})^2 = 0$ in $\mathbb{Q}(\sqrt{-p}) \subseteq K$ so -1 is a sum of three squares and so a sum of two squares (see Theorem 2.2) in $\mathbb{Q}(\sqrt{-p}) \subseteq K$. It follows that $s(K) \leq 2$. If $p \equiv 7(8)$, then $p = a_1^2 + a_2^2 + a_3^2 + a_4^2$ and we only get $s(K) \leq 4$.

The method fails if $p \equiv 1(4)$. In this case we proceed as follows (the proof also works for $p \equiv 3(4)$ again).

We know that $K/\mathbb{Q}$ is cyclic of degree $p-1$. Write $[K : \mathbb{Q}] = p-1 = 2^\gamma \cdot t$ where t is odd. Then by Galois theory we have a tower of fields

$$\mathbb{Q} = E_0 \subseteq E_1 \subseteq \ldots \subseteq E_\gamma \subseteq K$$

where $[E_{i+1} = E_i] = 2, [K; E_\gamma] = t$, all the E_i are normal over $\mathbb{Q}$ and $E_{i+1} = E_i(\sqrt{d_i}),\ \ d_i \in E_i^*$. View K as a subfield of $\mathbb{C}$ and let σ be the restriction of the complex conjugate to K. Clearly $K \not\subseteq \mathbb{R}$ since $-1 = \zeta + \zeta^2 + \ldots + \zeta^{p-1}$ is a sum of squares in K. Also σ has order 2. Further $F = \text{Fix}\ \sigma \subseteq \mathbb{R}$.

Now $\text{Gal}(K/E_{\gamma-1})$, a subset of $\text{Gal}(K/\mathbb{Q})$, has order $2t$ and so it has a subgroup of order 2. Indeed being a subgroup of $\mathrm{Gal}(K/\mathbf{Q})$, which is cyclic, this subgroup of order 2 is unique viz. $\langle\sigma\rangle$. Thus $\sigma \in \mathrm{Gal}(K/E_{\gamma-1})$ and so σ fixes $E_{\gamma-1}$ i.e. $E_{\gamma-1} \subset \mathbf{R}$.

We now claim that for $1 \le i < \gamma - 1, d_i$ is totally positive in E_i while $d_{\gamma-1}$ is totally negative in $E_{\gamma-1}$. First let $i < \gamma - 1$. Then $\sqrt{d_i} \in E_{i+1} \subset \mathbf{R}$, so $d_i > 0$ in the induced ordering. But each field involved is normal over $\mathbf{Q}$, so the same claim holds for every conjugate of $\sqrt{d_i}$ and since $E_{i+1} \subset \mathbf{R}$, it follows that every conjugate of d_i is positive i.e. d_i is positive in every ordering of E_i i.e. totally positive in E_i.

Now K is not formally real since -1 is a sum of squares in $K : -1 = \zeta + \ldots + \zeta^{p-1}$, i.e. $x_1^2 + \ldots + x_p^2$ represents 0 non-trivially in K. Hence by the Springer theorem (see later in this very chapter), it represents 0 non-trivially in E_γ as well because $[K : E_\gamma] = t$ is odd. Hence E_γ is not formally real, so there exists no embedding of E_γ in $\mathbf{R}$, i.e. no ordering of $E_{\gamma-1}$ can extend to $E_\gamma = E_{\gamma-1}(\sqrt{d_{\gamma-1}})$ i.e. $d_{\gamma-1}$ is negative in each ordering i.e. is totally negative.

Lemma 1. *Let $E = F(\sqrt{d})$ be a quadratic extension of fields where d is totally positive. Suppose F has the property that any sum of squares in F is already a sum of at most 4 squares in F. Then E also has this property.*

Proof. (Landau (1919); see [L7]). Let $\alpha \in E$ be a sum of squares in E (i.e. is totally positive in E). We have to show that α is a sum of 4 squares in E.

Write $\alpha = a + b\sqrt{d}$ $(a, b \in F)$. Since α is totally positive, so $\alpha, \overline{\alpha}$ are both positive in every ordering of E and every ordering of F extends to E.

Then $a^2 - db^2 = \alpha\overline{\alpha} = N\alpha$ is positive in each order of E (since $\alpha, \overline{\alpha}$ are) i.e. is totally positive. Hence by hypothesis $N\alpha = x_1^2 + x_2^2 + x_3^2 + x_4^2$ in F so $a^2 = db^2 + \sum x_j^2$ in F. Now for any order $<$ of F, since $d > 0$, we have $a^2 > \sum x_j^2 > x_1^2$ i.e. $a > x_1$. Also $\alpha + \overline{\alpha} = 2a$, so $2\alpha a = \alpha(2a) = \alpha(\alpha + \overline{\alpha}) = \alpha^2 + \alpha\overline{\alpha}$. Hence

$$\begin{aligned}(\alpha - x_1)^2 + x_2^2 + x_3^2 + x_4^2 &= \alpha^2 - 2\alpha x_1 + \sum_1^4 x_j^2 \\ &= \alpha^2 - 2\alpha x_1 + N\alpha \\ &= 2\alpha a - 2\alpha x_1 \\ &= 2\alpha(a - x_1).\end{aligned}$$

Since $a - x_1$ is totally positive it follows that

$$\begin{aligned}\alpha = \frac{(\alpha - x_1)^2 + x_2^2 + x_3^2 + x_4^2}{2(a - x_1)} &= \frac{\sum \text{four squares}}{\sum \text{four squares}} \\ &= \sum \text{four squares in } E. \qquad \square\end{aligned}$$

Now apply this lemma successively to the fields of our tower. By Lagrange's theorem, $\mathbf{Q}$ satisfies the property of the lemma so we see that $E_{\gamma-1}$ satisfies the property too. i.e. $-d_{\gamma-1} = x_1^2 + x_2^2 + x_3^2 + x_4^2$ in $E_{\gamma-1}$ (since $-d_{\gamma-1}$ is totally positive) i.e. $-1 = \sum_1^4 (x_j/\sqrt{d_{\gamma-1}})^2$ in $E_\gamma \subseteq K$; so $s(K) \leq 4$ as required.

So now we have proved that $s(K) \leq 4$ for the cyclotomic field $K = \mathbf{Q}(\zeta_p)$.

Lemma 2 (Chawla - 1969). *If the order d of 2 (mod p) is even, then $s(K) \leq 2$.*

Proof. Let $d = 2m$, then $2^{2m} \equiv 1(p)$ but $2^m \not\equiv 1(p)$. Since $(2^m)^2 \equiv 2^{2m} \equiv 1$ so $2^m \equiv -1 \pmod p$ i.e. $p | 2^m + 1$ and so $\zeta^{2^m+1} = 1$ or

$$\zeta^{2^m} = \zeta^{-1} \tag{3.7}$$

Now recall the identity

$$\begin{aligned}(1+x)(1+x^2)(1+x^4)\dots(1+x^{2^{m-1}}) &= 1 + x + x^2 + \dots + x^{2^m-1} \\ &= (x^{2^m} - 1)/(x-1).\end{aligned}$$

Putting $x = \zeta^2$, this gives

$$\begin{aligned}(1+\zeta^2)(1+\zeta^4)\dots(1+\zeta^{2^m}) &= ((\zeta^2)^{2^m} - 1)/(\zeta^2 - 1) \\ &= (\zeta^{-2} - 1)/(\zeta^2 - 1), \text{by (3.7)} \\ &= -\zeta^{-2}.\end{aligned}$$

Hence $-1 = \zeta^2(1+\zeta^2)(1+\zeta^4)\dots(1+\zeta^{2^m})$ is a sum of two squares in K by a repeated application of the 2-square identity. □

Note that if p is odd, then $s(K) \neq 1$; for otherwise $\sqrt{-1} \in \mathbf{Q}(\zeta_p)$, but $\mathbf{Q}(\sqrt{-1}, \zeta_p) = \mathbf{Q}(\zeta_{4p})$ so this cannot be. Alternatively we know that the unique quadratic field contained in $\mathbf{Q}(\zeta_p)$ is $\mathbf{Q}(\sqrt{\pm p})$ by the Gauss sum.

Thus, so far we have proved that if $K = \mathbf{Q}(\zeta_p), p$ an odd prime and $d =$ the order of 2 mod p, then $s(K) = 2$ if d is even while $s(K) = 2$ or 4 if d is odd.

For the remaining part of the theorem, we shall use the following (well known) result.

Lemma 3. *Let $\varphi_p(X) = X^{p-1} + \dots + X + 1$ be the cyclotomic polynomial-irreducible $/\mathbf{Q}$ and let d be the order of 2 (mod p). Then $\overline{\varphi}_p(X)$ (i.e. $\varphi_p(X)$ reduced mod 2 i.e. in $\mathsf{F}_2(X)$) factors as a product of $(p-1/d$ distinct irreducible polynomials, each of degree d. Indeed, more generally if $q \nmid n$, q prime, then $\overline{\varphi}_n(X)$ i.e. $\varphi_n(X)$ reduced mod q i.e. in $\mathsf{F}_q(X)$) factors as a product of $\varphi(n)/d$ distinct irreducible factors each of degree d, where d is the order of q (mod n).*

We can now prove the following

Lemma 4. *If the order d of 2 (mod p) is odd then $x^2 + y^2 + z^2$ does not represent 0 nontrivially in $K = \mathbf{Q}(\zeta)$ i.e. $s(K) > 2$.*

Proof. Suppose to the contrary. Then there exist $c_1, c_2, c_3 \in \mathbf{Q}(\zeta_p)$, not all 0, such that $c_1^2 + c_2^2 + c_3^2 = 0$. Clearing denominators, we may suppose that the $c_j \in \mathbf{Z}[\zeta_p]$. Choose $f_j(X) \in \mathbf{Z}[X]$ of degree at most $p-2$ with $f_j(\zeta) = c_j$ $(j = 1,2,3)$. Then, in $\mathbf{Z}[X]$

$$f_1^2(X) + f_2^2(X) + f_3^2(X) \equiv 0(\mathrm{mod}\ \varphi_p(X)). \tag{3.8}$$

Further, without loss of generality, f_1, f_2, f_3 are relatively prime in $\mathbf{Z}[X]$, for any common factor can be factored and omitted from (3.8) since $\varphi_p(X)$ is irreducible.

For some $h(X) \in \mathbf{Z}[X]$, write (3.8) as

$$f_1^2(X) + f_2^2(X) + f_3^2(X) = \varphi_p(X) \cdot h(X). \tag{3.9}$$

Comparing degrees, we get $p - 1 +$ deg $h \leq 2(p-2)$, i.e.

$$\deg h(X) \leq p - 3.$$

Now reduce (3.9) mod $2\mathbf{Z}[X]$ to get, in $\mathsf{F}_2[X]$,

$$(\overline{f}_1(X) + \overline{f}_2(X) + \overline{f}_3(X))^2 = \overline{\varphi}_p(X) \cdot \overline{h}(X)$$

(the cross terms on the left are all zero in F_2!). By Lemma 3, factor $\overline{\varphi}_p(X) = g_1(X)g_2(X)\dots g_r(X)$ into distinct irreducible factors in $\mathsf{F}_2[X]$ (all square-free).

If $\overline{f}_1(X) + \overline{f}_2(X) + \overline{f}_3(X) \neq 0$, then $\overline{\varphi}_p(X) \cdot \overline{h}(X)$ is a square and so $\overline{h}(X) = \overline{\varphi}_p(X) \cdot k^2(X)$ for some $k(X) \in \mathsf{F}_2[X]$. But then deg $h \geq$ deg $\varphi_p = p - 1$, which is a contradiction. Hence $\overline{f}_1 + \overline{f}_2 + \overline{f}_3 = 0$ and so $\overline{h} = 0$ in $\mathsf{F}_2(X)$ i.e.

$$\overline{f}_3 = -\overline{f}_1 - \overline{f}_2 = \overline{f}_1 + \overline{f}_2$$

(since $-1 = 1$) and $\overline{h} = 0$. In other words,

$$f_3 = f_1 + f_2 + 2g, h = 2h^*$$

for some $g(X), h^*(X) \in \mathbf{Z}[X]$. Plugging this back in (3.9), we get

$$f_1^2 + f_2^2 + f_3^4 + 2f_1f_2 + f_2^2 + 4g(f_1 + f_2) + 4g^2 = 2\varphi_p h^*,$$

i.e.

$$f_1^2 + f_1f_2 + f_2^2 + 2(f_1 + f_2 + g)g = \varphi_p h^* \text{ in } \mathbf{Z}[X].$$

Now again reduce mod 2 to get

$$\begin{aligned} \overline{f}_1^2 + \overline{f}_1\overline{f}_2 + \overline{f}_2^2 &= \overline{\varphi}_p\overline{h}^* \\ &= g_1g_2\dots g_r\overline{h}^* \text{ in } \mathsf{F}_2[X]. \end{aligned}$$

Let ϑ be a root of g_j. Since $\deg g_j = d$ so the field extension $L = \mathsf{F}_2(\vartheta)$ is of degree d over F_2 i.e. $L = \mathsf{F}_{2^d}$, the finite field of 2^d elements.

Let $a_1 = \overline{f}_1(\vartheta), a_2 = \overline{f}_2(\vartheta)$ ($\overline{f}_1(X), \overline{f}_2(X)$ evaluated at $X = \vartheta$ in L). Then $a_1^2+a_2^2+a_1a_2 = 0$ in L. If $a_1a_2 \neq 0$ then this gives $\left(\frac{a_1}{a_2}\right)^2+\left(\frac{a_1}{a_2}\right)+1 = 0$ in L i.e. $\frac{a_1}{a_2}$ is a roof of $X^2+X+1 = 0$ in L and this polynomial is irreducible $/\mathsf{F}_2$.

We thus get a tower of fields as shown. It follows that $2|d$, which is a contradiction, since d was supposed odd. Hence $a_1 = a_2 = 0$ ie. for every ϑ, a root of g_j (any j), we have $\overline{f}_1(\vartheta) = 0 = \overline{f}_2(\vartheta)$.

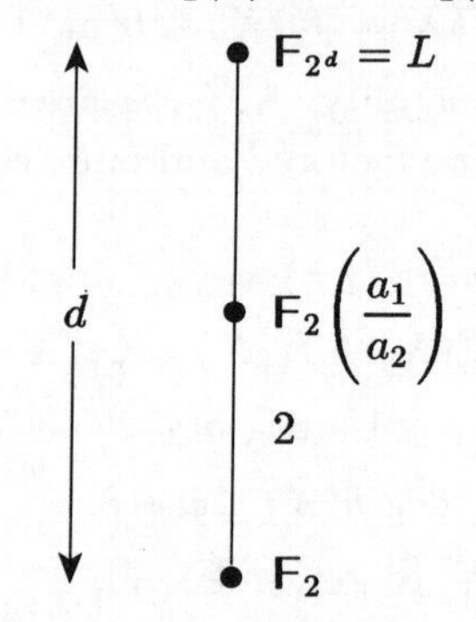

that is $g_j|\overline{f}_1, \overline{f}_2$ for all $j = 1, 2, \ldots, r$. By the uniqueness of factorization, we get $\overline{\varphi}_p|\overline{f}_1, \overline{f}_2$, but then $\deg f_1$ and $\deg f_2 \geq p-1$ which is again a contradiction. This finally proves everything. □

Remarks. (1) For $p \equiv 7(8)$, P. and S. Chawla [C6] determine the Stufe of $\mathbb{Q}(e^{2\pi i/p})$ by another method as follows.

That $s(K) \leq 4$ follows as before using the Gauss sums. Suppose to the contrary that $s(K) = 2$. The elements of K are polynomials in $\zeta = e^{2\pi i/p}$ of degree at most $p-1$ with rational coefficients. So let

$$-1 = f^2(\zeta) + g^2(\zeta) \tag{3.10}$$

Let $\sigma_r : \zeta \to \zeta^r$ $(r = 1, 2, \ldots, p-1)$ be the automorphisms of $K/\mathbb{Q}$. Apply σ_r to (3.10):

$$\begin{aligned} -1 &= f^2(\zeta^r) + g^2(\zeta^r) \\ &= a_r^2 + b_r^2, \end{aligned} \tag{3.11}$$

say. Here $a_r \neq 0$, otherwise $s(K) = 1$, which is false.

Let $R_p = \{r_1, r_2, \ldots, r_{(p-1)/2}\}$ be the set of all quadratic residues mod p and let

$$p_s = \sum_{r_1<\ldots<r_s} \frac{b_{r_1}}{a_{r_1}} \cdot \frac{b_{r_2}}{a_{r_2}} \cdots \frac{b_{r_s}}{a_{r_s}}$$

be the sth elementary symmetric function in b_r/a_r. Then

$$\begin{aligned}\prod_{r\in R_p}(a_r+ib_r)&=\prod_{r\in R_p}a_r\cdot\prod_{r\in R_p}(1+b_r/a_r)\\&=\prod_{r\in R_p}a_r((1-p_2+p_4-\ldots)+i(p_1-p_3+p_5-\ldots))\\&=A(\zeta)+B(\zeta),\end{aligned}\tag{3.12}$$

say. The second line is easy to see by considering, for example, the case $p=7$. Now for $r\in R_p$, the automorphisms σ_r keep $A(\zeta)$ and $B(\zeta)$ fixed; for as r runs through R_p so does r_1r for any $r_1\in R_p$ hence σ_r keeps the p_j and Πa_r fixed thus A,B fixed. Thus

$$A(\zeta^r)=A(\zeta),B(\zeta^r)=B(\zeta)\text{ for all }r\in R_p.$$

Now these σ_r form a subgroup Γ of order $(p-1)/2$ of $\mathrm{Gal}(K/\mathbb{Q})$ and so its fixed field L is a quadratic extension of $\mathbb{Q}$. This extension is $\mathbb{Q}(\sqrt{-p})$, for $\sqrt{-p}=\sum\zeta^r$ (a Gauss sum) is kept fixed by Γ and so this fixed field $L\supset Q(\sqrt{-p})$. But L is a quadratic extension of $\mathbb{Q}$ and so $L=\mathbb{Q}(\sqrt{-p})$.

Now since A,B are fixed under Γ, so $A,B\in\mathbb{Q}(\sqrt{-p})$; say $A=a+b\sqrt{-p},B=c+d\sqrt{-p},a,b,c,d\in\mathbb{Q}$. Now multiply the various equations (3.11) for all $r\in R_p$:

$$\begin{aligned}(-1)^{(p-1)/2}&=\prod_{r\in R_p}(a_r^2+b_r^2)=\prod_{r\in R_p}(a_r+ib_\gamma)\prod_{r\in R_p}(a_r-ib_r)\\&=(A+iB)(A-iB)\quad\text{(by (3.12))}\\&=A^2+B^2\\&=(a+b\sqrt{-p})^2+(c+d\sqrt{-p})^2.\end{aligned}$$

Equating reals and imaginaries, we get

$$-1=a^2+c^2-p(b^2+d^2)\text{ and}$$

$$ab+cd=0.$$

Hence $p=c^2/b^2+(b/(b^2+d^2))^2+(d/(b^2+d^2))^2$; i.e. p is a sum of three squares in $\mathbb{Q}$ which is a contradiction since $p\equiv 7(8)$. This completes the proof. □

(2) (S. Chawla [C5]). *Let* $K=\mathbb{Q}(e^{2\pi i/n})$, *where* $n\equiv 3(8)$ *is a positive integer, then* $s(K)=2$.

Proof. First note that if $n\equiv 3(8)$, then $\sqrt{-n}\in K$. For, writing $\zeta=e^{2\pi i/n}$, then the Gauss sum

$$\sum_{s=0}^{n-1}\zeta^{s^2}=i\sqrt{n}=\sqrt{-n},$$

since $n\equiv 3(8)$. Now write $n=t^2m$, where m is square-free. Then $m\equiv 3(8)$.

Further $\mathbb{Q}(\sqrt{-m}) = \mathbb{Q}(\sqrt{-n}) = K$, and by Theorem 3.2, $s(\mathbb{Q}(\sqrt{-m})) = 2$. It follows that $s(K) = 2$, since it is not 1. □

(3) *Indeed if n is any integer $n \geq 3$ and if there is an $m \equiv 3(8), m|n$, then $s(\mathbb{Q}(e^{2\pi i/n})) \leq 2$ (note that in the last, strict inequality can occur for example when $n = 12, m = 3$).*

For $\mathbb{Q}(e^{2\pi i/n}) \supseteq \mathbb{Q}(e^{2\pi i/m})$ and the Stufe of $\mathbb{Q}(e^{2\pi i/m}) = 2$.

For a general cyclotomic field $K^{(m)} = \mathbb{Q}(e^{2\pi i/m})$ if $2\|m$ then $K^{(m)} = K^{(m/2)}$ where $m/2$ is odd. If $4|m$ then $\sqrt{-1} \in K^{(m)}$ and so the Stufe of $K^{(m)} = 1$. We may therefore suppose m to be odd and at least 3.

Let p be any prime dividing m. Then $(e^{2\pi i/m})^{m/p} \in K^{(m)}$ i.e. $e^{2\pi i/p} \in K^{(m)}$ so $\mathbb{Q}(e^{2\pi i/p}) \leq K^{(m)}$. Since the Stufe of $\mathbb{Q}(e^{2\pi i/p}) \leq 4$ (Theorem 3.5), we see that the Stufe of $K^{(m)} \leq 4$ and so equals 2 or 4, since $i \notin K^{(m)}, m$ being supposed odd. The complete classification giving conditions under which the Stufe is 2 or 4 uses the Hasse-Minkowski theorem and we postpone its proof to Chapter 18.

Algebraic number fields

Let K be an algebraic number field i.e. a finite extension of the rationals $\mathbb{Q}$. Then $K = \mathbb{Q}(\alpha)$ and if $f(X) = \text{irr}(\alpha, \mathbb{Q})$ is of degree n, then $[K : \mathbb{Q}] = n$ and each element of K is a poynomial in α of degree at most $n-1$ with coefficients in $\mathbb{Q}$.

Suppose $f(X)$ has r real roots and s pairs of complex conjugate roots. If -1 is a sum of squares in K, say

$$-1 = \varphi_1^2(\alpha) + \varphi_2^2(\alpha) + \dots \tag{3.13}$$

and α' is a real root of $f(X)$, then applying the isomorphism $\alpha \to \alpha'$ to (3.13) gives

$$-1 = \varphi_1^2(\alpha') + \varphi_2^2(\alpha') + \dots > 0,$$

since α' is real, - a contradiction. Thus if -1 is to be a sum of squares in K, then all the roots of $f(X) = 0$ have to be non-real – in particular n is even. Such a K is called totally complex.

The converse is also true viz. that if all the roots of $f(X)$ are non-real (i.e. K is totally complex) then -1 is a sum of squares in K. Indeed -1 is a sum of at most four squares in K. This result again uses the Hasse-Minkowski theorem and we postpone its proof to Chapter 18. However, elementary proofs giving the result $s(K) \leq 4$, even for a given class of totally complex fields will always be desirable. The next class on the list would be the class of all totally complex quartic fields.

Our definitions tell us that a field K is totally complex iff K is not formally real. So if -1 is not a sum of squares in K, it does not mean necessarily that $K \subseteq \mathbf{R}$. For example let $K = \mathbf{Q}(\sqrt[3]{2} \cdot \omega)$ where $\omega = (-1 + \sqrt{-3})/2$. Then $K \not\subseteq \mathbf{R}$ but still -1 is not a sum of squares in K since $\mathrm{irr}((\sqrt[3]{2} \cdot \omega), Q) = X^3 - 2$ and this has one real zero viz. $\sqrt[3]{2}$.

We now look at a few very useful results applicable to general algebraic number fields - these will include Springer's theorem on odd degree extension fields. This we have already used during the proof of Theorem 3.5 and so a proof is no doubt due.

We start with the following

Theorem 3.6 (Pfister). *Let $L = K(\alpha)$ be an algebraic extension of fields and let $p(X) = \mathrm{irr}(\alpha, K)$. Suppose L is not formally real (i.e. is totally complex); then*

-1 is a sum of 2^{n-1} squares in L if and only if
$p(X)$ is a sum of 2^n squares in $K(X)$.

Proof. First let $p(X) = f_1^2(X) + \ldots + f_{2^n}^2(X)$, where by Cassels' lemma, we may suppose that $f_j(X) \in K[X]$. Put $X = \alpha$ to get $0 = f_1^2(\alpha) + \ldots + f_{2^n}^2(\alpha)$. Since $f_1(\alpha) \in L$, we see that $s(L) \leq 2^n - 1$ and being a power of 2, $s(L) \leq 2^{n-1}$ as required. To prove the converse proceed as follows. Let $l = \deg p(X) = [L : K] \geq 2$. By hypothesis

$$-1 = h_1^2(X) + \ldots + h_{2^{n-1}}^2(X)$$

where again by Cassels' lemma, $h_j(X) \in K[X]$, and $\deg\ h_j(X) \leq l - 1$. Then α is root of $1 + h_1^2(X) + \ldots + h_{2^{n-1}}^2(X)$ and so

$$1 + h_1^2(X) + \ldots + h_{2^{n-1}}^2(X) = p(X) \cdot q(X) \tag{3.14}$$

for some polynomial $q(X) \in K[X]$. We have then

$$\deg(p(X)) + \deg(q(X)) \leq 2 \max(\deg h_i(X)),$$

i.e.

$$\deg(q(X)) \leq 2(l-1) - l = l - 2.$$

We prove our result by induction on the degree l of $p(X)$. First for $l = 2, q(X)$ is a constant, say q_0. Then by (3.14)

$$1 + h_1^2(X) + \ldots + h_{2^{n-1}}^2(X) = q_0(X^l + \ldots,$$

say. Now equating the coefficients of the highest power of X gives q_0 to be a sum of at most 2^{n-1} squares in K, i.e. $q_0 \in G_{2^{n-1}}(K) \subset G_{2^n}(K(X))$. So again by (3.14) $p(X) \in G_{2^n}(K(X))$ as required, since $2^{n-1} + 1 \leq 2^n$.

Now let β be a root of $q(X)$ and let $q^*(X) = \mathrm{irr}\ (\beta, K) \in K[X]$, so that $q^*(X) | q(X)$ and

$$\deg\ q^* \leq\ \deg\ q \leq l - 2 < l.$$

Also $1 + h_1^2(\beta) + \ldots + h_{2^n-1}^2(\beta) = p(\beta)q(\beta) = 0$ i.e. -1 is a sum of 2^{n-1} squares in $K(\beta)$ and so $K(\beta)$ is totally complex and thus by the induction hypothesis $q^*(X)$ is a sum of 2^n squares in $K(X)$. This holds for all irreducible factors of $q(X)$ and so by the group property of $G_{2^n}(K(X)), q(X)$ itself is a sum of 2^n squares in $K(X)$. By (3.14), $p(X)$ is also. This completes the proof. □

Another promised elementary result is the

Theorem 3.7 (Springer). *Let $[K : F] = n$ be an extension of number fields, where n is odd. Then*

$$s(F) = s(K).$$

Proof. It is enough to prove that if

$$X_1^2 + \ldots + X_s^2 = 0 \tag{3.15}$$

is not solvable in F, then it is not solvable in K (i.e. that $s(F) \leq s(K)$ for, the reverse inequality is trivial). We use induction on n. For $n = 1$ the result is obvious. So let $n \geq 3$. Let $p(t) = \text{irr}\ (\alpha, F)$ where $K = F(\alpha)$. Suppose to the contrary that (3.15) is solvable in K. Let $X_j = g_j(t) \in F[t]$ be a solution of (3.15) where $\deg g_j \leq n - 1$, (don't forget elements of K are polynomials $f(\alpha) \in F[\alpha]$).

Then $0 = g_1^2(\alpha) + \ldots + g_s^2(\alpha)$ i.e $g_1^2(t) + \ldots + g_s^2(t)$ has $t = \alpha$ as a root and so $p(t)$ divides this polynomial, say

$$g_1^2(t) + \ldots + g_s^2(t) = p(t) \cdot h(t). \tag{3.16}$$

If an irreducible polynomial $f(t)$ divides all $g_j(t)$ say $g_j = fg'_j$, then $f^2(g_1'^2 + \ldots + g_s'^2) = ph$ where $\deg f \leq n - 1 < n$. Hence by the uniqueness of factorization $h = f^2 \cdot h'$ (since $f \not| p$). Cancelling f^2 we get $g_1'^2 + \ldots + g_s'^2 = ph'$. Thus, without loss of generality, $(g_1, \ldots, g_s) = 1$ i.e. there exist polynomials $h_1, \ldots, h_s$ such that

$$g_1(t) \cdot h_1(t) + \ldots + g_s(t) \cdot h_s(t) = 1.$$

It follows that the g_j cannot have a common root α in the algebraic closure C of K, for putting $t = \alpha$, this would then give $0 = 1$. Further $h(t) \neq 0$, otherwise $g_1^2(t) + \ldots + g_s^2(t) = 0$ which is false because equating the highest coefficient of t to 0 would give $a_1^2 + \ldots + a_r^2 = 0$ $(r \leq s)$, $a_j \in F$, which certainly is false by hypothesis. Now

$$\begin{aligned} \deg h &= 2 \max \deg g_j - \deg p \quad \text{(which is odd in the first place)} \\ &\leq 2(n-1) - n = n - 2. \end{aligned}$$

Let β be a root of $h(t)$ in $\mathbb{C}$. Put $t = \beta$ in (3.16) to get

$$g_1^2(\beta) + \ldots + g_s^2(\beta) = 0$$

i.e. $X_1^2 + \ldots + X_s^2$ represents 0 in $F(\beta)$. But $[F(\beta) : F] | \deg h(t)$ (which is odd) so $[F(\beta) : F]$ is odd and at most $\deg h \leq n - 2$. This contradicts our induction hypothesis. □

When the degree $[L : K]$ is even, $s(L)$ and $s(K)$ need not be equal (both supposed finite). However, we have the following.

Theorem 3.8. *Let K be formally real and let $d \in K^*$ be such that $d \in G_n(K)$ but $d \notin G_{n-1}(K)$. Suppose 2^k is the largest power of $2 \leq n$ (i.e. $2^k \leq n < 2^{k+1}$). Then $s(K(\sqrt{-d})) = 2^k$.*

Proof. Let $s = s(K(\sqrt{-d}))$, say $-1 = \sum_{j=1}^{s} \alpha_j^2$ $(\alpha_j \in K(\sqrt{-d}))$. Write $\alpha_j = b_j + \sqrt{-d} \cdot c_j$ $(b_j, c_j \in K)$ so that

$$-1 = \sum_{j=1}^{s} (b_j + c_j\sqrt{-d})^2.$$

Then

$$\sum_{j=1}^{s} b_j c_j = 0 \quad , \quad -1 = \sum_{j=1}^{s} b_j^2 - d \sum_{j=1}^{s} c_j^2.$$

Here $\sum c_j^2 \neq 0$, otherwise $-1 = \sum b_j^2$ $(b_j \in K)$, contradicting the formal reality of K.

It follows, from Theorem 2.3 (don't forget s is a Stufe, so a power of 2) that

$$\begin{aligned} d(\textstyle\sum c_j^2)^2 &= (\textstyle\sum c_j^2) + (\textstyle\sum c_j^2)(\textstyle\sum b_j^2) \\ &= \sum_{j=1}^{s} c_j^2 + \sum_{j=1}^{s} b_j c_j + w_2^2 + \ldots + w_s^2 \qquad (3.17) \\ &= \text{a sum of } 2s - 1 \text{ squares in } K, \end{aligned}$$

since $\sum b_j c_j = 0$.

Now in $K(\sqrt{-d})$, we have $(\sqrt{-d})^2 + d = 0$ i.e. $(\sqrt{-d})^2$ plus a sum of n squares equals 0 (since $d \in G_n(K)$) so $s(K(\sqrt{-d})) \leq n$. But $2^k \leq n$ and s is a power of 2 so $s | 2^k$. Were $s < 2^k$ (i.e. $s \leq 2^{k-1}$) then d would be a sum of $2s - 1$ squares in K (by (3.17)), in other words,

$$\begin{aligned} d &= \text{a sum of } 2^k - 1 \text{ squares in } K \\ &= \text{a sum of } n - 1 \text{ squares in } K. \end{aligned}$$

Thus $d \in G_{n-1}(K)$, contradicting our hypothesis. Hence $s = 2^k$. □

This result immediately leads to the following (cf. the proof of the second part of Theorem 2.2 just after Corollary 4 of Chapter. 2).

Corollary. *Let K be a formally real field and $L = K(X_1, \dots, X_n)$. Let $d = X_1^2 + \dots + X_n^2 \in L^*$. Suppose $2^k \leq n < 2^{k+1}$ and let $E = L(\sqrt{-d})$. Then $s(E) = 2^k$.* □

Exercises

1.

Definition. Let K be a field. If there exists a positive integer $p = p(K)$ such that any sum of squares in K is already a sum of p squares in K, then the least such p is called the *pythagoras number* (or the reduced height) of K. Prove

(i) $p(\mathbf{Q}) = 4$ (Lagrange's theorem) (see [L5] pp. 142-145)
(ii) $p(\mathbf{R}) = 1$
(iii) $p(K) \leq s(K) + 1$. (cf. (iii) on p. 15).

2.
(I) Let α be a zero of the irreducible quartic

$$X^4 + CX^2 + DX + E \quad (C, D, E \in \mathbf{Z}) \tag{$*$}$$

Prove that if $K = \mathbf{Q}(\alpha)$ is totally complex, then
(ii) Δ, the discriminant of $(*)$, is > 0
(iii) Either $C > 0$ or $E > C^2/4$ (see Burnside and Panton) *The Theory of Equations* Vol. I).
(II) Let $\alpha, \beta, \gamma, \delta$ be the zeros of $(*)$ (all complex), show that $\Delta = \lambda^2(\gamma - \delta)^2$ for some $\lambda \in \mathbf{Q}(\alpha, \beta)$.
(III) Show that we can select a second zero β of $(*)$ for which the field $\mathbf{Q}(\alpha, \beta)$ is either equal to K $(= \mathbf{Q}(\alpha))$ or is of degree 12 over $\mathbf{Q}$.
(IV) Prove that Stufe of $L \leq 4$.
(V) Deduce that Stufe of $K \leq 4$ (use Springer's theorem).

For a proof of this see [P7].

3. (i) Express -1 as a sum of two squares in $\mathbf{Q}(e^{2\pi i/5})$.
(ii) Express $x^4 + x^3 + x^2 + x + 1$ as a sum of four squares in $\mathbf{Q}[x]$. (We know by Theorem 3.6 that (i) and (ii) are equivalent.)

4. Do likewise with $\zeta = e^{2\pi i/7}$ and the polynomial irr $(\zeta, \mathbf{Q}) = x^6 + x^5 + x^4 + x^3 + x^2 + x + 1$.

5. Let $\zeta = e^{2\pi i/p}$ (p an odd prime). Then we know by Theorem 3.6 that Stufe $(Q(\zeta)) = 2$, respectively 4 iff $\operatorname{irr}(\zeta, \mathbb{Q}) = f(X) = X^{p-1} + \ldots + X + 1$ is a sum of four, respectively eight, squares in $\mathbb{Q}[X]$. In these two cases actually write $f(X)$ as a sum of four or eight squares in $\mathbb{Q}[X]$.

6. Show that if $s(K) = s$ and X is transcendental over K, then $s(K(X)) = s$ (see Theorem 11.8).

4

Hilbert's 17th problem and the function fields $\mathbb{R}(X)$, $\mathbb{Q}(X)$ and $\mathbb{R}(X, Y)$

In 1900, David Hilbert [H4]′ in his famous address at the International Congress of Mathematicians in Paris proposed as his 17th problem the following:

Hilbert's conjecture. *Let $f(X_1, \dots, X_n) \in \mathbb{R}(X_1, \dots, X_n)$. A necessary and sufficient condition that f is a sum of squares in $\mathbb{R}(X_1, \dots, X_n)$ is that f is positive definite (i.e. $f(a_1, \dots, a_n) \geq 0$ for all $a_1, \dots, a_n \in \mathbb{R}$ for which f is defined).*

A similar conjecture holds for $\mathbb{Q}(X_1, \dots, X_n)$. These conjectures were proved by Artin [A6] in 1927 for both $\mathbb{R}$ and $\mathbb{Q}$, but one still didn't know how many squares are needed for the representation. Some results were of course known when the number of variables $n = 1$ or 2. Let us first look at the field $\mathbb{R}$.

In $\mathbb{R}(X)$ two squares suffice:

Theorem 4.1. *Let $f(X) \in \mathbb{R}(X)$ be positive definite; then $f(X)$ is a sum of two squares.*

This had already been proved by Hilbert in 1893, as also was the next result.

Theorem 4.2 (Hilbert (1893)). *Let $f(X, Y) \in \mathbb{R}(X, Y)$ be positive definite; then $f(X, Y)$ is a sum of four squares.*

This was first proved by Hilbert [H4] and later again by Witt. We shall

give a proof due to Pfister [P5]. This proof has the advantage that it can be generalized to the case of n variables.

In 1966, James Ax (unpublished) proved that in $\mathbb{R}(X, Y, Z)$ eight squares suffice. This has now been proved by Pfister in a very elegant way (see Chapter 5). We state here the general theorem of Pfister for information.

Pfister's General Theorem (1967). *Let*

$$f(X_1, \ldots, X_n) \in \mathbb{R}(X_1, \ldots, X_n)$$

be positive definite. Then f is a sum of 2^n squares.

In terms of our earlier definition (Chapter 3, Exercises) of the pythagoras number $P(K)$ of a field K, this simply says that $P(\mathbb{R}(X_1, \ldots, X_n) \leq 2^n$.

We shall prove Theorem 4.2 first, which is merely a special case of Pfister's general theorem, since it brings out the mode of proof very neatly. The general case has a few further difficulties, which can, however, be surmounted fairly easily.

The next question is: *Is 2^n best possible?* In other words what is the true value of $P(\mathbb{R}(X_1, \ldots, X_n))$? In general, this would be very difficult to answer, even conjecture.

For $n = 1$ i.e. in $\mathbb{R}(X), 2$ is best possible that is there exist positive definite functions which are not squares in $R(X)$, e.g. $X^2 + d \ (d > 0)$.

For $n = 2$, 4 is indeed best possible, that is, $P(\mathbb{R}(X, Y)) = 4$. The function $1 + X^2Y^2(X^2 - 3) + X^2Y^4$ which we shall extensively use later, cannot be expressed as a sum of three squares in $\mathbb{R}(X, Y)$. This was proved by Cassels, Ellison and Pfister in 1971 [C3] in a most ad hoc way using elliptic curves. Thus the method is special for $n = 2$. This result was extended by Christie [C18].

We have remarked above that P_n (say) $= P(\mathbb{R}(X_1, \ldots, X_n)) \leq 2^n$. The function $1 + X_1^2 + \ldots X_n^2$ and Corollary 4 of Chapter 2 shows that $n+1 \leq P_n$. We have $P_1 = 2, P_2 = 4$. Even P_3 is not known exactly, we only know $4 \leq P_3 \leq 8$.

Let us look at some of the corresponding results for the function field $\mathbb{Q}(X_1, \ldots, X_n)$.

For $n = 1$, Landau [L4] showed that eight squares suffice:

Theorem 4.3. *Let $f(X) \in \mathbb{Q}(X)$ be a positive definite function. Then f is a sum of eight squares in $\mathbb{Q}(X)$.*

A proof of this will be given in the course of the proof of Theorem 4.2 of Pfister.

Is eight best possible? As for the case $\mathbb{R}$, if we let τ_n be the smallest natural number such that every sum of squares in $\mathbb{Q}(X_1, \ldots, X_n)$ is already a sum of τ_n squares, then Theorem 4.3 says that $\tau_1 \leq 8$. The function X^2+7 shows (see Corollary 3, Chapter 2) that $5 \leq \tau_1$. Pourchet [P6] has shown that $\tau_1 = 5$. We shall give a proof of Pourchet's theorem in a later chapter. Little is known regarding generalizations of these results. The latest big new result is that of Kazuya Kato and J.L. Colliot-Thélène [K1], where they prove that $\tau_2 \leq 8$. The proof is amazingly difficult and uses a sort of "higher level" class field theory, developed by Kato, K-theory and algebraic geometry. The appendix by Colliot-Thélène is quite readable. (Note that τ_n is simply $P(\mathbb{Q}(X_1, \ldots, X_n))$ in our earlier notation of pythagorian number.)

We now give a

Proof of Theorem 4.1. Without loss of generality, we may suppose that $f(X) \in \mathbb{R}[X]$; for otherwise if $f(X) = u(X)/v(X)(u, v \in \mathbb{R}[X]) = u(X)v(X)/v^2(X)$ and we merely look at $u(X).v(X)$. Further we may clearly suppose that $f(X)$ is square-free.

Now factorize $f(X)$ in $\mathbb{C}[X]$:

$$f(X) = \alpha(X - \alpha_1)(X - \overline{\alpha}_1)\ldots \tag{4.1}$$

Here there are no real roots for if $f(X) = (X - \rho) \cdot g(X)$ then for $X > \rho$ we must have $g(X) \geq 0$ since f is PSD while for $X < \rho$ we have $g(X) \leq 0$ again since f is PSD. It follows that $g(\rho) = 0$ contradicting the fact that $f(X)$ was square-free. Morever, by letting $X \to \infty$ we see that $\alpha > 0$ (because $f(X)$ is positive definite). It follows that $\alpha = b^2$ $(b \in \mathbb{R})$.

Now consider the polynomial $b(X - \alpha_1)(X - \alpha_2)\ldots$. This has complex coefficients and so may be written as $u(X) + iv(X)$ say, where $u, v \in \mathbb{R}[X]$. Then $b(X - \overline{\alpha}_1)(X - \overline{\alpha}_2)\ldots = u(X) - iv(X)$. Multiplying, we get $f(X) = u^2(X) + v^2(X)$ as required. □

Our aim now is to give Pfister's proof of Theorem 4.2.

An elementary lemma. *Let $f(X) \in K[X]$ be square-free. Then $K(X)/(f(X))$ is a direct sum of fields:*

$$K(\xi_1) \oplus K(\xi_2) \oplus \ldots \oplus K(\xi_s)$$

where $\xi_1, \xi_2, \ldots, \xi_s$ are roots of the distinct irreducible factors of $f(X)$ in $K[X]$ (one each).

Proof. Let $I = (f(X))$ and factorize I into prime ideals. Since $K[X]$ is a Euclidian domain these are maximal ideals, without common factors:

$$I = M_1 M_2 \ldots M_s, \quad (M_i, M_j) = 1.$$

Let $K(X) = R$. We claim that

$$R/I \cong R/M_1 \oplus R/M_2 \oplus \ldots \oplus R/M_s.$$

To see this, consider the map $\theta : R \to R/M_1 \oplus \ldots \oplus R/M_s$ given by $a \mapsto (a + M_1, \ldots, a + M_s)$. It is easily checked that ϑ is a ring homomorphism. To see that it is onto, we take a general element of the right side: $(a_1 + M_1, a_2 + M_2, \ldots, a_s + M_s)$. Then by the Chinese remainder theorem, there exists an $a \in R$ such that $a \equiv a_i$ (modulo M_i). It follows that

$$(a_1 + M_1, \ldots, a_s + M_s) = (a + M_1, \ldots, a + M_s)$$

and now a is the preimage of this general element under θ i.e. θ is onto.

What is ker θ? It is the set of those $a \in R$ for which $a + M_i = M_i$ for all i i.e. $a \in M_i$ for all i. Thus ker $\theta = M_1 \cap \ldots \cap M_s$. But these M_i are comaximal (i.e. maximal and coprime) so this intersection equals $M_1 \ldots M_s = I$. The result follows. □

Lemma 1. *Let K be a field with* char$K \neq 2$ *and suppose that the Stufe of every finite extension of K, which is not formally real, is at most 2^m. Then any element $f(X) \in K(X)$, which is a sum of squares in $K(X)$, is a sum of at most 2^{m+1} squares in $K(X)$, i.e. $P(K(X)) \leq 2^{m+1}$.*

Proof. Without loss of generality $f(X) \in K[X]$ for if

$$f(X) = u(X)/v(X) = u(X)v(X)/v^2(X),$$

we merely look at $u(X) \cdot v(X)$.

We use induction on the degree of $f(X)$. To start the induction, let deg $f = 0$ so that $f(X)$ is a constant, say c, given to be a sum of squares in $K(X)$ and we have to show that it is a sum of at most 2^{m+1} squares in $K(X)$.

First, since K is a finite extension of itself, it is not formally real, and its Stufe is at most 2^m (by hypothesis), so each element of K, in particular c, is a sum of at most $2^m + 1 \leq 2^{m+1}$ squares as required [cf.(iii) on p. 15].

So let K be formally real. We are given that c is a sum of squares in $K(X)$, so by Cassels' lemma, in $K[X]$, say

$$c = f_1^2(X) + \ldots + f_\gamma^2(X) \quad (fj(X) \in K[X]).$$

Here by equating the coefficients of the highest powers of X to zero, we contradict the formal reality of K, unless each $f_j(X)$ is a constant, say a_j, so that

$$c = a_1^2 + \ldots, a_\gamma^2 \quad (a_j \in K).$$

Now consider the field $K(\sqrt{-c})$. This is not formally real; indeed $0 = (\sqrt{-c})^2 + c = (\sqrt{-c})^2 + a_1^2 + \ldots + a_\gamma^2$. Hence the Stufe of $K(\sqrt{-c}) \leq 2^m$;

so we have, with $\alpha_j, \beta_j \in K$

$$0 = (\alpha_1 + \sqrt{-c}\beta_1)^2 + \ldots + (\alpha_{2^m+1} + \sqrt{-c}\beta_{2^m+1})^2.$$

Then $\sum \alpha_j^2 - c\sum \beta_j^2 = 0$ and $\sum \alpha_j \beta_j = 0$. The first relation gives $c = \sum \alpha_j^2 / \sum \beta_j^2$ where $\sum \beta_j^2$ (nor indeed $\sum \alpha_j^2$) can be zero, since K is formally real. Now $\sum \alpha_j^2, \sum \beta_j^2 \in G_{2^m+1} \subset G_{2^{m+1}}$. This latter is a group, so $c \in G_{2^{m+1}}$.

This starts the induction. Now suppose the result is true for polynomials of degree less than N and let $f(X)$ have degree N. If $f(X)$ has a square factor, say $g^2(X)$, then $f(X)/g^2(X) \in K[X]$ and is a sum of squares since $f(X)$ is and has degree less than N. Hence by the induction hypothesis, it is a sum of at most 2^{m+1} squares and then so is clearly $f(X)$.

So suppose $f(X)$ has no square factor. We first look at the special case when $f(X)$ is irreducible in $K[X]$, this is only to get a feel of the general case.

Since $f(X)$ is irreducible, it is prime and so maximal.

Thus $K[X]/(f(X))$ is a field; call it $K(\xi)$ where $f(\xi) = 0$. Thus $K(\xi)$ is a finite extension of K and it is not formally real, for, in the relation

$$f(X) = h_1^2(X) + \ldots + h_\gamma^2(X),$$

(don't forget $f(X)$ is given to be a sum of squares) put $X = \xi : 0 = h_1^2(\xi) + \ldots + h_\gamma^2(\xi)$. Here, not all the $h_j(\xi)$ are zero for otherwise $f(X)|h_j(X)$ (all j) and then $f^2(X)|\sum h_j^2(X) = f(X)$, which is false. Thus

$$0 = h_1^2(\xi) + \ldots + h_\gamma^2(\xi)$$

is a non-trivial representation of 0 as a sum of squares in $K[\xi]$, so $K(\xi)$ is not formally real. Thus by hypothesis, the Stufe of $K(\xi)$ is at most 2^m, i.e. there exists a solution of

$$a_1^2(\xi) + \ldots + a_{2^m+1}^2(\xi) = 0$$

where $a_j(\xi) \in K(\xi)$ are polynomials in ξ of degree less than $\deg f$. Thus $a_1^2(X) + \ldots + a_{2^m+1}^2(X)$ is a polynomial in $K[X]$ which vanishes at $X = \xi$ and so is divisible by $\mathrm{irr}(\xi, K) = f(X)$:

$$a_1^2(X) + \ldots, + a_{2^m+1}^2(X) = f(X)g(X). \tag{4.2}$$

Then $\deg f + \deg g \leq 2 \max_j (\deg a_j(X)) < 2 \deg f$ so $\deg g < \deg f$. Now $f(X)$ is by hypothesis, a sum of squares (and we have to show that it is a sum of at most 2^{m+1} squares) and so a sum of 2^T squares for sufficiently large $T(\geq m)$. Since G_{2^T} is a group, by Corollary 2, Chapter 2, it follows by (4.2) that $g(X)$ is also a sum of squares ($g(X) \in G_{2^T}$). Now $\deg g < \deg f$ so by the induction hypothesis $g(X)$ is a sum of at most 2^{m+1} squares i.e. $g(x) \in G_{2^{m+1}}$. So again by (4.2), $f(X) \in G_{2^{m+1}}$, since $2^m + 1 \leq 2^{m+1}$. This completes the proof of the case when $f(X)$ is irreducible.

So now suppose $f(X)$ is general, but square-free. Then $K[X]/(f(X))$ is a direct sum of fields:

$$K(\xi_1) \oplus \ldots \oplus K(\xi_s)$$

where $\xi_1, \ldots, \xi_s$ are roots, one each, of the irreducible factors of $f(X)$ in $K[X]$. Our aim now is to get a relation like (4.2). Then as before we should be through. Let $f(X) = f_1(X) \ldots f_s(X)$. Then

$$K[X]/(f(X)) \cong K[X]/(f_1(X)) \oplus \ldots \oplus K[X]/(f_s(X)).$$

Here again, none of the $K(\xi_i)$ are formally real, for in the relation $f(X) = h_1^2(X) + \ldots + h_\gamma^2(X)$, put $X = \xi_1$ say. Since ξ_1 is a root of $f_1(X), f(\xi) = 0$, hence

$$0 = h_1^2(\xi_1) + \ldots + h_\gamma^2(\xi_1).$$

As before if all $h_i(\xi_1) = 0$ then $f_1^2 | f(X)$ which is false since $f(X)$ was supposed square-free. So $K(\xi_1)$ is not formally real and similarly none of the $K(\xi_i)$ are formally real.

We have, in each of the fields $K(\xi_i) = K[X]/(f_i(X))$,

$$0 = \sum_{j=1}^{2^m+1} h_{ij}^2(\xi_1) \quad (i = 1, 2, \ldots, s)$$

since the Stufe of $K(\xi_i)$ is at most 2^m by hypothesis, where the $h_{ij}(X)$ are polynomials (clear the denominators if any). Then we have the congruences

$$\sum_{j=1}^{2^m+1} h_{ij}^2(X) \equiv 0 \bmod f_i(X) \ (i = 1, 2, \ldots, s).$$

By the Chinese remainder theorem, find $g_j(X)$ such that

$$g_j(X) \equiv h_{ij}(X) \text{mod } f_i(X), \ j = 1, 2, \ldots, 2^m + 1.$$

Written out in full to see clearly:

$$h_{11}^2(X) + h_{12}^2(X) + \ldots + h_{1\ 2^m+1}^2(X) \equiv 0 \bmod f_1(X)$$
$$h_{21}^2(X) + h_{22}^2(X) + \ldots + h_{2\ 2^m+1}^2(X) \equiv 0 \bmod f_2(X)$$

....

Now find

$$g_1(X) \equiv h_{11}(X) \bmod f_1(X)$$
$$h_{21}(X) \bmod f_2(X)$$

....

$$g_2(X) \equiv h_{12}(X) \bmod f_1(X)$$
$$h_{22}(X) \bmod f_2(X)$$

....

then

$$g_1^2(X) + \ldots + g_{2^m+1}^2(X) \equiv h_{11}^2(X) + h_{22}^2(X) + \ldots \bmod f_1(X)$$
$$\equiv h_{21}^2(X) + h_{22}^2(X) + \ldots \bmod f_2(X)$$
$$\ldots\ldots\ldots\ldots\ldots\ldots\ldots\ldots\ldots\ldots$$

and so $g_1^2 + \ldots + g_{2^m+1}^2(X) \equiv 0 \bmod f(X)$, since $f_1, f_2 \ldots$ are all coprime. This gives a relation like (4.2) and so completes the proof of Lemma 1. □

As an immediate consequence of this lemma we can prove Theorem 4.3 of Landau as follows:

Proof of Theorem 4.3. Let $K = \mathbb{Q}$. Then by Theorem 18.3 the Stufe of every finite extension of $\mathbb{Q}$, which is not formally real is at most $4 = 2^2$, so by Lemma 1, $p(\mathbb{Q}(X)) \le 2^{2+1} = 8$. □

Lemma 2. *Let L be a finite extension of $\mathbb{R}(X)$, which is not formally real. Then any quadratic form in $L(i)$ ($i^2 = -1$) in at least three variables represents 0 non-trivially.*

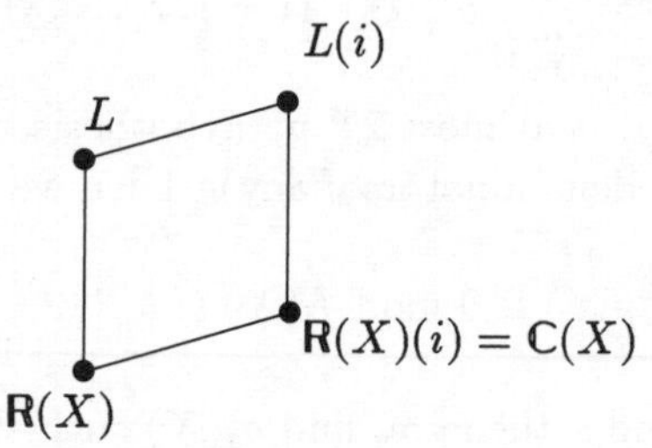

Proof. $L(i)$ is a finite extension of $\mathbb{C}(X)$ and so $L(i) = \mathbb{C}(X, \xi)$, where ξ satisfies the monic polynomial $\xi^\gamma + a_1(X)\xi^{\gamma-1} + \ldots + a_\gamma(X) = 0$, so that $a_j(X) \in \mathbb{C}[X]$ (rather than $\mathbb{C}(X)$). Let the quadratic form be

$$A_1X_1^2 + A_2X_2^2 + A_3X_3^2 \tag{4.3}$$

(without loss of generality, diagonal, but in any case we shall be applying the lemma to a diagonal form). Then $A_1, A_2, A_3 \in L(i) = \mathbb{C}(X, \xi)$. Clearing denominators in (4.3), we may suppose that the $A_j \in \mathbb{C}[X, \xi]$ and let us look for a solution in $\mathbb{C}[X, \xi]$; so let

$$(\text{for } i = 1, 2, 3)\ X_i = Y_0^{(i)} + Y_1^{(i)}\xi + Y_2^{(i)}\xi^2 + \ldots + Y_{\gamma-1}^{(i)}\xi^{\gamma-1}$$

where $Y_j^{(i)} \in \mathbb{C}[X]$ and where deg $Y_j^{(i)}$(in X) is at most N say, to be determined later. Each $Y_j^{(i)}$, being a polynomial in X of degree at most N, has $N+1$ coefficients, so all together there are $(N+1)\cdot\gamma\cdot 3$ coefficients since j takes γ values, and i takes three values, $1, 2, 3$. Now substitute for the X_i

in (4.3):

$$0 = A_1 \left\{ \sum_{j=0}^{\gamma-1} Y_j^{(1)} \xi^j \right\}^2 + A_2 \left\{ \sum_{j=0}^{\gamma-1} Y_j^{(2)} \xi^j \right\}^2 + A_3 \left\{ \sum_{j=0}^{\gamma-1} Y_j^{(3)} \xi^j \right\}^2 \tag{4.4}$$

Now the defining polynomial for ξ implies that ξ^k $(k \geq \gamma)$ is a polynomial in ξ of degree $\gamma - 1$ with coefficients in $\mathbb{C}[X]$, say

$$\xi^k = a_{\gamma-1}^{(k)}(X)\xi^{\gamma-1} + \ldots + a_0^{(k)}(X).$$

Substituting this in (4.4), we get $\sum_{j=0}^{\gamma-1} T_j(X)\xi^j = 0$. Now let $c_1 = \max \deg a_j^{(k)}(X), c_2 = \max \deg A_i(X)$. Then

$$\deg T_j(X) \leq 2N + c_1 + c_2 = 2N + C,$$

say. Now (4.4) will become zero if we can make each $T_j(X)$ individually zero. Since $\deg T_j(X) \leq 2N+C$, so each T_j involves at most $2N+C+1$ coefficients to be made zero i.e. there are at most $(2N+C+1)\gamma$ equations to be satisfied. The number of unknowns are $3\gamma(N+1)$, (these are actually homogeneous quadratic equations). Since $\mathbb{C}$ is algebraically closed and $3\gamma(N+1) > (2N+C+1)\gamma$ which is at least the number of equations, so there is a non-trivial solution. □

Lemma 3. *Suppose $a, b, c \in L$ with L as above. Then $X^2 + aY^2 = b^2 + c^2$ is solvable in L i.e. the form $X^2 + aY^2$ represents $b^2 + c^2$ in L.*

Proof. If $i \in L$ then by Lemma 2 the forms $X^2 + aY^2 - (b \pm ic)Z^2$ represent 0 non-trivially. If, in this representation, $Z = 0$ then $X^2 + aY^2$ represents 0 nontrivially in L and so represents all elements (see Theorem 11.4) of L, in particular $b \pm ic$. If $Z \neq 0$ then dividing throughout by Z we see that $X^2 + aY^2$ represents $b \pm ic$ again. So $b \pm ic$ is always represented by $X^2 + aY^2$, say,

$$\left.\begin{aligned} X_1^2 + aY_1^2 &= b + ic \\ X_2^2 + aY_2^2 &= b - ic \end{aligned}\right\}, X_1, Y_1, X_2, Y_2 \in L = L(i).$$

Multiplying, we get $(X_1X_2 + aY_1Y_2)^2 + a(X_1Y_2 - X_2Y_1)^2 = b^2 + c^2$. □

If $i \notin L$, we see as above that there exist $\xi, \eta \in L(i)$ such that:

$$\xi^2 + a\eta^2 = b + ic$$

and so also

$$\overline{\xi}^2 + a\overline{\eta}^2 = b - ic.$$

Multiplying, we get $(\xi\overline{\xi} - a\eta\overline{\eta})^2 + a(\xi\overline{\eta} + \overline{\xi}\eta)^2 = b^2 + c^2$. Here $\xi\overline{\xi} - a\eta\overline{\eta}$ and $\xi\overline{\eta} + \overline{\xi}\eta$ are both elements of L since they are invariant under "bar". Thus $X^2 + aY^2 = b^2 + c^2$ has been solved in L.

Lemma 4. *If $e_1, e_2, e_3, e_4 \in L$ then $e_1^2 + e_2^2 + e_3^2 + e_4^2$ is a sum of three squares in L.*

Proof. Put $a = e_3^2 + e_4^2$, $b = e_1$, $c = e_2$ in Lemma 3. Then there exists a solution u, v say of $u^2 + v^2(e_3^2 + e_4^2) = e_1^2 + e_2^2$. Then

$$\begin{aligned} e_1^2 + e_2^2 + e_3^2 + e_4^2 &= u^2 + v^2(e_3^2 + e_4^2) + e_3^2 + e_4^2 \\ &= u^2 + (v^2 + 1)(e_3^2 + e_4^2) \\ &= u^2 + (ve_3 + e_4)^2 + (ve_4 - e_3)^2 \end{aligned}$$

which is a sum of three squares. It now follows that the Stufe of L is at most 2 for if s is the Stufe so that $-1 = e_1^2 + \ldots + e_s^2$, then by Lemma 4, we can successively reduce the right side to a sum of three squares. Hence $s \leq 3$; but s is a power of 2 (see Theorem 2.2) so $s \leq 2$. □

And now finally we come to the

Proof of Theorem 4.2. Let $K = \mathbf{R}(X)$. By the above, the Stufe of every finite extension of $\mathbf{R}(X)$, which is not formally real is at most 2. Hence by Lemma 1, any element of $\mathbf{R}(X, Y)$, which is a sum of squares, is a sum of at most $2^{1+1} = 4$ squares. □

The fields considered above have either been $\mathbf{R}$ or $\mathbf{Q}$. For a general field K we could formulate Hilbert's 17th problem on similar lines. We give a quick survey of what is known for this general set up, together with all the historical developments and complete sets of references. One should note that what is true for the real numbers $\mathbf{R}$ is almost always valid for real closed fields R. For a detailed account of this and some other references, the reader is referred to [P3], [P4].

Let K be any field and let

$$f(X_1, \ldots, X_n) = \frac{g(X_1, \ldots, X_n)}{h(X_1, \ldots, X_n)} \in K(X_1, \ldots, X_n)$$

be a rational function in n variables with coefficients in K. We call f positive definite if for all $a_1, \ldots, a_n \in K$ for which $h(a_1, \ldots, a_n) \neq 0$, we have $f(a_1, \ldots, a_n) \geq 0$ for all orderings $>$ of K. If K has no ordering, then every f is positive definite. Hilbert's 17th problem can now be formulated in three parts:

P1 *Let $K = R$ be a real closed field. Does*

$$f(X_1, \ldots, X_n) \in R(X_1, \ldots, X_n),$$

f positive definite, imply f is a sum of r squares in $R(X_1, \ldots, X_n)$ for some natural number r? The converse to this is of course trivial.

P2 *Let K be an arbitrary field. Does*

$$f(X_1, \ldots, X_n) \in K(X_1, \ldots, X_n),$$

f positive definite, imply f is a sum of r squares in $K(X_1, \ldots, X_n)$ for some natural number r?

P3 *If* P1 *respectively* P2 *is true, is there an upper bound on r depending only on n (the number of variables $X_1, \ldots, X_n$) and on K but not on f?*

The answers are the following:
P1 is true (Artin [A.6] - 1927).
P2 is in general, false (counterexample due to Dubois [D3], (1967)).
P3 is true for $K = R$, a real closed field (Pfister, [P2], 1967) but unsolved for other fields K where P2 holds.

The investigation of positive definite functions began with Hilbert in the year 1888, when he first proved, [H3], the negative result that if $f(X_1, \ldots, X_n) \in \mathbb{R}[X_1, \ldots, X_n]$ is a positive definite polynomial, then f need not be a sum of squares of polynomials in $\mathbb{R}[X_1, \ldots, X_n]$, except for $n = 1$. Hilbert's proof was geometric and did not explicitly yield any such polynomials. The first explicit example was given by Motzkin [M2] 1966 viz. the polynomial.

$$f(X, Y) = 1 + X^4Y^2 + X^2Y^4 - 3X^2Y^2$$

which we have already met previously in this very connection. Later R.M. Robinson [R8] produced other examples.

In 1893, Hilbert [H4] succeded in proving the case $n = 2$ of P1; indeed one can deduce from his proof the stronger result that four squares suffice for $n = 2$. This deduction was carried out by Landau [L4] in 1906, who at the same time looked at the case $K = \mathbb{Q}$, $n = 1$ of P2. Landau's result for $K = \mathbb{Q}$, $n = 1$ we have already looked at earlier. The subsequent improvement due to Pourchet [P6] will be taken up later in Chapter 17.

In 1927, Artin [A6] proved P1 in general and P2 under the condition that K has only Archimedean orderings (for instance if K is an algebraic number field). In contrast to Hilbert's proof of P1 for the case $n = 2$, which was essentially constructive, Artin's proof relies heavily on Artin-Schrier theory of formally real fields and also on Sturm's theorem.

The first constructive proof of P1 for $K = \mathbb{R}$ was given by Habicht [H1], in 1940, under the extra condition that f is strictly poisitive, say $f \geq 1$. In 1955, A. Robinson [R7] proved P1 by using lower predicate calculus and the model completion of $\mathbb{R}$. In 1956 Kreisel [K4] gave a constructive proof for arbitrary $f \geq 0$ and showed that there is an upper bound on the number r of squares, depending only on n and the degree of f (but not on the coefficients of f). Finally, in 1972, Knebusch [K2] gave a proof of Artin's theorem P1 that avoids Sturm's theorem.

We have already said that P2 is in general, false. An example was given by Dubois [D3], in 1967, in which he constructs a field K with only one ordering (which is, of course, non-Archimedean) and a positive definite polynomial $f(X) \in K[X]$, which is not a sum of squares in $K(X)$. Thereby it was shown that some condition of the field K, such as

K is real closed or

All orderings of K are Archimedean etc.

is necessary if P2 is to be true. We give here an

Elementary exposition of Dubois' example.

Let $\mathbb{Q}$ be the rational field and t an indeterminate over $\mathbb{Q}$. We order $\mathbb{Q}(t)$ as follows:

Let $h(t) = f(t)/g(t) \in \mathbb{Q}(t)$. Write $h(t)$ as $h(t) = r \cdot t^s \cdot f_1(t)/g_1(t)$, where f_1, g_1 have constant term equal to 1 and $s \in \mathbb{Z}$.

Now say $h(t) > 0$ iff $r > 0$. It is easy to check that this defines an order in $\mathbb{Q}(t)$. In this order, t satisfies the following properties (and then we say t is infinitesimal):

(1) $t > 0$.

(2) for any $r \in \mathbb{Q}$, $r > 0$, we have $r > t$.

Indeed the properties (1), (2) imply that the rational function $h(t) = f(t)/g(t) = r \cdot t^s \cdot f_1(t)/g_1(t)$, written as above is greater than 0 iff $r > 0$ for

$$h(t) = r \cdot t^s f_1(t) \cdot g_1(t)/g_1^2(t).$$

Here $t^s > 0$, $g_1^2(t) > 0$ (all squares positive). We claim that $f_1(t) \cdot g_1(t) = 1 + a_1 t + \ldots + a_j t^j > 0$, so $h(t) > 0$ iff $r > 0$ as required.

To prove the claim, we have indeed

$$a_0 + a_1 t + \ldots + a_j t^j > 0 \text{ iff } a_0 > 0.$$

The general term $a_l t^l < \frac{1}{j}$; this is clear if $a_l < 0$, while if $a_l > 0$ we have $t^l < t$ (since $t < 1$) which is less than any positive rational and so less than $\frac{1}{ja_l}$. Thus $a_1 t + \ldots + a_j t^j < \frac{1}{j} + \ldots + \frac{1}{j} = 1$ and $1 + a_1 t + \ldots + a_j t^j > 0$. □

Now we show that the order $>$ defined above on $\mathbb{Q}(t)$, is non-archimedean. For this we must produce an element $\alpha > 0$ in $\mathbb{Q}(t)$ for which no integer n satisfies $n > \alpha$. Take $\alpha = 1/t$ so that $\alpha > 0$ certainly. Now $n > \frac{1}{t}$ (> 0) for some $n \Rightarrow t > 1/n$ which contradicts (2) above as required.

Now let K be a real closure of $\mathbb{Q}(t)$ and let F be a field over $\mathbb{Q}(t)$ consisting of elements of K obtained from $\mathbb{Q}(t)$ by means of a finite sequence of rational operations and extraction of square roots of elements of $\mathbb{Q}(t)$, exactly as in ruler and compass constructions. Thus,

$$\text{every positive element of } F \text{ has a square root in } F. \qquad (*)$$

This already implies that F has a unique order for (*) is equivalent to saying that for any element α of F

$$\begin{aligned} \textit{Either} \quad \alpha &= \text{ a square in } F \\ \textit{or} \quad -\alpha &= \text{ a square in } F \end{aligned}$$

(for one of $\pm\alpha > 0$).

Let $P = \{\alpha \in F | \alpha \text{ is a square }\}$. Then $P + P \subset P$ for if $\alpha, \beta \in P$, say $\alpha = x^2$, $\beta = y^2$, then $\alpha + \beta = x^2 + y^2$ and if this not in P, then it in $-P$, say $= -z^2$ so $x^2 + y^2 + z^2 = 0$; so F is not formally real, which is false, since $F \subset K$ (real and closed).

Also $P \cdot P \subset P$ clearly. So P is an order in F. If now P^* is any order of F, then $P^* \supset P$ (since all orders contain squares), but P is an order, so no strictly bigger P^* can be an order, hence $P^* = P$.

Thus F has a unique order which is non-archimedean.

Now let $f(X) = (X^3 - t)^2 - t^3 \in F[X]$. This $f(X)$ is not a sum of squares in $F(X)$, indeed not even in $K(X)$, for $f(t^{1/3}) = -t^3 < 0$.

We claim that $f(X)$ is a positive definite function on F i.e. that $F(\alpha) \geq 0$ for all $\alpha \in F$.

The easiest way to show this is to work in a larger field, which naturally includes F. Define $F_n = \mathbb{R}((t_n))$ to be the field of formal power series in t_n over $\mathbb{R}$. Thus the elements of $\mathbb{R}((t))$ are of the form

$$a_N t^N + a_{N+1} t^{N+1} + \ldots + \ldots \qquad (n \in \mathbb{Z}).$$

(Note that N can be negative. Thus these are the Laurent expansions). We embed F_{n-1} in F_n by putting

$$t_n^2 = t_{n-1} \quad (n = 1, 2, \ldots) \text{ and put } t_0 = t.$$

Thus $\mathbb{R}((t)) = \mathbb{R}((t_0)) \subset \mathbb{R}((t_1)) \subset \mathbb{R}((t_2)) \subset \ldots$ where $t_0 = t_1^2, t_1 = t_2^2, t_2 = t_3^2, \ldots$. In fact t_1 is algebraic $/\mathbb{R}((t_0))$ (it satisfies $t_0 = t_1^2$) and indeed

$$\mathbb{R}((t_0))(t_1) = \mathbb{R}((t_1))$$

and so on at each stage; for an element of the left side is of the form

$$A(t_0) + t_1 B(t_0)$$

and writing $t_0 = t_1^2$, this is an element of the right side. Conversely for any element of the right side, we separate out terms with even powers of t_1 and odd powers of t_1 and write it as $E(t_1) + O(t_1) = E(t_1) + t_1$ (even powers of t_1). Now put $t_0 = t_1^2$ throughout; so this equals $A(t_0) + t_1 B(t_0)$, an element of the left side as required.

Next let

$$\mathsf{F} = \bigcup_{n=0}^{\infty} F_n.$$

Now all the F_n have a unique order inherited from $\mathbb{R}$ by the convention that t_n is positive but infinitesimal. Indeed any element of $\mathbb{R}((t_n))$ looks like

$$a_N t_n^N + a_{N+1} t_n^{N+1} + \ldots + a_0 + a_1 t_n + a_2 t_n^2 + \ldots$$
$$= a_N t_n^N (1 + b_1 t_n + b_2 t_n^2 + \ldots) > 0 \text{ iff } a_N > 0.$$

This gives the unique order. Hence F has a unique order.

We now show that F is closed under the taking of square roots of positive elements. So let $A \in \mathsf{F}$, $A > 0$. Then $A \in$ some F_n; say $A = r t_n^s (1 + a_1 t_n + a_2 t_n^2 + \ldots)$, where since $A > 0$ we have $r > 0$. Here if s is even, $\sqrt{t_n^s} \in F_n$ itself while if s is odd, then

$$\sqrt{t_n^s} = \sqrt{t_n} \cdot t_n^{(s-1)/2} = t_{n+1} \cdot t_n^{(s-1)/2} \in F_{n+1} F_n \subset \mathsf{F}.$$

Also, since $r > 0$, $\sqrt{r} \in \mathbb{R}$. It remains to show that $1 + a_1 t_n + a_2 t_n^2 + \ldots$ has a square root in F. Write t for t_n. We shall show that it has a square root in F_n i.e. we wish to solve

$$(1 + b_1 t + b_2 t^2 + \ldots)^2 = 1 + a_1 t + a_2 t^2 + \ldots \quad (b_j \in \mathbb{R}).$$

We equate coefficients of t^j:

For $j = 1, 2b_1 = a_1$ so $b_1 = a/2 \in \mathbb{R}$.

For $j = 2, b_1^2 + 2b_2 = a_2$; since $b_1$is known, so is b_2 and it is real.

Now suppose $b_1, b_2, \ldots b_n$ have been found. Now equate coefficients of t^{n+1} to get

$$b_{n+1} + b_1 b_n + b_2 b_{n-1} + \ldots + b_n b_1 + b_{n+1} = a_{n+1}.$$

Since $b_1, b_2, \ldots, b_n$ are all known, so b_{n+1} can be determined (since the characteristic does not equal 2). □

Now identify naturally F as a subfield of F. This is possible since F is constructed from $\mathbb{Q}(t) = \mathbb{Q}(t_0)$ by extraction of square roots of positive elements and so is clearly contained in F.

Finally identify t_n with the 2^nth roots of t viz. $t_1 = \sqrt{t_0} = \sqrt{t}, t_2 = \sqrt{t_1} = \sqrt[4]{t}, t_3 = \sqrt[8]{t}$ etc. $\ldots, t_n = 2^n$th roots of t as required. With this identification, every non-zero element x of F is of the form (note that $x \in F_n$ for some n)

$$x = r t_n^M (1 + a_1 t_n + a_2 t_n^2 + \ldots)$$
$$= r t^{M/2^n} (1 + a_1 t^{1/2^n} + a_2 t^{2/2^n} + \ldots)$$
$$= r t^m \ + \text{higher powers of } t \cdot b\text{'s},$$

meaning t^ρ, where $\rho > m$ $(\rho, m \in \mathbb{Q})$, and m has degree 2^n. Here $r \neq 0$.

Now in the relevant order rt^m dominates the rest of the terms, so we may write

$$x = rt^m + \text{smaller terms},\ r \neq 0,\ m \in \mathbb{Q},$$

and m has denominator 2^s.

We shall indeed show that the polynomial

$$f(X) = (X^3 - t)^2 - t^3$$

is positive definite in F and so clearly in F.

Take any $x \in \mathsf{F}$ of the above form and put $X = x$. Then the left side becomes

$$\begin{aligned}(rt^m &+ \text{smaller terms }))^3 - t^2 - t^3 \\ &= (r^3t^{3m} + \text{smaller terms } - t)^2 - t^3.\end{aligned}$$

Case 1. $3m < 1$. Then t^{3m} dominates the bracket so it equals $r^2t^{6m}+$ smaller terms greater than t^3 since $6m < 3$ (indeed $6m < 2$).

Case 2. $3m > 1$. Then $-t$ dominates the bracket so it equals t^2 plus smaller terms greater than t^3 again.

Finally note that $3m \neq 1$, otherwise $m = \frac{1}{3}$, which is false, as m has denominator a power of 2. This completes the proof. □

Going back to the question in P3; the first important step (after Artin's proof of P1) was taken by Tsen [T2] in 1936 with his very general result on quasi-algebraically closed fields. His paper was forgotten during the war and his results had to be rediscovered by Lang [L6] before they reached the mathematical community of the western world. P3 was finally solved for real closed fields by Pfister [P2], the number of squares needed being 2^n. We have already looked at the proof of this result for the field R of real numbers. The same result is valid for any real closed field:

$$n + 1 \leq P(R(X_1, \ldots, X_n)) \leq 2^n$$

where $P(K)$ is the pythagoras number of $K = R(X_1, \ldots, X_n)$.

For a general field K, P3 is still an open problem. Other open problems in this topic are

Problem 1. *What is the true value of $P(R(X_1, \ldots, X_n))$, where R is real closed?*

Problem 2. *Is $P(\mathbb{Q}(X_1, \ldots, X_n))$ bounded by some function of n?*

A good guess would be $P(\mathbb{Q}(X_1, \ldots, X_n)) \leq 2^n + 3$. This is true for $n = 0$ (Lagrange's theorem) and for $n = 1$ (Pourchet's theorem). Pfister asked if 2^{n+2} would serve as an upper bound. Here again for $n = 0, 1$ the bound works.

Problem 3. *If K is a field with $P(K) < \infty$, is it true that $P(K(X)) < \infty$?*

5

Positive definite functions and sums of squares in $R(X_1,\dots,X_m)$.

One of the aims of this chapter in the following beautiful result of Pfister:

Theorem 5.1. *Let R be a real closed field. Then any positive definite function in $R(X_1,\dots,X_m)$ is a sum of at most 2^m squares in $R(X_1,\dots,X_m)$. i.e. $P(R(X_1,\dots,X_m))\leq 2^m$.*

During the proof of this result we shall need to develop the beautiful new idea of the generalization of the identities (1.1), (1.2), (1.3) and (2.1) to similar more general diagonal forms

$$X_1^2+aX_2^2, X_1^2+aX_2^2+bX_3^2+baX_4^2, \text{ etc.}$$

This generalization is ultimate in the sense that the relevant properties are satisfied by no further generalizations. All this will not be required in the proof of Theorem 5.1 but the topic is so beautiful that we shall pursue it (as much as is possible in the spirit of this book) in Chapter 12.

Theorem 5.1 will follow from the following two results:

Theorem 5.2. *Let K be a field with* char$K\neq 2$ *and the property that the Stufe of every algebraic extension of K, which is not formally real, is at most 2^m. Then any sum of squares in $K(X)$ is already a sum of at most 2^{m+1} squares in $K(X)$; i.e. $p(K(X))\leq 2^{m+1}$.*

Remark. This result is the content of Lemma 1 of Chapter 4; it is trivial if K is not formally real for then, K being an algebraic extension of itself,

$s(K) \leq 2^m$; so for any polynomial $f(X) \in K[X]$, we have, as required,

$$f = \left(\frac{f+1}{2}\right)^2 - \left(\frac{f-1}{2}\right)^2 = \text{a sum of } 2^m + 1 \text{ squares}$$
$$\leq \text{ a sum of } 2^{m+1} \text{ squares.}$$

So we may suppose K to be formally real.

Theorem 5.3. *Let R be a real closed field and L a non-formally real algebraic extension of $R(X_1, \ldots, X_m)$. Then $s(L) \leq 2^m$.*

Deduction of Theorem 5.1 from Theorems 5.2 and 5.3. Take

$$K = R(X_1, \ldots, X_{m-1})$$

in Theorem 5.2. By Theorem 5.3 the Stufe of every non-formally real algebraic extension L of K is $\leq 2^{m-1}$. So by Theorem 5.2, $p(K(X_m)) \leq 2^m$ i.e.

$$p(R(X_1, \ldots, X_{m-1}, X_m)) \leq 2^m.$$

□

Although a proof of Theorem 5.2 was given in Chapter 4 we give here a different one.

Second proof of Theorem 5.2. Let $f(X)$ be such a sum of squares in $K(X)$. It is enough to take $f(X) \in K[X]$ for if $f = g/h$ then $fh^2 = gh \in K[X]$ and knowing the result in $K[X]$ gives fh^2 to be a sum of 2^n squares in $K(X)$, hence so is f as required.

The result is trivial if $f = 0$. So let $f = a_0 + a_1X + \ldots + a_lX^l$ $(a_j \in K,\ a_l \neq 0)$. We first show that l is even and that a_l is a sum of squares in K. By hypothesis there exist $g_i(X), h(X) \in K[X]$ such that

$$fh^2 = g_1^2 + \ldots + g_t^2 \tag{5.1}$$

Let

$$h(X) = c_0 + c_1X + \ldots + c_mX^m \quad (c_m \neq 0)$$
$$g_i(X) = b_{i0} + b_{i1}X + \ldots + b_{ir}X^r \quad (\text{at least one } b_{ir} \neq 0),$$

and let $r = \max_{1 \leq i \leq t} (\deg g_i)$. On comparing degrees in (5.1) we see that $l + 2m = 2r$. Thus l is even. (Note that if $l + 2m < 2r$ then equating the coefficient of X^{2r} on both sides of (5.1) gives $\Sigma b_{ir}^2 = 0$ $(b_{ir} \in K)$, a contradiction since K is formally real). (5.1) gives $a_lc_m^2 = \sum_{i=1}^t b_{ir}^2$ so a_l is a sum of squares in K as required.

Write $l = 2k$. If $k = 0$ then $f(X) = a_0$ which is a sum of squares in K. We show that it is a sum of at most 2^n squares in K: indeed $X^2 + a_0$ is irreducible over K for if it factors as $(X+\sqrt{-a_0})(X-\sqrt{-a_0})$ then $\sqrt{-a_0} \in K$; say $\sqrt{-a_0} = a \in K$ i.e. $-a^2 = a_0$, a sum of squares in K, so -1 is a sum of squares in K which is false as K is formally real.

Let α be a root of $X^2 + a_0$ and let $L = K(\alpha)$, so that in L, $\alpha^2 + a_0 = 0$. But a_0 is a sum of squares, so

$$-\alpha^2 = \text{ a sum of squares in } L$$

$$\text{i.e. } -1 = \text{ a sum of squares in } L$$

Thus by hypothesis

$$-1 = \text{ a sum of } 2^{n-1} \text{ squares in } L$$

$$= \sum_{i=1}^{2^{n-1}} (b_i + \alpha c_i)^2 \quad (b_i, c_i \in K).$$

Then $1 + \sum_{i=1}^{2^{n-1}} b_i^2 = -\alpha^2 \sum_{i=1}^{2^{n-1}} c_i^2 = a_0 \sum_{i=1}^{2^{n-1}} c_i^2$. Here $\Sigma c_i^2 \neq 0$ for otherwise $-1 = \Sigma b_i^2$ in K, which is false. Then $f = a_0 = \frac{1+\Sigma b_i^2}{\Sigma l_i^2} \in G_{2^n}$ as required.

Now let $k > 0$. Then $f(X) = a_0 + a_1 X + \ldots + a_{2k} X^{2k}$. Here, as shown above, a_{2k} is a sum of squares in K and so, as for a_0, it is a sum of 2^n squares in K. We claim that without loss of generality f may be taken monic; for $f = a_{2k} \cdot f/a_{2k}$ where f/a_{2k} is monic. Here $f(X)$ and a_{2k} are sums of squares in $K(X)$ so f/a_{2k} is a sum of squares. If we have proved the result for monic polynomials, then f/a_{2k} is a sum of at most 2^n squares in $K(X)$ and so is a_{2k} and therefore so is f. Thus without loss of generality f is monic and also clearly square-free (just divide out by the square factor). Now making a common denominator we have

$$f(X) = (g_1^2(X) + \ldots + g_t^2(X))/h^2(X) \qquad (5.2)$$

Let α be a root of $f(X) = 0$ and let $p(X) = \operatorname{irr}(\alpha, K)$. Then there exists an index i such that $g_i(\alpha) \neq 0$, for otherwise $p(X)|g_i(X)$ for all i and so $p^2(X)|f^2(X)$ by (5.2).

Thus putting $X = \alpha$ in (5.2) we get

$$0 = f(\alpha) \cdot h^2(\alpha) = \sum_{i=1}^{t} g_i^2(\alpha).$$

So $L = K(\alpha)$ is not formally real, hence by hypothesis $s(L) \leq 2^{m-1}$ and so by Theorem 3.5, $p(X)$ is a sum of 2^n squares in $K(X)$. This being so for each irreducible factor $p(X)$ of $f(X)$, we see that $f(X)$ is a sum of 2^n squares in $K(X)$. □

To prove Theorem 5.3, we need to generalize to 2^n variables the quadratic form $X^2 + aY^2$ (which we have already met in Chapter 4), which satisfies the following curious identity:

$$(X_1^2 + aX_2^2)(Y_1^2 + aY_2^2) = (X_1Y_1 + aX_2Y_2)^2 + a(X_1Y_2 - X_2Y_1)^2 \qquad (5.3)$$

Compare this with equation (1.1). The generalization of this form to four variables is

$$\begin{aligned}\Phi(X_1, X_2, X_3, X_4) &= X_1^2 + aX_2^2 + b(X_3^2 + aX_4^2)\\ &= X_1^2 + aX_2^2 + bX_3^2 + abX_4^2.\end{aligned}$$

This satisfies an identity similar to (1.2), one which we shall be using extensively in the proof of Pourchet's theorem (Chapter 17).

In general let $n = 2^m$ and let

$$\begin{aligned}\Phi(\mathbf{X}) &= \Phi(X_1, \ldots, X_n)\\ &= (X_1^2 + a_1X_2^2) + a_2(X_3^2 + a_1X_4^2)\\ &\quad + a_3(X_5^2 + a_1X_6^2 + a_2(X_7^2 + a_1X_8^2)) + \ldots\\ &\quad + a_m(X_{2^{m-1}+1}^2 + a_1X_{2^{m-1}+2}^2 + \ldots + a_1a_2\ldots a_{m-1}X_{2^m}^2),\end{aligned}$$

i.e.

$$\begin{aligned}\Phi(X_1, \ldots, X_{2^m}) &= \Phi(X_1, \ldots, X_{2^{m-1}})\\ &\quad + a_m\Phi(X_{2^{m-1}+1}, \ldots, X_{2^m}).\end{aligned} \qquad (5.4)$$

This is the so-called m-fold Pfister form.

To prove Theorem 5.3, we need the following

Theorem 5.4. *Let L be a non-formally real algebraic extension of $R(X_1, \ldots, X_m)$ (R a real closed field). Then the Pfister form Φ represents the element $b^2 + c^2$ in L for any $b, c \in L^*$, ($n = 2^m$ of course).*

From Theorem 5.4, it is easy to deduce Theorem 5.3 as follows.

Consider the sum $E = e_1^2 + \ldots + e_{2^{m+1}}^2$ ($e_j \in L$) of 2^{m+1} squares in L. Put $b = e_1$, $c = e_2$,

$$\begin{aligned}a_1 &= e_3^2 + e_4^2,\\ a_2 &= e_5^2 + e_6^2 + e_7^2 + e_8^2,\\ &\ldots\ldots\ldots\ldots\ldots\ldots\\ a_m &= e_{2^m+1}^2 + e_{2^m+2}^2 + \ldots + e_{2^{m+1}}^2.\end{aligned}$$

By Theorem 5.4, $b^2 + c^2 = \Phi(X_1, \ldots, X_{2^m})$ is solvable in L say

$$b^2 + c^2 = \Phi(\tau_1, \tau_2, \ldots, \tau_{2^m}) \ (\tau_j \in L).$$

Then

$$\begin{aligned}
E &= b^2 + c^2 + a_1 + a_2 + \ldots + a_m \\
&= \Phi(\tau_1, \ldots, \tau_{2^m}) + a_1 + \ldots + a_m \\
&= \tau_1^2 + a_1\tau_2^2 + a_2(\tau_3^2 + a_1\tau_4^2) + a_3(\tau_5^2 + a_1\tau_6^2 + a_2\tau_7^2 + a_1a_2\tau_8^2) \\
&\quad + \ldots + \ldots + a_1 + a_2 + \ldots + a_m \\
&= \tau_1^2 + a_1(1 + \tau_2^2) + a_2(1 + \tau_3^2 + a_1\tau_4^2) + \\
&\qquad a_3(1 + \tau_5^2 + a_1\tau_6^2 + a_2\tau_7^2 + a_1a_2\tau_8^2) + \ldots \\
&= \tau_1^2 + (e_3^2 + e_4^2)(1 + \tau_2^2) + (e_5^2 + e_6^2 + e_7^2 + e_8^2)(1 + \tau_3^2 + a_1\tau_4^2) + \ldots.
\end{aligned}$$

Now this is the sum of one square plus two squares plus four squares and so on up to 2^m squares, by the 2-, 4-, ... square identities. So the right side equals the sum of $2^{m+1} - 1$ squares. Thus any sum of 2^{m+1} squares (viz. E) is a sum of $2^{m+1} - 1$ squares.

If now $-1 = e_1^2 + \ldots e_s^2$ and $s \geq 2^{m+1}$, then this may be successively reduced to a sum of $2^{m+1} - 1$ squares; so $s(L) < 2^{m+1} \leq 2^m$ (being a power of 2). □

It remains to prove Theorem 5.4. For this we need to develop further properties of the Pfister form Φ.

The first result we prove is an identity similar to equation (2.1), which is satisfied by this Pfister form Φ. We have the following

Theorem 5.5. *Let $n = 2^m$ and let $\Phi(X_1, \ldots, X_{2^m})$ be inductively defined by*

$$\Phi(X_1, X_2) = X_1^2 + aX_2^2$$

and (5.4) above. Then $\Phi(\mathbf{X})\Phi(\mathbf{Y}) = \Phi(\mathbf{Z})$ where the components Z_j of $\mathbf{Z} = (Z_1, \ldots, Z_{2^m})$ are linear functions of the Y_j, with coefficients in $K(X_1, \ldots, X_{2^m})$, where char$k \neq 2$.

Proof. We use introduction on m. The result is true for $n = 2$ by (5.3) above. So suppose the result is true for m and let us try to prove it for $m + 1$ i.e.

$$\Phi(X_1, \ldots, X_{2^{m+1}})\Phi(Y_1, \ldots, Y_{2^{m+1}}) = \Phi(Z_1, \ldots, Z_{2^{m+1}}).$$

Write $\mathbf{Z} = (Z_1, \ldots, Z_{2^{m+1}}) = A\mathbf{Y}$ for some suitable matrix $A = (a_{ji})$ with $a_{ji} \in K(X_1, \ldots, X_{2^m})$. Write $\Phi = \mathbf{X}' A_{2^m \times 2^m} \mathbf{X} = \mathbf{X}' A_m \mathbf{X}$, say, where the matrix A_m of Φ looks like the following:

$$A_1 = \begin{pmatrix} 1 & 0 \\ 0 & a_1 \end{pmatrix},$$

$$A_2 = \begin{pmatrix} 1 & 0 & 0 & 0 \\ 0 & a_1 & 0 & 0 \\ 0 & 0 & a_2 & 0 \\ 0 & 0 & 0 & a_2a_1 \end{pmatrix} = \begin{pmatrix} A_1 & 0 \\ 0 & a_2A_1 \end{pmatrix}, \ldots, \ldots$$

etc. until

$$A_m = \begin{pmatrix} A_{m-1} & 0 \\ 0 & a_mA_{m-1} \end{pmatrix}.$$

So $\Phi(X_1, \ldots, X_{2^m}) = \mathbf{X}'A_m\mathbf{X}$ and we have to prove $\mathbf{X}'A_m\mathbf{X}\mathbf{Y}'A_m\mathbf{Y} = \mathbf{Z}'A_m\mathbf{Z}$ ($\mathbf{Z} = T\mathbf{Y}$), i.e. $\mathbf{Y}'(\mathbf{X}'A_m\mathbf{X}A_m - T'A_mT)\mathbf{Y} = 0$ (since $\mathbf{X}'A_m\mathbf{X}$ is a scalar) and as this is true for all $\mathbf{Y}$, we have

$$\mathbf{X}'A_m\mathbf{X}A_m = T'A_mT,$$

or $\Phi(X_1, \ldots, X_{2^m})A_m = T'A_mT$ for a suitable T. So assume this for m (remember $n = 2^m$) for the two sets of variables $X_1, \ldots, X_{2^m}$ and $X_{2^m+1}, \ldots, X_{2^{m+1}}$. We thus have, by the induction hypothesis, two matrices (the same matrix actually, just different variables as entries) $T^{(1)}, T^{(2)}$, say, such that

$$\Phi(X_1, \ldots, X_{2^m})A_m = T^{(1)'}A_mT^{(1)}$$

and

$$\Phi(X_{2^m+1}, \ldots, X_{2^{m+1}})A_m = T^{(2)'}A_mT^{(2)}.$$

We are required to prove that

$$\Phi(X_1, \ldots, X_{2^{m+1}})A_{m+1} = T'A_{m+1}T$$

for a suitable matrix T. We try

$$T = \begin{pmatrix} T^{(1)} & a_{m+1}T^{(2)} \\ T^{(2)} & M \end{pmatrix}$$

where M is to be determined. We want

$$\Phi(X_1, \ldots, X_{2^{m+1}}) \begin{pmatrix} A_m & 0 \\ 0 & a_{m+1}A_m \end{pmatrix}$$

$$= \begin{pmatrix} T^{(1)} & a_{m+1}T^{(2)} \\ T^{(2)} & M \end{pmatrix} \begin{pmatrix} A_m & 0 \\ 0 & a_{m+1}A_m \end{pmatrix} \begin{pmatrix} T^{(1)} & a_{m+1}T^{(2)} \\ T^{(2)} & M \end{pmatrix}$$

$$= \begin{pmatrix} B & C \\ D & E \end{pmatrix},$$

where $B = T^{(1)'}A_mT^{(1)} + a_{m+1}T^{(2)'}A_mT^{(2)}$, $C = a_{m+1}(T^{(1)'}A_mT^{(2)} + T^{(2)'}A_mM)$, $D = a_{m+1}(T^{(2)'}A_mT^{(1)} + M'A_mT^{(2)})$, and $E = a_{m+1}^2T^{(2)'}A_mT^{(2)} + a_{m+1}M'A_mM$. Now, by the induction hypothesis, $B = \Phi(X_1, \ldots, X_{2^m})A_m + a_{m+1}\Phi(X_{2^m+1}, \ldots, X_{2^{m+1}})A_m$. By choosing $M = -A_m^{-1}T^{(2)'^{-1}}T^{(1)'}A_mT^{(2)}$, it follows that $C = D = 0$. Finally we can check that $E = A_{m+1}B$, and we are done. This completes the proof of Theorem 5.5. □

Now let $B(\mathbf{X},\mathbf{Y})$ be the bilinear form associated with the quadratic form Φ i.e.

$$\Phi(\mathbf{X}+\mathbf{Y}) = \Phi(\mathbf{X}) + \Phi(\mathbf{Y}) + 2B(\mathbf{X},\mathbf{Y})$$

so that $B(\mathbf{X},\mathbf{Y}) = \mathbf{X}A_m\mathbf{Y}'$. Write $\mathbf{e} = (1,0,0,\ldots,0)$ ($2^m = n$ coordinates). We have the following

Lemma 1. $B(\mathbf{Z},\mathbf{e}) = B(\mathbf{X},\mathbf{Y})$.

Proof. For $m = 1$ we have: The left side equals

$$(Z_1, Z_2)\begin{pmatrix} 1 & 0 \\ 0 & a_1 \end{pmatrix}\begin{pmatrix} 1 \\ 0 \end{pmatrix} = (Z_1, Z_2)\begin{pmatrix} 1 \\ 0 \end{pmatrix} = Z_1;$$

and the right side equals

$$(X_1, X_2)\begin{pmatrix} 1 & 0 \\ 0 & a_1 \end{pmatrix}\begin{pmatrix} Y_1 \\ Y_2 \end{pmatrix} = (X_1, X_2)\begin{pmatrix} Y_1 \\ a_1 \ Y_2 \end{pmatrix} = X_1Y_1 + a_1X_2Y_2$$

and these are equal by (5.3).

Now use induction on m. We are required to prove that $\mathbf{Z}A_m\mathbf{e}' = \mathbf{X}A_m\mathbf{Y}'$ i.e. that

$$(Z_1, \ldots, Z_{2^m})\begin{pmatrix} 1 & & & & \\ & a_1 & & 0 & \\ & & a_2 & & \\ & & & a_1a_2 & \\ & 0 & & & \ddots \end{pmatrix}\begin{pmatrix} 1 \\ 0 \\ 0 \\ \vdots \\ 0 \end{pmatrix}$$

$$= (X_1, \ldots, X_{2^m})\begin{pmatrix} 1 & & 0 \\ & a_1 & \\ 0 & & \ddots \end{pmatrix}\begin{pmatrix} Y_1 \\ Y_2 \\ \vdots \\ Y_{2^m}. \end{pmatrix}.$$

In other words

$$(Z_1, \ldots, Z_{2^m})\begin{pmatrix} 1 \\ 0 \\ \vdots \\ 0 \end{pmatrix} = (X_1, \ldots X_{2^m})\begin{pmatrix} Y_1 \\ a_1Y_2 \\ a_2Y_3 \\ a_2a_1Y_4 \\ \vdots \end{pmatrix}$$

i.e. $Z_1 = X_1Y_1 + a_1X_2Y_2 + a_2X_3Y_3 + a_1a_2X_4Y_4 + \ldots + a_1a_2\ldots a_mX_{2^m}Y_{2^m}$. So suppose this is true - our induction hypothesis for m.

Then $(Z_1)_{m+1}$ is the first coordinate of TY, where we have

$$TY = \begin{pmatrix} T^{(1)} & a_{m+1}T^{(2)} \\ T^{(2)} & M \end{pmatrix}\begin{pmatrix} Y_1 \\ \vdots \\ Y_{2^{m+1}} \end{pmatrix}.$$

Then, using the induction hypothesis,

$$\begin{aligned}
(Z_1)_{m+1} &= T^{(1)}\begin{pmatrix} Y_1 \\ \vdots \\ Y_{2^m} \end{pmatrix} + a_{m+1}T^{(2)}\begin{pmatrix} Y_{2^m+1} \\ \vdots \\ Y_{2^{m+1}} \end{pmatrix} \\
&= (X_1Y_1 + a_1X_2Y_2 + \ldots + a_1a_2\ldots a_mX_{2^m}Y_{2^m}) \\
&\quad + a_{m+1}(X_{2^m+1}Y_{2^m+1} + a_1X_{2^m+2}Y_{2^m+2} + \ldots \\
&\quad + a_1a_2\ldots a_mX_{2^{m+1}}Y_{2^{m+1}}) \\
&= \mathbf{X}A_{m+1}\mathbf{Y}' = B(\mathbf{X},\mathbf{Y})
\end{aligned}$$

as required. □

Lemma 2. *Let $n = 2^m$. Then*

$$\Phi(\mathbf{Y})\Phi(\mathbf{Z}/\Phi(\mathbf{Y}) + i\mathbf{e}) = \Phi(\mathbf{X} + i\mathbf{Y}).$$

Proof. The left hand side is, using $\Phi(\lambda\mathbf{X}) = \lambda^2\Phi(\mathbf{X})$, $B(\lambda\mathbf{X},\mu\mathbf{Y}) = \lambda\mu B(\mathbf{X},\mathbf{Y})$,

$$\begin{aligned}
&\Phi(\mathbf{Y})(\Phi(\mathbf{Z}/\Phi(\mathbf{Y})) + \Phi(i\mathbf{e}) + 2B(\mathbf{Z}/\Phi(\mathbf{Y}), i\mathbf{e}) \\
&= \Phi(\mathbf{Y})\left(\Phi(\mathbf{Z})/(\Phi(\mathbf{Y}))^2 - 1 + \frac{2i}{\Phi(\mathbf{Y})}B(\mathbf{Z},\mathbf{e})\right) \\
&= \Phi(\mathbf{Y})\left(\frac{\Phi(\mathbf{X})\Phi(\mathbf{Y})}{(\Phi(\mathbf{Y}))^2} - 1 + \frac{2i\ B(\mathbf{X},\mathbf{Y})}{\Phi(\mathbf{Y})}\right), \\
&= \Phi(\mathbf{X}) - \Phi(\mathbf{Y}) + 2i\ B(\mathbf{X},\mathbf{Y}) \\
&= \Phi(\mathbf{X} + i\mathbf{Y}),
\end{aligned}$$

since $\Phi(\mathbf{Z}) = \Phi(\mathbf{X})\cdot\Phi(\mathbf{Y})$ by Theorem 5.5. □

We now require a generalization of Lemma 2 of Chapter 4, due to Tsen and Lang. Indeed their result is much more general and we give here only the relevant special case required for our purpose.

Theorem 5.6 (Tsen-Lang). *Let R be a real closed field and let L be a finite extension of $R(Z_1,\ldots,Z_m)$ which is not formally real. Then any quadratic form over $L(i)$ ($i^2 = -1$) in at least $2^m + 1$ variables represents zero non-trivially in $L(i)$.*

Using this result (a proof follows soon) enables us to prove Theorem 5.4.

Proof of Theorem 5.4. If Φ represents 0, it is universal (see Theorem 11.4) and so represents $b^2 + c^2$. If not, consider the form $\Phi - (b + ic)Z^2$ over $L(i)$, which has more than 2^m variables. This then represents 0 over $L(i)$ by the

Tsen-Lang theorem. Choose a representation with $Z \neq 0$ and so (dividing by Z) with $Z = 1$. That is Φ represents $b+ic$ in $L(i)$, say, $\Phi(\mathbf{X}+i\mathbf{Y}) = b+ic$ and so also $\Phi(\mathbf{X} - i\mathbf{Y}) = b - ic$. Multiplying and using Lemma 2, gives

$$\Phi^2(\mathbf{Y}) \cdot \Phi(\mathbf{Z}/\Phi(\mathbf{Y}) + i\mathbf{e}) \cdot \Phi(\mathbf{Z}/\Phi(\mathbf{Y}) - i\mathbf{e}) = b^2 + c^2 \qquad (5.5)$$

We shall show that the left side here equals $\Phi(\lambda)$ for some $\lambda \in L$. Let $\mathbf{Z}/(\Phi(\mathbf{Y}) = (u_1, u_2, \ldots, u_{2^m}) = (u_1, \mathbf{v})$, say. Then $\mathbf{Z}/\Phi(\mathbf{Y})+i\mathbf{e} = (u_1 + i, \mathbf{v})$ and $\mathbf{Z}/\Phi(\mathbf{Y}) - i\mathbf{e} = (u_1 - i, \mathbf{v})$; also for any vector $\mathbf{X}$, $\Phi(\mathbf{X}) = X_1^2 + g(X_2^2, \ldots, X_{2^m})$. Using this we now find that the left hand side of (5.5) above equals

$$\begin{aligned}
&\Phi^2(\mathbf{Y})\{(u_1 + i)^2 + g(\mathbf{v})\}\{(u_1 - i)^2 + g(\mathbf{v})\} \\
&\quad = \Phi^2(\mathbf{Y})\{(u_1^2 + 1)^2 + (g(\mathbf{v}))^2 + 2(u_1^2 - 1)g(\underline{v})\} \\
&\quad = \Phi^2(\mathbf{Y})\{(u_1^2 + 1 - g(\mathbf{v}))^2 + 4u_1^2 g(\mathbf{v})\} \\
&\quad = \Phi^2(\mathbf{Y}) \cdot \Phi(u_1^2 + 1 - g(\mathbf{v}), 2u_1\mathbf{v}).
\end{aligned}$$

Now using Theorem (5.5), we have $\Phi(\mathbf{X})\cdot\Phi(\mathbf{Y}) = \Phi(\mathbf{Z})$, so the above equals $\Phi(\lambda)$ $(\lambda \in L)$ as required. This finally completes the proof of Theorem 5.1. □

We now give the proof of the Tsen-Lang theorem; see [T2], [L6], [R5], [G2]. Indeed we prove the following more general result.

Theorem 5.6′ (Tsen-Lang). *Let L be a finite (and so algebraic) extension of $K = \mathbb{C}(Z_1, \ldots, Z_m)$ and let $f_j(X_1, \ldots, X_n)$ $(1 \leq j \leq r)$ be r quadratic forms in the n variables $X_1, \ldots, X_n$ over L. If $n \geq r \cdot 2^m + 1$ then the f_j's have a common non-trivial zero in L.*

Remarks.

1. Actually much stronger results than Theorem 5.6′ can be proved (see [L6]).
2. Theorem 5.6 follows on taking $r = 1$ in Theorem 5.6′. Since we only need Theorem 5.6 why do we not adapt the proof of Theorem 5.6′ for the special case $r = 1$ and thus simplify matters? The answer will be clear in the proof of Theorem 5.6.
3. We shall prove Theorem 5.6′ for $\mathbb{C} = \mathbb{R}(i)$; however, the proof works for any real closed field R, (so that $C = R(i)$ is algebraically closed).

We need the following

Lemma 3. *Let $f_j(X_1, \ldots, X_n)$ $(1 \leq j \leq r)$ be defined over K. If $n \geq r \cdot 2^m + 1$ then the f_j have a non-trivial zero in K.*

Proof. We use induction on m. For $m = 0$, the result reads: If $n \geq r + 1$

then the r quadratic forms $f_j(X_1, \ldots, X_n)$ have a common non-trivial zero in K. The proof of this is immediate since $\mathbb{C}$ is algebraically closed and there are more X's than there are f's; so just solve.

Now suppose the lemma is true for $m-1$; we prove it for m. On clearing denominators we may suppose without loss of generality that all the $f_j(X_1, \ldots, X_n) \in \mathbb{C}[Z_1, \ldots, Z_m]$. Let t be the largest of the degrees in Z_m of the coefficients in $f_1, \ldots, f_r$, considered as polynomials in Z_m with coefficients in $\mathbb{C}[Z_1, \ldots, Z_{m-1}]$. We wish to make each $f_j(X_1, \ldots, X_n) = 0$ $(1 \le j \le r)$ for suitable $X_j \in \mathbb{C}(Z_1, \ldots, Z_m)$, not all zero. Take

$$X_j = a_{0j} + a_{1j}Z_m + a_{2j}Z_m^2 + \ldots + a_{sj}Z_m^s \quad (1 \le j \le n) \tag{5.6}$$

where the $a_{ij} \in \mathbb{C}(Z_1, \ldots, Z_{m-1})$ are to be determined; we take the same s (large) in each X_j. Insert these values of X_j in each

$$f_k(X_1, \ldots, X_n) = 0$$

to get equations of the type

$$f_i(X_1, \ldots, X_n) = \sum_{p=0}^{2s+t} A_p^{(i)} Z_m^p \quad (1 \le i \le r) \tag{5.7}$$

where the $A_p^{(i)}$ are quadratic forms in the $n(s+1)$ variables a_{ij} appearing in (5.6) with coefficients in $\mathbb{C}(Z_1 \ldots, Z_{m-1})$. For example suppose, without loss of generality, $f_1 = \alpha_1 X_1^2 + \ldots + \alpha_n X_n^2$ to be diagonal, where α_j are polynomials in Z_m of degree at most t with coefficients in $\mathbb{C}(Z_1, \ldots, Z_{m-1})$. Then

$$f_1 = (b_{01} + b_{11}Z_m + \ldots + b_{t1}Z_m^t)(a_{01} + a_{11}Z_m + \ldots + a_{s1}Z_m^s)^2 + \ldots$$

and on expanding and rewriting as a polynomial in Z_m we get what we want since the highest power of Z_m is $2s+t$. Now remembering that there is also the coefficient of Z_m^0, we see that there are $(2s+t+1)r$ quadratic forms $A_p^{(i)}$ in (5.7) in the $n(s+1)$ variables a_{ij} of (5.6). By our induction hypothesis these have a non-trivial zero in $\mathbb{C}(Z_n, \ldots, Z_{m-1})$ if

$$n(s+1) > (2s+t+1) \cdot r \cdot 2^{m-1}$$

i.e. if

$$(n - r \cdot 2^m)s > (t+1) \cdot r \cdot 2^{m-1} - n,$$

and by choosing s sufficiently large, this inequality can be satisfied. In other words not all a_{ij} equal zero i.e. not all X_j equal zero, as required. □

Proof of Theorem 5.6′. Let L be of degree t over K and let $w_1, \ldots, w_t$ be a basis of L/K. If $\alpha \in L$ then

$$\alpha = \sum_{i=1}^{t} a_i w_i \quad (a_i \in K).$$

Put

$$X_k = \sum_{p=1}^{t} U_{kp} w_p \quad (U_{kp} \in K,\ 1 \le k \le n). \tag{5.8}$$

Then as in the lemma

$$f_i(X_1, \ldots, X_n) = \sum_{p=1}^{t} B_p^{(i)} w_p \quad (1 \le i \le r) \tag{5.9}$$

where the $B_p^{(i)}$ are quadratic forms in the nt variables U_{kp} with coefficients in K. The number of forms is rt so by the lemma they have a non-trivial zero in K if $nt > r \cdot t \cdot 2^m$ which is the hypothesis. Thus there exists U_{kp} not all 0 satisfying (5.8) i.e. X_k not all zero satisfying (5.9) in L. □

Exercises

1. Show that if F is a field of Stufe s, then

$$s \le P(F) \le s + 1.$$

(Hint: say $-1 = a_1^2 + \cdots + a_s^2$ $(a_j \in F)$, then each a in F can be written as

$$a = \left(\frac{a+1}{2}\right)^2 - \left(\frac{a-1}{2}\right)^2 \tag{5.10}$$

$$= \left(\frac{a+1}{2}\right)^2 + a \text{ sum of } s \text{ squares}.$$

2. Let A be a ring with 1 and let the Stufe of A (defined as for fields) be $s(A) = s$. Show that

$$s \le P(A) \le s + 2.$$

(Hint: let $a \in A$ be a sum of squares, say,

$$a = c_1^2 + \cdots c_n^2.$$

Put $x = c_1 + \cdots + c_n + 1$. Show that $a - x^2$ is of the form $(y+1)^2 - y^2$ (for a suitable y). Then

$$a = x^2 + (y+1)^2 + a \text{ sum of } s \text{ squares}.$$

3. Show that if *either* 2 is a unit in A *or* $s(A)$ is even, then $P(A) \le s + 1$.

If 2 is a unit in A, the identity (5.10) above gives $P(A) \le s + 1$. So let $s(A)$ be even. We have

$$a = \text{a sum of squares}$$

and

$$-1 = \text{a sum of squares}$$

So $-a$ = a sum of squares, and by the proof of Exercise 1, $-a = \alpha^2 + \beta^2 - \gamma^2$ for some $\alpha, \beta, \gamma \in A$, giving $\gamma^2 - a = \alpha^2 + \beta^2$. Also $-1 = z_1^2 + \cdots + z_s^2$ (say). Multiplying we get using the 2-identity $s/2$ times,

$$\begin{aligned} a - \gamma^2 &= (\alpha^2 + \beta^2)(z_1^2 + \cdots + z_s^2) \\ &= \text{a sum of squares.} \end{aligned}$$

Hence a is a sum of $s+1$ squares giving $P(A) \leq s+1$)

4. Prove that $P(K(X)) = s+1$ where s is the Stufe of K and X is transcendental over K (see Theorem 16.3). (By Exercise 6 of Chapter. 3 and Exercise 1 above, we have $s \leq P(K(X)) \leq s+1$ where $s = s(K) = s(K(X))$. If $P = s$, then $X = \left(\frac{X+1}{2}\right)^2 - \left(\frac{X-1}{2}\right)^2$ would be a sum of s squares in $K(X)$. Then $X = f_1^2 + \cdots + f_s^2$, $f_j \in K[X]$ by Cassels' lemma of Chapter 2. Now equate coefficients of the highest power of X on the right side to zero to contradict $s(K) = s$.

5. Let K be an algebraic number field. Show that $P(K) \neq 1$. For otherwise $m = 1^2 + \cdots + 1^2$ (m terms), a sum of squares in K and so $= \alpha^2$ ($\alpha \in K$) since $P(K) = 1$; i.e. $\sqrt{m} = \alpha \in K$ for all $m \in \mathbb{N}$ giving $[K : \mathbb{Q}] = \infty$; a contradiction.

6. Let K be an algebraic number field which is not formally real (in particular $i \notin K$). Suppose there is an odd rational prime p which stays prime in $K(i)$. Show that $P(K) \neq 2$.
Hint: suppose to the contrary that $P(K) = 2$. Since $p = 1^2 + \cdots + 1^2$ is an SOS, $p = \alpha^2 + \beta^2$ ($\alpha, \beta \in K$). Clearing denominators find the least positive rational integer t such that

$$\begin{aligned} t^2 p &= x^2 + y^2 \qquad (*) \\ &= (x+iy)(x-iy) \ \text{ in } K[i], \end{aligned}$$

where $x, y \in [K]$, the ring of integers of K. But p is prime in $K[i]$, so, without loss of generality, $p | x + iy$, and $\sigma p | \sigma(x+iy)$, where σ is the automorphism $i \to -i$ of $K(i)/K$. That is $p | x - iy$, since $x, y, p \in K$. Thus $p | x + iy$ and $x - iy$, so $p | x$ and y, giving a contradiction to the minimality of t in $(*)$).

6

Introduction to Hilbert's theorem (1888) in the Ring $\mathbb{R}[X_1, X_2, \ldots, X_n]$

In Chapters 4 and 5 we looked at the positive semi-definite (PSD) functions and sums of squares (SOS) in the function field $\mathsf{R}(X_1, X_2, \ldots, X_n)$. In the present chapter and the following three, we consider the same problem in the ring $\mathsf{R}[X_1, X_2, \ldots, X_n]$. As we shall see, the situation is more complicated and more interesting as it leads to a number of unanswered questions and opens up many avenues for research.

Already in 1888, Hilbert [H3] had studied the question of whether a PSD real form is always a SOS of other real forms. By a form, we mean of course, a homogeneous polynomial. Denote by $\mathcal{P}_{n,m}$ the class of all real forms $F(X_1, \ldots, X_n)$ in the n real variables $X_1, \ldots, X_n$ and of degree m (the n-ary m-ics) which are PSD (i.e. $F(a_1, a_2, \ldots, a_n) \geq 0$ for all $a_j \in \mathsf{R}$). We let $\Sigma_{n,m}$ be the class of all n-ary m-ics which can be written as sums of squares of other forms, in fact of n-ary $\frac{m}{2}$-ics, so that m is necessarily even. We have clearly $\Sigma_{n,m} \subseteq \mathcal{P}_{n,m}$. Hilbert asked:

$$\textit{For what pairs } (n, m) \textit{ is } \mathcal{P}_{n,m} \textit{ equal to } \Sigma_{n,m}? \qquad (6.1)$$

Hilbert solved the problem completely, his answer being the following.

Theorem 6.1.
(a) $\mathcal{P}_{n,m} = \Sigma_{n,m}$ *if*

(i) $n = 2$, *all* m *(even)* ≥ 2,
(ii) *all* $n \geq 2$, $m = 2$,
(iii) $n = 3$, $m = 4$, *and indeed*
(iv) $f \in \mathcal{P}_{3,4}$ *implies* f *is a sum of* 3 *squares*

(b) *In all other cases* $\Sigma_{n,m} \subset$ *(strictly)* $\mathcal{P}_{n,m}$.

Before we give a proof of this theorem, we make a few remarks about its historical development.

$m \backslash n$	2	3	4	5	...
2	√	√	√	√	...
4	√	√	×	×	...
6	√	×	×	×	...
8	√	×	×	×	...
⋮	⋮	⋮	⋮	⋮	⋱

Since m is even, Theorem 6.1 may be summarized in the above chart, where a tick (√) denotes an affirmative answer to (6.1) whereas a cross (×) denotes a negative one. Hilbert had already shown that in the two basic cases $(3,6)$ and $(4,4)$, it is possible to construct, in principle, forms in $\mathcal{P}-\Sigma$. Why these two cases are referred to as the basic cases will be explained in the proof of the theorem. Hilbert's method was rather complicated and elaborate and did not lend itself to a really practical construction. As a result, no explicit example was obtained in [H3] and indeed no such example appeared in print in the 80 years following Hilbert's paper.

Following through Hilbert's method, W.S. Ellison (unpublished) worked out a very complicated ternary sextic which was PSD but not a SOS i.e. was contained in $\mathcal{P}_{3,6} - \Sigma_{3,6}$.

Independently of Hilbert's method, T.S. Motzkin [M2] showed that the Motzkin polynomial (form)

$$M(x,y,z) = z^6 + x^4y^2 + x^2y^4 - 3x^2y^2z^2$$

(or as a non-homogeneous polynomial, the really simple looking thing $x^2y^2(x^2+y^2-3)+1$) belongs to $\mathcal{P}_{3,6} - \Sigma_{3,6}$. This was, therefore, the first form to appear in print as an example to prove b) of Theorem 6.1.

Using a drastic simplification of Hilbert's method and independently of Motzkin, R.M. Robinson [R8] constructed new examples of forms in $\mathcal{P}_{4,4} - \Sigma_{4,4}$ and also of forms in $\mathcal{P}_{3,6} - \Sigma_{3,6}$. More such examples were obtained by M.D. Choi [C7].

The central idea in Hilbert's proof is to associate a ternary quartic curve in the complex projective plane and then to use the classically well-developed theory of algebraic curves. To quote Choi and Lam, "this beautiful piece of work, albeit a remarkable testament to the mathematical prowess and insight of the young Hilbert, is unfortunately not easy to read". This is so not only because of the century old terminology, but also because of the use of various non-trivial facts such as dimension counts, Max Noether's lemma and the like that are drawn from the theory of the algebraic curves. We shall therefore give two proofs of Hilbert's theorem; one due to Choi and

Lam [C9] which uses only the rudiments of real analysis and the second, a modern simplified version of Hilbert's proof for which my grateful thanks are due to Professor J.W.S. Cassels. The Choi-Lam proof, however, only deals with (iii) of Theorem 6.1.

Proof of (a) (i). Let the (real) form be

$$f(X,Y) = a_m X^m + a_{m-1}X^{m-1}Y + \ldots + a_0 Y^m \quad (a_j \in \mathbf{R}).$$

On dividing throughout by Y^m and letting $X/Y \to X$, we may suppose that f is the polynomial

$$f(X) = a_m X^m + a_{m-1}X^{m-1} + \ldots + a_0.$$

We are given that f is PSD, so by Hilbert's theorem in $\mathbf{R}(X)$ (Theorem 4.1), we know that

$$f(X) = u^2(X) + v^2(X) \quad (u(X), v(X) \in \mathbf{R}(X)).$$

But then by Cassels' lemma (see Chapter 2)

$$f(X) = \alpha^2(X) + \beta^2(X) \quad (\alpha(X), \beta(X) \in \mathbf{R}[X]).$$

Thus $f(X) \in \Sigma$ i.e. $\mathcal{P} \subseteq \Sigma$ so $\mathcal{P} = \Sigma$ □

We can revert back to the form $f(X,Y)$ by noting that m being even, Y^m is a square.

Proof of (a) (ii). Let the form be $f(X_1, X_2, \ldots, X_n)$ ($n \geq 2$, $m = 2$, so that f is a quadratic form) which is PSD. On completing squares we get

$$f = a_1 f_1^2 + \ldots + a_r f_r^2 - b_1 g_1^2 - \ldots - b_t g_t^2$$

$(r + t \leq n,\ a_j, b_j > 0)$. But f is PSD; hence the b_j are all 0 and feeding the a_j into the f_j we get

$$f = (\sqrt{a_1} f_1)^2 + \ldots + (\sqrt{a_r} f_r)^2 \quad (r \leq n)$$

□

Note that here n squares suffice. Thus every PSD quadratic form is a sum of at most n squares of linear forms (linear since $\frac{m}{2} = 1$).

For the proof of (b) of Theorem 6.1, we first note the very important fact that if we can produce examples of forms in

(i) $\mathcal{P}_{3,6} - \Sigma_{3,6}$ and

(ii) $\mathcal{P}_{4,4} - \Sigma_{4,4}$

then we can complete the general case very easily as follows:

If $F(X_1, X_2, \ldots . X_n)$ is a form of degree m which is PSD, so that $m = 2d$ say, but not a SOS, then we can construct from it forms of higher degree which are PSD but not SOS; indeed the form

$$X_1^2 F(X_1, X_2, \ldots, X_n)$$

is clearly PSD; also it is not a SOS; for if say

$$X_1^2 F = \sum_{j=1}^{k} f_j^2, \tag{6.2}$$

then putting $X_1 = 0$ in this gives

$$0 = f_1^2(0, X_2, \ldots, X_n) + f_2^2(0, X_2, \ldots, X_n) + \ldots.$$

It follows that $f_j(0, X_2, \ldots, X_n) = 0$ for all $X_2, \ldots X_n$ (since each $f_j^2 \geq 0$ for all $X_2, \ldots, X_n$) i.e. that $f_j(0, X_2, \ldots, X_n) \equiv 0$ i.e. $f_j(X_1, X_2, \ldots, X_n)$ vanishes when $X_1 = 0$ and so it splits a factor X_1:

$$f_j = X_1 \cdot g_1(X_1, X_2, \ldots, X_n).$$

Hence we may divide out by X_1^2 in (6.2) to get $F = g_1^2 + \ldots, +g_k^2$ giving a contradiction.

Similarly we may regard F (PSD but not a SOS) as a form in the variables $X_1, X_2, \ldots, X_n, X_{n+1}$ and we easily see that F is certainly PSD and also not a SOS of forms in $X_1, X_2, \ldots, X_n, X_{n+1}$. For if say

$$F(X_1, X_2, \ldots, X_n) = \sum_{j=1}^{k} f_j^2(X_1, \ldots, X_n, X_{n+1}),$$

then writing f_j as a polynomial in $X_{n+1}(= x$ say), we get

$$F = \sum_{j=1}^{k} (a_{0j} + a_{1j}x + \ldots + a_{r_j j}x^{r_j})^2$$

where the $a_{ij} \in \mathbb{R}[X_1, \ldots, X_n]$ $(1 \leq i \leq r_j,\ 1 \leq j \leq k)$. Putting $x = 0$, we get

$$F(X_1, \ldots, X_n) = \sum_{j=1}^{k} a_{0j}^2(X_1, \ldots X_n),$$

which is a SOS of forms in $\mathbb{R}[X_1, \ldots, X_n]$ - a contradiction.

The above remarks may be combined to give the following

Theorem 6.2. *If* $F(X_1, \ldots, X_n) \in \mathcal{P}_{n,m} - \Sigma_{n,m}$, *then*

(i) $X_1^{2i}\, F(X_1, \ldots, X_n) \in \mathcal{P}_{n,m+2i} - \Sigma_{n,m+2i}$

(ii) $F(X_1, \ldots, X_n) \in \mathcal{P}_{n+s,m} - \Sigma_{n+s,m}$. (all $s \geq 1$).

Thus to prove (b) of Theorem 6.1, we need only produce examples of (i) ternary sextics (ii) quaternary quartics which are PSD but not SOS. In future these two cases will therefore be referred to as the "*basic*" cases

We now give examples in these basic cases to prove Theorem 6.1, part (b).

(1) The quaternary quartic $q = (X, Y, Z, W) =$

$$W^4 + X^2Y^2 + Y^2Z^2 + Z^2X^2 - 4XYZW$$

is PSD but not a SOS (of quadratic forms necessarily) i.e.

$$q \in \mathcal{P}_{4,4} - \Sigma_{4,4}.$$

(2) The ternary sextics

$$S(X,Y,Z) = X^4Y^2 + Y^4Z^2 + Z^4X^2 - 3X^2Y^2Z^2$$

and

$$M(X,Y,Z) = Z^6 + X^4Y^2 + X^2Y^4 - 3X^2Y^2Z^2$$

are both contained in $\mathcal{P}_{3,6} - \Sigma_{3,6}$.

That they are all PSD follows by the arithmetic-geometric mean inequality (AGI) applied respectively to the non-negative quantities

(i) $$W^4, X^2Y^2, Y^2Z^2, Z^2X^2$$

(ii) $$X^4Y^2, Y^4Z^2, Z^4X^2$$

and

(iii) $$X^4Y^2, Y^4X^2, Z^6.$$

As for not being SOS, there are two methods of tackling the problem and it is worth while giving the proof using either of the methods. The first one compares coefficients and in general turns out to be simpler in principle as well as in practice. The second one is Hilbert's original method, or rather Robinson's simplified version of it. For this latter method one needs to determine the zero set $\mathfrak{S}(f)$ of the form f in question. This in itself is a very interesting study and leads to some striking results which we shall look at in the following chapters.

Let us deal with the quaternary quartic q first:

Method 1 to show that q is not SOS. Suppose to the contrary that

$$q = \sum q_j^2, \quad q_j \in \mathbf{R}[X,Y,Z,W] \tag{6.3}$$

The q_j are quadratic forms and so involve the following monomials:

$$\begin{array}{cccc} \underline{X^2} & XY & XZ & \underline{XW} \\ & \underline{Y^2} & YX & \underline{YW} \\ & & \underline{Z^2} & \underline{TW} \\ & & & W^2 \end{array}$$

But the underlined ones can not occur in any q_j: for first notice that X^2, Y^2, Z^2 can not occur because if some of the q_j did have say X^2 (say a_jX^2 appears in q_j) then equating coefficients of X^4 on both sides of (6.3) we get $a_1^2 + a_2^2 + \ldots = 0$, which is false since $\mathbf{R}$ is formally real. Now we can

easily deduce that XW, YW, ZW also do not occur in any of the q_j (again since q is free of X^2W^2, Y^2W^2, Z^2W^2).

Thus each q_j involves only the monomials W^2, XY, YZ, ZX. But now there is no way of getting the term (monomial) $XYZW$ from $\sum q_j^2$. □

Indeed this argument works over any formally real field.

Method 2 to show that q is not a SOS. We first have to determine the zero set $\mathfrak{S}(q)$ of q. Since we know exactly when the arithmetic mean equals the geometric mean we easily see that

$$\mathfrak{S}(q) = \{(1,0,0,0),(0,1,0,0),(0,0,1,0),(1,1,1,1),$$
$$(1,-1,-1,1),(-1,1,-1,1),(-1,-1,1,1)\}.$$

Indeed the arithmetic mean of $a_1, a_2, \ldots, a_n$ is equal to their geometric mean if and only if the non-zero a_j are all equal to each other. Hence $q(X,Y,Z,W) = 0$ if and only if

$$\begin{aligned}
\text{either } X^2Y^2 &= Y^2Z^2, Z^2X^2 = W^4 = 0\\
\text{or } Y^2Z^2 &= Z^2X^2, X^2Y^2 = W^4 = 0\\
\text{or } Z^2X^2 &= X^2Y^2, Y^2Z^2 = W^4 = 0\\
\text{OR } \quad X^2Y^2 &= Y^2Z^2 = Z^2X^2, W^4 = 0\\
\text{or } X^2Y^2 &= Y^2Z^2 = W^4, Z^2X^2 = 0\\
\text{or } X^2Y^2 &= Z^2X^2 = W^4, Y^2Z^2 = 0\\
\text{or } Z^2X^2 &= Y^2Z^2 = W^4, X^2Y^2 = 0\\
\text{OR } \quad X^2Y^2 &= Y^2Z^2 = Z^2X^2 = W^4.
\end{aligned}$$

The last set gives the four points

$$(1,-1,-1,1),(-1,1,-1,1),(-1,-1,1,1) \text{ and } (1,1,1,1).$$

The others are got using the remaining equations. We always view $\mathfrak{S}(q)$ projectively since we have homogenized all our polynomials into forms. Thus $(0,0,0,0)$ is not to be regarded as a zero since it is not a projective point. Further $(0,0,0,\delta)$ for example is the same as $(0,0,0,1)$ for all $\delta \neq 0$.

Now it is an easy exercise to show that if q_i is a quadratic form in X, Y, Z, W which vanishes on all the above seven points, then

$$q_i = a_i(XY - ZW) + b_i(XZ - YW) + c_i(XW - YZ)$$

and conversely (trivially). This may be checked as follows: Let

$$q_i = \alpha X^2 + \beta Y^2 + \gamma Z^2 + \delta W^2 + aXY + bXZ + cXW + dYZ + eYW + fZW;$$

then $(1,0,0,0) \in \mathfrak{S}(q_1)$ implies $0 = \alpha$, similarly $0 = \beta = \gamma$. Also $(1,-1,-1,1)$

$\in \mathfrak{S}(q_1)$ gives

$$0 = \delta - a - b + c + d - e - f \quad (i)$$

and similarly $$0 = \delta - a + b - c - d + e - f \quad (ii)$$

and $$0 = \delta + a - b - c - d - e + f \quad (iii)$$

and $$0 = \delta + a + b + c + d + e + f \quad (iv)$$

These imply $\delta = 0,\ a + f = b + e = c + d = 0$ □

So now if $q = \sum q_i^2$, then each q_i must also vanish on the seven points. But each such q_i also vanishes on the eighth point $(0,0,0,1)$. Hence q must vanish on $(0,0,0,1)$; but it does not. Hence $q \neq \sum q_i^2$ as required. □

This completes the quaternary quartic case using both the methods. We next look at the two ternary sextics; S denotes the Robinson ternary sextic and M the Motzkin ternary sextic; the symbol $R(X,Y,Z)$ is reserved for another beautiful even symmetric ternary sextic of Robinson viz.

$$\begin{aligned} R(X,Y,Z) = X^6 + Y^6 + Z^6 & \\ & - (X^4Y^2 + Y^4Z^2 + Z^4X^2 + X^2Y^4 + Y^2Z^4 \\ & + Z^2X^4) + 3X^2Y^2Z^2, \end{aligned}$$

which is also PSD but not a SOS. We shall come to this example later in a different context.

Suppose to the contrary that $M = \sum q_i^2$, where the q_i are ternary cubics. Each q_i involves the following monomials:

$$\begin{array}{ccccccc} & & & \underline{X^3} & & & \\ & & X^2Y & & \underline{X^2Z} & & \\ & \underline{XY^2} & & XYZ & & XZ^2 & \\ \underline{Y^3} & & Y^2Z & & \underline{YZ^2} & & \underline{Z^3} \end{array}$$

Here the underlined ones can not occur in any q_i (as for the quaternary quartic q, first eliminate X^3, Y^3, Z^3; then eliminate XY^2, YZ^2, ZX^2). Thus

$$q_i = a_iX^2Y + b_iY^2Z + c_iZ^2X + d_iXYZ.$$

But then in $\sum q_i^2$ the term $X^2Y^2Z^2$ has coefficient $\sum d_i^2$ which is at least 0, whereas in M, this coefficient is -3; giving a contradiction.

In case of S, the situation is similar: In the equation $S = \sum q_i^2$, the q_i cannot contain the terms X^3, Y^3, Z^3 and so also not the terms XY^2, YZ^2, ZX^2, giving

$$q_i = a_iX^2Y + b_iY^2Z + c_iZ^2X + d_iXYZ.$$

As before the term $X^2Y^2Z^2$ has coefficient -3 on one side and $\sum d_i^2$ on the other side - a contradiction.

It is surprising that the Hilbert-Robinson method fails to work for either of these two ternary sextics S, M. The zeros of S are

$$\mathfrak{S}(S) = \{(1,0,0);(0,1,0);(0,0,1);(1,1,1);\\ (-1,1,1);(1,-1,1);(1,1,-1)\}.$$

However one may show that the set of all cubics vanishing on $\mathfrak{S}(S)$ do not vanish on an eighth point; hence the method fails for S and similarly it fails for M; where

$$\mathfrak{S}(M) = \{(1,0,0);(0,1,0);(1,1,1)\\ (-1,1,1);(1,-1,1);(1,1,-1)\}.$$

So much for (b) of Theorem 6.1. It remains to show (iii) and (iv) of part (a). Before we do this however, we shall list some easy facts concerning forms and polynomials. These may be omitted by readers conversant with this topic.

(1) Forms of degree m in n variables may always be replaced by polynomials of degree m in $n-1$ variables. For example let

$$f(X,Y,Z) = 3X^2YZ^2 + 4XY^4 + Z^3Y^2 + Z^5$$

be a form of degree five. Take the variable Z say; then

$$f(X,Y,Z) = Z^5\left(\frac{3X^2}{Z^2}\cdot\frac{Y}{Z} + \frac{4X}{Z}\cdot\frac{Y^4}{Z^4} + \frac{Y^2}{Z^2} + 1\right)$$

and f may be replaced by the polynomial

$$\varphi(X,Y) = 3X^2Y + 4XY^4 + Y^2 + 1$$

where $X \to X/Z,\ Y \to Y/Z$. Conversely $\varphi(X,Y)$ may be homogenized by the substitution $X \to X/Z,\ Y \to Y/Z$ and $f(X,Y,Z)$ may be recovered.

(2) Suppose $F(X_1, X_2, \ldots, X_n)$ is a SOS:

$$F = \sum_{j=1}^{k} f_j^2(X_1, X_2, \ldots, X_n) \tag{6.4}$$

where F, f_j are polynomials (or forms), F being of degree m. The highest degree terms on the right side can not cancel out (as $\mathbf{R}$ is formally real), so they must exactly match the ones on the left side. Hence F has even degree: $m = 2d$. Further each f_j has degree at most d and at least one f_j has degree exactly d. If F is a form of degree $m(=2d)$, then each f_j is a form of degree $\frac{m}{2} = d$.

(3) The number of coefficients in an n-ary m-ic polynomial F is $\binom{m+n}{n}$ where as, if F is a form, then this number is $\binom{m+n-1}{n-1}$. This is best seen by

considering examples with small m, n and then using induction:

(i) $m = 2d = 2,\ n = 2,$

$$F(X,Y) = a_1X^2 + a_2XY + a_3Y^2 + a_4X + a_5Y + a_6;$$

and there are six coefficients. We have $\binom{m+n}{n} = \binom{4}{2} = 6$ as required. The corresponding form is $a_1X^2 + a_2XY + a_3Y^2$ and there are three coefficients. We have, as required,

$$\binom{m+n-1}{n-1} = \binom{2+1}{1} = 3.$$

(ii) $m = 2,\ n = 3$

$$F(X,Y,Z) = a_1X^2 + a_2Y^2 + a_3Z^2 + a_4XY + a_5XZ + a_6YZ + a_7X + a_8Y + a_9Z + a_{10}$$

and

$$\binom{m+n}{n} = \binom{15}{3} = 10.$$

The corresponding form is

$$a_1X^2 + a_2Y^2 + a_3Z^2 + a_4XY + a_5YZ + a_6ZX$$

and, as required,

$$\binom{m+n+1}{n-1} = \binom{4}{2} = 6.$$

(4) In equation (6.4), the number of f_j on the right hand side can always be made at most $\binom{m+n}{n}$, i.e. if a representation of F as in (6.4) is possible with k large, we can reduce the number of terms to $\binom{m+n}{n}$.

Proof. The set of all real polynomials of degree m in n variables form a vector space of dimension $\binom{m+n}{n}$, a basis being the various monomials. For example if $m = 2 = n$, the dimension is six, a basis being $1, X, Y, X^2, Y^2, XY$.

Thus if $k > \binom{m+n}{n}$; then the k polynomials $f_1^2, f_2^2, \ldots, f_k^2$ are linearly dependent; say

$$\alpha_1 f_1^2 + \ldots + \alpha_k f_k^2 = 0 \quad (\alpha_j \in \mathbf{R}).$$

Let $\alpha = \max|\alpha_j|$, where by renaming, we may suppose that $\alpha = \pm\alpha_k$. Then

$$f_k^2 = \beta_1 f_1^2 + \ldots + \beta_{k-1} f_{k-1}^2\ , \quad |\beta_j| = \left|\frac{\alpha_j}{\alpha_k}\right| \leq 1.$$

Substituting this value of f_k^2 in (6.4), we reduce k by 1; but introduce coefficients in front of the remaining terms:

$$F = f_1^2 + \ldots + f_{k-1}^2 + \beta_1^2 f_1^2 + \ldots + \beta_{k-1} f_{k-1}^2$$
$$(1+\beta_1)f_1^2 + \ldots + (1+\beta_{k-1})f_{k-1}^2\ , \text{ where } |\beta_j| \leq 1;$$

so each of the $1 + \beta_j \geq 0$. Then each of $1 + \beta_j$ may be absorbed in f_j to give $F = g_1^2 + \ldots + g_{k-1}^2$. The process continues until k reduces to $\binom{m+n}{n}$ as required.

(5) Let $F(X_1, \ldots, X_n)$ be a given polynomial of degree m and suppose

$$|F(X_1, \ldots, X_n)| \leq k \qquad (*)$$

for all $(X_1, \ldots, X_n)$ belonging to a set $S \subseteq \mathbf{R}^n$ of points with the property that S^0 (S interior) is non-empty. Then we can compute a bound for the coefficients of F by repeated use of the Lagrange interpolation formula as follows: The formula merely tells us that if $f(x)$ is a polynomial of degree n with coefficients in a field K such that $f(x)$ assumes given values $f(\alpha_1), \ldots, f(\alpha_{n+1})$ at the different points $\alpha_1, \ldots, \alpha_{n+1}$, then $f(x)$ is uniquely determined, by the formula ([W4], p. 66)

$$f(x) = \sum \frac{f(\alpha_i)(x - \alpha_0)\ldots(x - \alpha_{i-1})(x - \alpha_{i+1})\ldots(x - \alpha_n)}{(\alpha_i - \alpha_0)\ldots(\alpha_i - \alpha_{i-1})(\alpha_i - \alpha_{i+1})\ldots(\alpha_i - \alpha_n)}.$$

Now regard F as a polynomial in X_n with coefficients in the field $\mathbf{R}(X_1, \ldots, X_{n-1})$; then the condition $(*)$ determines all the coefficients as bounded. Proceeding by induction we get what was stated.

Thus if k is small, then the coefficients of F will be small.

(6) Suppose we have a convergent sequence of polynomials of degree at most $2d$:

$$F_k(X_1, \ldots, X_n) \to F(X_1, \ldots, X_n) \text{ as } k \to \infty.$$

This could be taken as any one of

(i) pointwise convergence
(ii) uniform convergence on bounded sets
(iii) convergence of the coefficients of F_n.

since they are all equivalent because after all the F_n are well behaved polynomials.

Now suppose each F_k is a SOS:

$$F_k(X_1, \ldots, X_n) = f_{k1}^2(X_1, \ldots, X_n) + f_{k2}^2(X_1, \ldots, X_n) + \ldots + f_{ks}^2(X_1, \ldots, X_n).$$

Here, by (4) above, we can suppose $s \leq \binom{2d+n}{n}$ and so independent of k. All of the polynomials f_{kj} are of degree $\leq d$ and they are uniformly bounded on any bounded set. Hence by (5) above, their coefficients are uniformly bounded; so we can choose a sequence of values of k for which all the coefficients of the polynomials f_{kj} $(j = 1, 2, \ldots, s)$ will approach limits i.e. (iii) in (6) holds. The limiting polynomials will give a representation of $F(X_1, \ldots, X_n)$ as a SOS of real polynomials.

(7) Regard a real polynomial of degree $m = 2d$ in n variables as a point in the $\binom{n+2d}{n}$-dimensional coefficient space V. Then

(I) The PSD polynomials are a closed convex cone $\mathcal{C}_1$ in V:

(a) closed since clearly the limit of a sequence of PSD polynomials is PSD.

(b) convex since f_1, f_2 PSD implies $\lambda f_1 + (1-\lambda) f_2$ is PSD.

(c) cone since f PSD, $\lambda > 0$ implies λf PSD.

(II) The polynomials which are SOS of real polynomials will also form a closed convex cone $\mathcal{C}_2$ (which in general is a proper subset of $\mathcal{C}_1$). That $\mathcal{C}_2$ is a convex cone is trivial as in (I). That it is closed follows by (6) above since the limiting polynomial of SOS is itself a SOS as shown in (6).

(III) Polynomials which are SOS of real binomials (including monomials) will form a still smaller closed convex cone $\mathcal{C}_3$ (check this). And finally,

(IV) Polynomials which are SOS of monomials will give the smallest (so far) closed convex cone $\mathcal{C}_4$.

Thus we have the following inclusions:

$\mathcal{C}_4$	$\subset$	$\mathcal{C}_3$	$\subset$	$\mathcal{C}_2$	$\subset$	$\mathcal{C}_1$	$\subset$	V
SOS of monomials		SOS of monomials and binomials		all SOS		PSD		all polynomials of degree at most $2d$

This last cone $\mathcal{C}_4$ lies in an $\binom{n+d}{n}$ - dimensional subspace of V and so being in a proper subspace of V, it has no interior points.

(V) The strictly positive definite polynomials from the open convex cone $\mathcal{C}_1^0$ and it is easy to see that $\overline{\mathcal{C}_1^0} = \mathcal{C}_1$.

Remark 1. The reader may consult R.M. Robinson's paper [R8] for all these results.

Remark 2. There is an appendix at the end of the book on convex sets.

Exercises

The main theorem of Appendix 1 shows how important extremal forms are. We have neither developed any techniques nor proved any results in our text in this direction. An excellent example is provided by Choi, Lam and Reznick [C11] where all extremal forms for the PSD symmetric ternary sextics are determined and the interested reader is advised to look at this paper in detail. The object of these exercises is to develop techniques where by the three forms S, M, Q defined in this chapter are proved extremal. We denote the set of all extremal forms of $\mathcal{P}_{n,m}$ by $\mathcal{E}(\mathcal{P}_{n,m})$ and of $\Sigma_{n,m}$ by $\mathcal{E}(\Sigma_{n,m})$.

1. Suppose $(n,m) \neq (3,4)$, $(2,m)$, $(n,2)$ so that $\Sigma_{n,m} \subsetneq \mathcal{P}_{n,m}$. Let $f \in \mathcal{P}_{n,m} - \Sigma_{n,m}$ and write (using the theorem of Appendix 1) $f = f_1 + \cdots + f_r$ where $f_j \in \mathcal{E}(\mathcal{P}_{n,m})$. Show that not all f_j are in $\Sigma_{n,m/2}$.

2. Show that if a form $F \in \mathcal{E}(\mathcal{P}_{n,m})$, then (i) $X^{2i}F \in \mathcal{E}(\mathcal{P}_{n,m+2i})$ (all $i \geq 0$),
(ii) $F \in \mathcal{E}(\mathcal{P}_{n+j,m})$ (all $j \geq 0$).

Remark. Thus if we have an example of a form in $\mathcal{E}(\mathcal{P}_{3,6})$ and one in $\mathcal{E}(\mathcal{P}_{4,4})$, then we get examples of forms in $\mathcal{E}(\mathcal{P}_{n,m})$ for each pair $(n,m) \neq (3,4)$, $n \geq 3$, $m \geq 4$.

3. Suppose $F \in \mathcal{P}_{3,6}$ does not contain any terms in X^2Y^4, Y^2Z^4, Z^2X^4 and suppose F vanishes on the zero set $\mathfrak{S}(S)$ of the Robinson ternary sextic S given after Theorem 6.2. By following the hints given below prove that $F = \alpha S$ for some real number α.

(i) Show that F cannot contain X^6, Y^6, Z^6 (use the fact that F vanishes on $(1,0,0), (0,1,0), (0,0,1)$).

(ii) Show that F is free of $X^5Y, X^5Z, Y^5Z, Y^5X, Z^5X, Z^5Y$ (use F is PSD). Hence write F as
$$F(X,Y,Z) = X^4(aY^2 + bYZ) + X^3(cY^3 + \cdots) \qquad (*)$$

(iii) Using $f \geq 0$, show that $b = 0$. Thus F is free of X^4YZ and similarly of Y^4ZX, Z^4XY.

(iv) Noting that all the terms in $(*)$, except aX^4Y^2 and cX^3Y^3 are divisible by Z, show by considering the relation $F(X,Y,0) \geq 0$, that $c = 0$. Thus F is free of X^3Y^3 and similarly of Y^3Z^3, Z^3X^3. Hence
$$\begin{aligned} F = \{&\alpha X^4Y^2 + \beta Y^4Z^2 + \gamma Z^4X^2 - 3\varepsilon X^2Y^2Z^2\} \\ &+ X^2YZ(\lambda Y^2 + \lambda' Z^2) + Y^2ZX(\mu Z^2 + \mu' X^2) \qquad (*)' \\ &+ Z^2XY(\nu X^2 + \nu' Y^2) = F^* + \text{ the rest,} \end{aligned}$$
where F^* is the curly bracket term. Now let $f_1(X,Y) = F(X,Y,Y)$, $f_2(X,Y) = F(X,Y,-Y)$.

(v) Using $\mathfrak{S}(S) \subseteq \mathfrak{S}(F)$ show that $f_1(1,0) = f_1(1,\pm 1) = 0$. Deduce that f_1 is divisible by $Y(X^2 - Y^2)$. Hence using $f_1 \geq 0$, show that $f_1(X,Y) = a_1Y^2(X^2-Y^2)^2$ and similarly $f_2(X,Y) = a_2Y^2(X^2-Y^2)^2$.

Finally using $(*)'$ prove that $\lambda, \mu, \nu, \lambda', \mu', \nu'$ are all zero and that $\alpha = \beta, -2\alpha = \gamma - 3\varepsilon$. Show further, using symmetry, that $\alpha = \beta = \gamma = \varepsilon$. It now follows that $F = \alpha S$ as required.

4. Show by following the steps given below that $S(X,Y,Z) \in \mathcal{E}(\mathcal{P}_{3,6})$.

(i) Suppose $S \geq F \in \mathcal{P}_{3,6}$. Show that $\mathfrak{S}(S) \subseteq \mathfrak{S}(F)$.

(ii) Suppose $F = aX^2Y^4 + bY^2Z^4 + cZ^2X^4 +$ the rest.

Using $F \in \mathcal{P}_{3,6}$, show that F is free of X^5, Y^5, Z^5. Hence using the leading coefficient argument show that a, b, c are non-negative.

Applying the same argument to $S - F$ (which is non-negative), instead of to F, show that a, b, c are all non-positive. Hence using Exercise 3, show that $F = \alpha S$.

(iii) Deduce that S is extremal i.e. that if $S = F_1 + F_2$ $(F_1, F_2 \in \mathcal{P}_{3,6})$ then $F_1 = \alpha S, F_2 = \beta S$ $(\alpha + \beta = 1)$.

Likewise for the Motzkin ternary sextic we have

5. Suppose $F \in \mathcal{P}_{3,6}$ does not contain terms in $X^4Z^2, Y^4Z^2, X^2Z^4, Y^2Z^4$ and suppose F vanishes on $\mathfrak{S}(M)$ (see p.76). Show that $F = \alpha M$ for some $\alpha \in \mathbf{R}$. Deduce that $M \in \mathcal{E}(\mathcal{P}_{3,6})$.

For the Robinson quaternary quartic Q, we have

6. Suppose $F \in \mathcal{P}_{4,4}$ does not contain terms in X^2W^2, Y^2W^2, Z^2W^2 and suppose F vanishes on $\mathfrak{S}(Q)$ (see page 76). Show that $F = \alpha Q$ $(\alpha \in \mathbf{R})$ and deduce that $Q \in \mathcal{E}(\mathcal{P}_{4,4})$.

As it happens, it is possible to deduce the extremeness of M and of Q from that of S. This is done in the following

7. Prove that $S \in \mathcal{E}(\mathcal{P}_{3,6}) \Rightarrow M \in \mathcal{E}(\mathcal{P}_{3,6})$ along the following lines:

(i) Verify that $M(X^2, YZ, XZ) = X^4Z^2S(X, Y, Z)$.

(ii) Suppose $M \geq F \in \mathcal{P}_{3,6}$, then

$$X^4Z^2S(X, Y, Z) = M(X^2, YZ, XZ) \geq F(X^2, YZ, XZ) \geq 0.$$

But X^4Z^2S is extremal, by Exercise 2; so

$$F(X^2, YZ, XZ) = \alpha M|X^2, YZ, XZ), \quad (\alpha \in \mathbf{R}).$$

Replacing Y by XY/Z this becomes

$$X^6F(X, Y, Z) = F(X^2, XY, XZ) = \alpha M(X^2, XY, XZ)$$
$$= \alpha X^6 M(X, Y, Z).$$

8. Show similarly that $M \in \mathcal{E}(\mathcal{P}_{3,6}) \Rightarrow Q \in \mathcal{E}(\mathcal{P}_{4,4})$ (see [C11] p. 9).

7

The two proofs of Hilbert's main theorem; Hilbert's own and the other of Choi and Lam.

The main part of Hilbert's Theorem 6.1 is the following

Theorem 7.1. (Hilbert, 1888) *Every PSD ternary quartic is a sum of squares of ternary quadratics and indeed three squares always suffice.*

We plan to give two proofs in this chapter. The first proof, due to Choi and Lam [C9], uses arguments from elementary analysis - and also makes use of the Krein-Milman theorem, which is a popular tool of functional analysts. This proof only shows the first part of the theorem, i.e. that $\mathcal{P}_{3,4} = \sum_{3,4}$.

The second proof is Hilbert's original. To give the Choi-Lam proof we need the following.

Lemma 1. *Let $T(X, Y, Z) \in \mathcal{P}_{3,4}$. Then there exists a quadratic form $q(X, Y, Z)$ ($\neq 0$) such that $T \geq q^2$, where by $T \geq q^2$ we mean of course that the form $T - q^2 \geq 0$ i.e. is PSD.*

Proof. Let $\mathfrak{S}(T)$ denote the set of zeros of T.

Case 1: $\mathfrak{S}(T) = \emptyset$. Consider the positive continuous function

$$\varphi(X, Y, Z) = T(X, Y, Z)/(X^2 + Y^2 + Z^2)^2$$

defined for all $(X, Y, Z) \neq (0, 0, 0)$. On the unit sphere S^2 (a compact set) let $\mu = \inf \varphi \geq 0$. By compactness of S, μ is attained, i.e. $\mu = \varphi(\alpha, \beta, \gamma)$ for some $(\alpha, \beta, \gamma) \in S^2$; so $\mu \neq 0$ since $\mathfrak{S}(T) = \emptyset$. Thus

$$T(X, Y, Z) \geq \mu(X^2 + Y^2 + Z^2)^2 \quad \text{on } S^2 \tag{7.1}$$

We claim that (7.1) holds for all the points of $\mathbf{R}^3$; for let $(\alpha, \beta, \gamma) \neq (0,0,0)$ be any point of $\mathbf{R}^3$ and let $N = \sqrt{(\alpha^2+\beta^2+\gamma^2)}$. Then $(\alpha/N, \beta/N, \gamma/N) \in S^2$ and so

$$T\left(\frac{\alpha}{N}, \frac{\beta}{N}, \frac{\gamma}{N}\right) \geq \mu\left(\frac{\alpha^2}{N^2} + \frac{\beta^2}{N^2} + \frac{\gamma^2}{N^2}\right)^2,$$

i.e. $T(\alpha,\beta,\gamma) \geq \mu(\alpha^2+\beta^2+\gamma^2)^2$ as claimed. For the point $(0,0,0)$, (7.1) is true trivially. Since μ is > 0, we see that (7.1) gives, as claimed,

$$T(X,Y,Z) \geq (\sqrt{\mu}(X^2+Y^2+Z^2))^2.$$

Case 2: $|\mathfrak{S}(T)| = 1$. By changing coordinates, we suppose without loss of generality that $T(1,0,0) = 0$. Write T as a polynomial in X:

$$T(X,Y,Z) =$$
$$AX^4 + X^3(\alpha_1 Y + \alpha_2 Z) + X^2 \cdot f(Y,Z) + 2Xg(Y,Z) + h(Y,Z),$$

where f, g, h are quadratic, cubic, and quartic forms respectively. Since $T(1,0,0) = 0$, we get $A = 0$. Further, for any fixed Y, Z, the X^3 term could be made negative and to dominate the rest by choosing X large enough, thus making $T(X,Y,Z) < 0$, a contradiction. Hence $\alpha_1 = \alpha_2 = 0$ so

$$T(X,Y,Z) = X^2 f + 2Xg + h.$$

Moreover $f \geq 0$, $h \geq 0$, since if say, f were not PSD, then we could find $(Y,Z) = (\alpha,\beta)$ such that $f(\alpha,\beta) < 0$ and choose X large enough to make $T < 0$, a contradiction. Similarly if $h(\alpha,\beta) < 0$, choose $X = \delta$ so small that $T(\delta,\alpha,\beta)$ stays negative, again a contradiction. Now

$$fT = (Xf+g)^2 + (fh - g^2) \tag{7.2}$$

Here $fh - g^2 \geq 0$ for otherwise there exist $(Y,Z) = (\alpha,\beta)$ such that $fh - g^2 < 0$ at (α,β). Then

$$(Xf+g)^2 = (X(\lambda Y^2 + \mu YZ + \nu Z^2) + (\vartheta_1 Y^3 + \vartheta_2 Z^3 + \varphi_1 Y^2 Z + \varphi_2 YZ^2))^2.$$

Now choose $X = \delta$ to make the right side equal to zero when $(Y,Z) = (\alpha,\beta)$. Then at (δ,α,β),

$$(Xf+g)^2 = 0$$

and so at (δ,α,β) (7.2) gives

$$f(\alpha,\beta)\cdot T(\delta,\alpha,\beta) = (fh-g^2)_{(\alpha,\beta)} < 0$$

which is a contradiction since f and T are both non-negative.

Now we consider two subcases:

Subcase 1: f is of rank 1. Then on completing squares, $f = f_1^2$, where f_1 is a linear form, say $f_1 = \alpha Y + \beta Z$. The zero set of f_1 is just $\left(\frac{\beta}{\alpha}, 1\right) = (\beta, \alpha)$ (projectively). Plugging this into $fh - g^2$ we see that $(fh-g^2)_{(\beta,\alpha)} =$

$-(g(\beta,\alpha_1))^2 < 0$ which is a contradiction unless $g(\beta,\alpha) = 0$ i.e. $f_1 | g$, say $g = f_1 g_1$. Then (7.2) gives

$$\begin{aligned} fT &\geq (Xf+g)^2 \quad (\text{since } fh - g^2 \geq 0) \\ &= (Xf_1^2 + f_1 g_1)^2 \quad (\text{since } f = f_1^2, g = f_1 g_1) \\ &= f_1^2 (Xf_1 + g_1)^2 \\ &= f(Xf_1 + g_1)^2. \end{aligned}$$

Hence $T \geq (Xf_1 + g_1)^2$ as required.

Subcase 2: f is of rank 2. So that $f = f_1^2 + f_2^2$, with f_1, f_2 linear in Y, Z and so $\mathfrak{S}(f_1)$ and $\mathfrak{S}(f_2)$ are singleton points. Now f_1, f_2 cannot be simultaneously zero otherwise they would be multiples of each other and so rank$f$ would be 1. Hence $f > 0$. Also $fh - g^2 > 0$ (we have already shown that $fh - g^2 \geq 0$); for if (b, c) were a zero of $fh - g^2$, then it could be completed to a zero $(-g(b,c)/f(b,c), b, c)$ of T (check). But T has only one zero $(1,0,0)$ so $(b,c) = (0,0)$, which is not admissible and $fh - g^2$ has no zero i.e. $fh - g^2 > 0$ as claimed.

Hence now $(fh - g^2)/f^3 \geq \mu > 0$ on the unit circle S^1 ($\mu = \inf(fh - g^2)/f^3)$) just as in Case 1 and so $fh - g^2 \geq \mu f^3$ everywhere. But then by (7.2), $fT \geq fh - g^2 \geq \mu f^3$ giving $T \geq (\sqrt{\mu} f)^2$ □

Case 3: $|\mathfrak{S}(T)| \geq 2$. Here without loss of generality we may arrange the coordinate system so that $(1,0,0)$ and $(0,1,0)$ are two of the zeros of T. As in Case 2, T is of degree at most 2 in X as well as in Y and so it is easy to check that

$$T(X,Y,Z) = X^2 f(Y,Z) + 2XZg(Y,Z) + Z^2 h(Y,Z),$$

where f, g, h are quadratic forms as in Case 2. Start with

$$T = \Sigma^3 \alpha X^4 + \Sigma^6 \beta X^3 Y + \Sigma^3 \gamma X^2 YZ + \Sigma^3 \delta X^2 Y^2.$$

where the power in Σ indicates the number of terms in the summation.

$$\text{So} \quad fT = (Xf + Zg)^2 + Z^2(fh - g^2) \tag{7.3}$$

and here $fh - g^2 \geq 0$ for if $fh - g^2 < 0$ at $(Y, Z) = (\alpha, \beta)$ then taking $X = \delta = -\beta g(\alpha,\beta)/f(\alpha,\beta)$ (don't forget $f > 0$) we see that

$$\begin{aligned} (fT)_{(\delta,\alpha,\beta)} &= (\delta f + \beta g(\alpha,\beta))^2 + \beta^2(fh - g^2) \\ &= 0 + \beta^2(fh - g^2) < 0, \text{ a contradiction.} \end{aligned}$$

If f (or h) is of rank 1, we get the desired result by an argument similar to the Subcase 1 of Case 2. Hence we may further suppose f and h to be of rank 2 i.e. $f > 0$, $h > 0$ (again as before). We again consider 2 subcases:

Subcase 1: $fh - g^2$ has (b, c) say, as a non-trivial zero. Let:

$$\alpha = -g(b,c)/f(b,c),$$

and

$$\begin{aligned}T_1(X,Y,Z) &= T(X+\alpha Z,Y,Z)\\ &= X^2f+2XZ(g+\alpha f)+Z^2(h+2\alpha g+\alpha^2 f).\end{aligned}$$

Then

$$\begin{aligned}&\{h(Y,Z)+2\alpha g(Y,Z)+\alpha^2 f(Y,Z)\}_{(b,c)}\\ &\qquad = h+2g(-g/f)+f\cdot g^2/f^2\end{aligned}$$

substituting the value of α

$$\begin{aligned}&= h-g^2/f\\ &= \left(\frac{hf-g^2}{f^2}\right)_{(b,c)} = 0.\end{aligned}$$

Thus rank $(h+2\alpha g+\alpha^2 f)\leq 1$; thus, as in Subcase 1 of Case 2, T_1 is not less than the square of a quadratic form and hence so is T, as required.

Subcase 2: $fh-g^2>0$. Then as in Subcase 2 of Case 2, we have

$$\frac{fh-g^2}{(Y^2+Z^2)f}\geq \mu>0$$

on the circle S^1 and so $fh-g^2\geq \mu(Y^2+Z^2)f$ everywhere. Hence

$$\begin{aligned}fT &\geq Z^2(fh-g^2)\quad \text{(by (7.3))}\\ &\geq \mu Z^2(Y^2+Z^2)f\end{aligned}$$

giving, as required,

$$\begin{aligned}T &\geq (\sqrt{\mu}ZY)^2+(\sqrt{\mu}Z^2)^2\\ &\geq (\sqrt{\mu}Z^2)^2.\end{aligned}$$

This completes the proof of the lemma. □

We now use the required version of the Krein-Milman theorem; we give the statement here and say something about it in the appendix on convex sets.

Theorem. *Let $\mathcal{E}=\mathcal{E}(\mathcal{P}_{n,m})$ be the set of all extremal PSD forms; then $\mathcal{E}$ spans $\mathcal{P}_{n,m}$ i.e. every form in $P_{n,m}$ is a finite sum of forms in $\mathcal{E}$.*

Proof of Theorem 7.1. Let $T\in\mathcal{P}_{3,4}$ and write

$$T=T_1+T_2+\ldots+T_k$$

where the T_j are extremal forms of $\mathcal{P}_{3,4}$ i.e. $T_j\in\mathcal{E}(\mathcal{P}_{3,4})$. Apply Lemma 1 to each T_j; we see that $T_j\geq q_j^2$ i.e. $T_j-q_j^2$ is PSD $=q_j'$ say or

$$T_j=q_j^2+q_j'\quad (q_j^2,q_j'\in\mathcal{P}_{3,4}).$$

But T_j is extremal, so this is the trivial decomposition i.e. $T_j = q_j^2$. So now $T = q_1^2 + \ldots + q_k^2$, a SOS. □

Hilbert's Proof. We now give Hilbert's proof of Theorem 7.1.

Let $\mathcal{F}$ be the class of all real ternary quartics F. A typical F has shape

$$F(X,Y,Z) = a_1X^4 + a_2Y^4 + a_3Z^4 + a_4X^3Y + a_5X^3Z + a_6XY^3 + a_7XZ^3 + a_8Y^3Z + a_9YZ^3 + a_{10}X^2Y^2 + a_{11}Y^2Z^2 + a_{12}X^2Z^2 + a_{13}X^2YZ + a_{14}XY^2Z + a_{15}XYZ^2.$$

Using the 15 a_j as coordinates, we represent F as a point in $\mathbf{R}^{15}$ so that the PSD forms constitute the closed convex cone $\mathcal{P} = \mathcal{P}_{3,4}$. The strictly positive definite forms are the open convex cone $\mathcal{P}^0$. Clearly $\overline{\mathcal{P}^0} = \mathcal{P}$.

Now let $\mathcal{A}$ be the set of all ternary quartics F which can be written as

$$F = f^2 + g^2 + h^2 \tag{7.4}$$

where f, g, h are real quadratic forms. Since F is PSD we have $\mathcal{A} \subset \mathcal{P}$.

Again let $\mathcal{B}$ be the set of all ternary quartics F which can be written as (7.4) with the additional condition that f, g, h have no non-trivial (i.e. $\neq (0,0,0)$) common zero, real or complex. This is Hilbert's key idea. Then we have the following

Lemma 2. *$\mathcal{B}$ is open.*

Proof. Using the coefficients as coordinates, we represent an ordered triple (f, g, h) of ternary quadratic forms by a point in $\mathbf{R}^{18}$ (f, g, h have shape $a_1X^2 + a_2Y^2 + a_3Z^2 + a_4XY + a_5YZ + a_6ZX$). The map

$$(f, g, h) \to F = f^2 + g^2 + h^2 \tag{7.5}$$

is an algebraic map $\mathbf{R}^{18} \to \mathbf{R}^{15}$. Consider this map in the neighbourhood of (f_0, g_0, h_0), where f_0, g_0, h_0 have no common (real or complex) non-trivial zero, so that $F_0 = f_0^2 + g_0^2 + h_0^2$ belongs to $\mathcal{B}$. We want to show that each F in a sufficiently small neighbourhood of F_0 is also in $\mathcal{B}$; so let u, v, w be ternary quadratics and let δ be small. Then $(f_0 + u\delta)^2 + (g_0 + v\delta)^2 + (h_0 + w\delta)^2$ belongs to a sufficiently small neighbourhood of F_0. It equals

$$F_0 + 2(uf_0 + vg_0 + wh_0)\delta + 0(\delta^2).$$

Consider the so-called tangent map of (7.5):

$$(u, v, w) \xrightarrow{\Phi} 2(uf_0 + vg_0 + wh_0). \tag{7.6}$$

Here again, Φ is a linear map from $\mathbf{R}^{18}$ of triples $(u, v, w) \to \mathbf{R}^{15}$ of ternary quartics $2(uf_0 + vg_0 + wh_0)$. What is the kernel of Φ? It can be shown (see Appendix 1 to this chapter) that under the condition that f_0, g_0, h_0 have no common zero, we have $uf_0 + vg_0 + wh_0 = 0$ if and only if there exist constants λ, μ, ν such that

$$\left.\begin{aligned} u &= \nu g_0 - \mu h_0 \\ v &= \lambda h_0 - \nu f_0 \\ w &= \mu f_0 - \lambda g_0 \end{aligned}\right\} \tag{7.7}$$

Thus

$$(u, v, w) \in \ker\ \Phi \text{ if and only if } uf_0 + vg_0 + wh_0 = 0$$

i.e. if and only if (7.7) holds, which gives a condition on u, v, w viz. if

$$f_0 = a_1X^2 + a_2Y^2 + a_3Z^2 + a_4XY + a_5YZ + a_6ZX$$
$$g_0 = b_1X^2 + b_2Y^2 + b_3Z^2 + b_4XY + b_5YZ + b_6ZX$$
$$h_0 = c_1X^2 + c_2Y^2 + c_3Z^2 + c_4XY + c_5YZ + c_6ZX,$$

then a basis for ker Φ is

$$\underline{v}_1 = (b_1, b_2, b_3, \ldots, b_6, -a_1, -a_2, \ldots, -a_6, 0, 0, 0, 0, 0, 0)$$
$$\underline{v}_2 = (-c_1, -c_2, \ldots, -c_6, 0, 0, 0, 0, 0, 0, a_1, a_2, \ldots, a_6)$$
$$\underline{v}_3 = (0, 0, 0, 0, 0, 0, c_1, c_2, \ldots, c_6, -b_1, -b_2, \ldots, -b_6)$$

since $(u, v, w) \in$ ker Φ iff (7.7) holds, i.e. iff (check this)

$$(u, v, w) = \nu\underline{v}_1 + \lambda\underline{v}_2 + \mu\underline{v}_3.$$

Thus ker Φ has dimension 3. If follows that

$$\begin{aligned} \text{dim (image)} &= \text{ dim (domain) - dim (kernel)} \\ &= 18 - 3 = 15; \end{aligned}$$

but this is just the dimension of the image space. Thus Φ is onto. Hence by the implicit function theorem, (for some explanation see Appendix 2 to this chapter) every F in some neighbourhood of F_0 can be written as $f^2 + g^2 + h^2$ with f, g, h having no common (real or complex) zero. So $F \in \mathcal{B}$ □

What is the closure of $\mathcal{B}$? We have

Lemma 3. *$\overline{\mathcal{B}} \subset \mathcal{A}$ and further if $F \in \overline{\mathcal{B}}$ and $F \notin \mathcal{B}$ then either $F \notin \mathcal{P}^0$ or the curve $F(X, Y, Z) = 0$ in the complex projective plane has at least two (real or complex) double points.*

Proof. $\mathcal{B} \subset \mathcal{A}$ trivially and $\mathcal{A}$ is closed so $\overline{\mathcal{B}} \subset \mathcal{A}$. Further since $F \in \overline{\mathcal{B}}$ ($\subset \mathcal{A}$) and $F \notin \mathcal{B}$, so

$$F = f^2 + g^2 + h^2$$

where f, g, h *have* a non-trivial common zero $\underline{\mathcal{H}} = (\alpha, \beta, \gamma)$, possibly complex, so that $F(\alpha, \beta, \gamma) = 0$. If $\underline{\mathcal{H}}$ is real then F vanishes at the real point (α, β, γ), so is *not* strictly positive definite i.e. $F \notin \mathcal{P}^0$ as required.

Otherwise $\mathcal{H}$ is distinct from its complex conjugate $\overline{\mathcal{H}}$ i.e. represents a different point in the complex projective plane. Then both $\mathcal{H} = (\alpha, \beta, \gamma)$ and $\overline{\mathcal{H}} = (\overline{\alpha}, \overline{\beta}, \overline{\gamma})$ give double points of $F(X, Y, Z) = 0$ since either point is a zero of f, g, h, so a double zero of f^2, g^2, h^2 i.e. a double zero of $f^2 + g^2 + h^2$ i.e. a double point of $F(X, Y, Z) = 0$ as required. □

The property of having at least two double points imposes two algebraic conditions on the coefficients of F. Hence the set $\mathcal{E}$ of real F with at least two double points has codimension at least two. We can now prove Hilbert's main

Theorem 7.1. $\mathcal{P} = \mathcal{A}$.

Proof. Clearly $\mathcal{A} \subset \mathcal{P}$. To prove that $\mathcal{P} \subset \mathcal{A}$ let $F_1 \in \mathcal{P}$ and we wish to prove that $F_1 \in \mathcal{A}$. Since $\mathcal{E}$ has codimension two, we can chose $F_0 \in \mathcal{B}$ such that the line segment

$$F_t = (1 - t)F_0 + tF_1 \quad (0 \leq t \leq 1)$$

does not meet $\mathcal{E}$, except possibly at F_1. Since $F_0 \in \mathcal{B} \subset \mathcal{P}^0$ and $F_1 \in \mathcal{P}$ so $F_t \in \mathcal{P}^0$ for $0 \leq t < 1$ ($\mathcal{P}, \mathcal{P}^0$ are convex). By Lemma 3, F_t can be on the boundary of $\mathcal{B}$ (i.e. in $\overline{\mathcal{B}} - \mathcal{B}^0 = \overline{\mathcal{B}} - \mathcal{B}$ (since $\mathcal{B}$ is open)) at worst only for $t = 1$ (for suppose $F_t \in \overline{\mathcal{B}}, F_t \notin \mathcal{B}^0 (= \mathcal{B})$; then by Lemma 3, since $F_t \in \mathcal{P}^0$, so the curve $F_t = 0$ has at least two double points i.e. $F_t \in \mathcal{E}$ so $t = 1$). Hence $F_t \in \mathcal{B}$ for $0 \leq t < 1$ and so $F_1 \in \overline{\mathcal{B}} \subseteq \mathcal{A}$ as required.

Appendix 1

We prove the following result used in Hilbert's proof of the theorem.

Lemma 4. *Let f, g, h; u, v, w be quadratic forms in X, Y, Z and suppose that*

$$uf + vg + wh = 0 \tag{1}$$

and that f, g, h have no common non-trivial zero (over the algebraic closure). Then there are constants $\lambda, \mu, \nu \in \mathbb{C}$ such that

$$u = \nu g - \mu h,$$
$$v = \lambda h - \nu f,$$
$$w = \mu f - \lambda g.$$

Proof. It is enough to show that w is a linear combination of f, g with coefficients in $\mathbb{C}$; for once this is done, i.e. once we have proved that

$$\begin{aligned} u &= \alpha_2 g + \alpha_3 h \\ v &= \beta_1 f + \beta_3 h \\ w &= \gamma_1 f + \gamma_2 g. \end{aligned}$$

(1) gives $(\beta_3 + \gamma_2)gh + (\alpha_3 + \gamma_1)hf + (\alpha_2 + \beta_1)fg = 0$. Now substitute for (X, Y, Z) a zero of f which is not a zero of g or of h and we get $\beta_3 + \gamma_2 = 0$ and similarly $\alpha_3 + \gamma_1 = 0$ and $\alpha_2 + \beta_1 = 0$ giving what is required.

It is clear that f, g have only finitely many common zeros (the four intersections of the conics $f = 0, g = 0$) and so after a linear change of coordinate system, we may suppose that the line $X = 0$ (the side of the triangle of reference) does not pass through any common non-trivial zero of f, g.

We now use Hilbert's Nullstellensatz (see Van der Waerden, *Modern Algebra* Vol.II, pp. 5-6): if $\varphi \in K[X_1, \ldots, X_\alpha]$ is a polynomial which vanishes at all the zeros common to the polynomials $f_1, f_2, \ldots, f_\beta$; then for some integer n, we have

$$f^n \equiv 0(f_1, \ldots, f_\beta)$$

i.e. f^n is a linear combination of $f_1, \ldots, f_\beta$ in $K[X_1, \ldots, X_\alpha]$.

Apply this to $\varphi = X \in K[X, Y, Z]$, which vanishes at all the zeros common to f, g, h (vacuously as there are no such zeros). So by the Nullstellensatz

$$X^n = rf + sg + th \tag{2}$$

for some integer $n \geq 2$ and forms r, s, t of degree $n - 2$ in X, Y, Z.

Multiplying (1) by t, (2) by w and subtracting, we get $(rw - ut)t + (sw - vt)g = wX^n$ i.e. say

$$bf + cg = wX^n \tag{3}$$

where b, c are forms of degree n.

Now suppose n is minimal in (3). If $n = 0$, we are done. If $n > 0$, we get a contradiction as follows:

Put $X = 0$ in (3) to get

$$b_0 f_0 + c_0 g_0 = 0 \tag{4}$$

where $b(0, Y, Z) = b_0$ etc. Now by our choice of the line $X = 0$ we see that f_0, g_0 have no common zero and by (4) if we substitute for Y, Z a zero η, ζ of f_0, we see that $f_0 | c_0$ in $K[X, Y, Z]$; say $-c_0/f_0 = d = b_0/g_0$ (similarly), i.e.

$$b_0 = dg_0, \; c_0 = -df_0 \text{ for some } d \in K[X, Y, Z] \tag{5}$$

Now by replacing b, c in (3) by $b - dg, \; c + df$, we may suppose that b, c vanish at $X = 0$ (by (5)) i.e. that $X | b, c$. But then we may cancel an X in (3) to get a smaller n for which (3) holds giving the required contradiction. □

Appendix 2

The standard implicit function theorem, as given say in W. Rudin's *Principles of Mathematical Analysis* is as follows:

Theorem (Implicit function theorem) *Let $F_j(\mathbf{X},\mathbf{Y})$ $(1 \le j \le n)$ be n continuously differentiable functions of the $m+n$ variables*

$$\mathbf{X} = (X_1, \ldots, X_m),\ \mathbf{Y} = (Y_1, \ldots, Y_n),$$

defined in some neighbourhood of $(\mathbf{X},\mathbf{Y}) = (\mathbf{a},\mathbf{b})$. Suppose that $F_j(\mathbf{a},\mathbf{b}) = 0$ and that at $(\mathbf{a},\mathbf{b})$ the $n \times n$ matrix $\left(\frac{\partial F_j}{\partial Y_k}\right)$ is non-singular. Then in some neighbourhood of $\mathbf{X} = \mathbf{a}$, there are n continuously differentiable functions $G_k(\mathbf{X})$ $(1 \le k \le n)$ such that $F_j(\mathbf{X}, G_1(\mathbf{X}), \ldots, G_n(\mathbf{X})) = 0$ $(1 \le k \le n)$ and $(G_1(\mathbf{a}), \ldots, G_n(\mathbf{a})) = \mathbf{b}$.

What we need is the following

Corollary. *Let $F_j(\mathbf{X},\mathbf{Z})$ $(1 \le j \le n)$ be n continuously differentiable functions of the $m+N$ variables $\mathbf{X} = (X_1, \ldots, X_m)$, $\mathbf{Z} = (Z_1, \ldots, Z_N)$ in some neighbourhood of $(\mathbf{X},\mathbf{Z}) = (\mathbf{a},\mathbf{c})$, where $N \ge n$. Suppose that $F_j(\mathbf{a},\mathbf{c}) = 0$ $(1 \le j \le n)$ and that the rank of the $n \times N$ matrix $M = \left(\frac{\partial F_j}{\partial Z_k}\right)_{\mathbf{X}=\mathbf{a},\mathbf{Z}=\mathbf{c}}$ $(1 \le j \le n, 1 \le k \le N)$ is n. Then there are continuously differentiable functions $G_k(\mathbf{X})$ $(1 \le k \le N)$ such that $F_j(\mathbf{X}, G_1(\mathbf{X}), \ldots, G_N(\mathbf{X})) = 0$ $(1 \le j \le n)$ and $(G_1(\mathbf{a}), \ldots, G_N(\mathbf{a})) = \mathbf{c}$.*

Proof. Since the rank of M is n we can select a non-singular $n \times n$ minor. On renumbering the variables $Z_1, \ldots, Z_N$, we may suppose without loss of generality, that the $n \times n$ matrix $\left(\frac{\partial F_j}{\partial Z_k}\right)_{\mathbf{X}=\mathbf{a},\mathbf{Z}=\mathbf{c}}$ $(1 \le j \le n, 1 \le k \le n)$ is non-singular. Then we put $G_k(\mathbf{X}) = c_k$ $(n < k \le N)$. The theorem applies to the functions $F_j^*(\mathbf{X},\mathbf{Y}) = F_j(\mathbf{X},\mathbf{Y}, c_{n+1}, \ldots, c_N)$ where $\mathbf{Y} = (Z_1, \ldots, Z_n), (c_1, \ldots, c_n) = \mathbf{b}$.

In the application to Hilbert's theorem, we have $m = n = 15, N = 18$. The $\mathbf{X} = (X_1, \ldots, X_{15})$ are the coefficients of the quadratic form. The $\mathbf{Z} = (Z_1, \ldots, Z_{18})$ are the $3 \times 6 = 18$ coefficients of the quadratic forms. The equations $F_j(\mathbf{X},\mathbf{Z}) = 0$ are just the statement that the quartic form is the sum of the squares of the three quadratic forms. Finally the statement that the rank of the matrix M is 15 is just the result proved in Appendix 1 above.

8

Theorems of Reznick and of Choi, Lam and Reznick [R4], [C12]

Hilbert's conjecture for the function field $K_n = \mathbb{R}(X_1, \ldots, X_n)$ and the subsequent proof of the conjecture given by Artin partly closes the topic so to speak; this result followed by Pfister's theorem almost completely closes the topic as it were in the sense that the only question that remains regarding this is the exact value of the Pythagoras number $P(K_n)$ of K_n and it would be a very difficult invariant to determine in general.

For the ring $\mathbb{R}[X_1, \ldots, X_n]$, we have now seen that the set $\Delta_{n,m} = \mathcal{P}_{n,m} - \Sigma_{n,m}$ is non-empty except for the pairs $(n, m) = (2, m \text{ (even)}), (n, 2), (3, 4)$ and it is precisely this fact that makes life so interesting and leads to so many natural questions, mainly about the nature of the set $\Delta_{n,m}$. For many of these questions it is enough to consider the two basic examples $\Delta_{3,6}$ and $\Delta_{4,4}$, answers in the general case being easy corollaries in most cases to the answers in $\Delta_{3,6}$ and $\Delta_{4,4}$. Hence we shall mainly concentrate our attention on these two basic examples and indeed on $\Delta_{3,6}$ as the other case $\Delta_{4,4}$ usually turns out to be a little more complicated; but similar in principle to the $\Delta_{3,6}$ case.

Reznick's theorem on the "simplest" elements in $\Delta_{3,6}$.

The simplicity of Motzkin's and Robinson's examples of forms belonging to $\Delta_{3,6}$ and to $\Delta_{4,4}$ contrasts with the complexity of Hilbert's original method of constructing such forms. However, there may conceivably be still simpler examples of forms belonging to $\Delta_{3,6}$ or $\Delta_{4,4}$. By "simpler" we mean of course ones with fewer number of terms (monomials). Our first striking result is that such examples of ternary sextics involve at least four monomials.

Theorem 8.1 (Bruce Reznick). *Suppose the ternary sextic*

$$f(X_1, X_2, X_3) \in \Delta_{3,6} = \mathcal{P}_{3,6} - \Sigma_{3,6};$$

then f involves at least four terms.

Remark 1. In [R4], Reznick determines all the four-term extremal forms in $\Delta_{n,m}$ for $n = 3$, $m \leq 12$ and gives a method of constructing such "simplest" extremal forms for other values of n and m.

Remark 2. Although we shall give a proof in $\Delta_{3,6}$ we could have just as easily worked in the general case up to a point. However, the examples are best understood and visualized for small values of n and m.

General definitions and notations. Let

$$p(X_1, X_2, \ldots, X_n) = \sum_{i=1}^{k} a_i X_1^{r_{i1}} X_2^{r_{i2}} \ldots X_n^{r_{in}}$$

be a form of degree m ($m = 2d$) so that

$$\sum_{j=1}^{k} r_{ij} = m \quad (i = 1, 2, \ldots, k) \tag{8.1}$$

We write $\underline{X} = X_1 X_2 \ldots X_n$ (*not* an n-vector; just a notation) and $\underline{r}_i = (r_{i1}, r_{i2}, \ldots, r_{in})$ (an n-vector in $\mathbf{R}^n$, indeed in $\mathbf{Z}^n$), where we suppose $\underline{r}_i$ to be distinct n-tuples (i.e. if $\underline{r}_{i_1} = \underline{r}_{i_2}$, then the two corresponding terms can be joined up into a single term $(a_{i_1} + a_{i_2})\underline{X}^{r_{i_1}}$) and we suppose also that $a_i \neq 0$ (i.e. if $a_i = 0$, we simply omit the term corresponding to a_i). Then $p(X_1, \ldots, X_n)$ is abbreviated to

$$p(X_1, X_2, \ldots, X_n) = \sum_{i=1}^{k} a_i \underline{X}^{\underline{r}_i} \tag{8.2}$$

Example 1. $n = 3$, $m = 6$ and take the variables as X, Y, Z. We write the terms of $p(X, Y, Z)$ in lexicographical order:

$$\begin{aligned} p(X, Y, Z) &= a_1 X^6 + a_2 X^5 Y + a_3 X^5 Z + a_4 X^4 Y^2 + a_5 X^4 Y Z \\ &\quad + a_6 X^4 Z^2 + a_7 X^3 Y^3 + \ldots + a_{28} Z^6 \\ &= \sum_{i=1}^{28} a_i \underline{v}^{\underline{r}_i}, \end{aligned}$$

where

$$\underline{v} = XYZ$$

and

$$\underline{r}_1 = (6, 0, 0),\ \underline{r}_2 = (5, 1, 0),\ \underline{r}_3 = (5, 0, 1), \ldots, \underline{r}_{28} = (0, 0, 6).$$

We see that the sum of the coordinates of each $\underline{r}_i$ is $6 : r_{i1} + r_{i2} + r_{i3} = 6$ for all i (verifying (8.1)).

Definitions.

(i) $C(p)$, the cage of p, is the convex hull of the $\underline{r}_i$'s. Then $C(p) \subset \mathbf{R}^3$; indeed all the $\underline{r}_i$ lie in the plane $X + Y + Z = 6$ for our example.

(ii) $F(p) = C(p) \cap \mathbf{Z}^n$. This is the set of all lattice points in $C(p)$.

(iii) $E(p)$ = set of all extreme points of $C(p)$ so that $E(p) \subset$ points defined by the $\underline{r}_i$; all are therefore lattice points.

In general $C(p) \subset (n-1)$-dimensional hyperplane but if some of the a_i are 0, then $C(p)$ could lie in a smaller dimensional hyperplane; for example in the ternary sextic case $C(p) \subset$ the plane $X + Y + Z = 6$, but if for example,

$$p(X,Y,Z) = a_1X^6 + a_2X^5Y + a_{11}X^2Y^4 + a_{22}Y^6$$

then $C(p)$ is spanned by the vectors $(6,0,0), (5,1,0), (2,4,0)$ and $(0,6,0)$ and so all lie on the line

$$X + Y + Z = 6, \quad Z = 0.$$

Example 2. $n = 2$, $m = 2$, $p(X,Y) = a_1X^2 + a_2XY + a_3Y^2$, $\underline{r}_1 = (2,0)$, $\underline{r}_2 = (1,1)$, $\underline{r}_3 = (0,2)$. All the $\underline{r}_i$ lie on the line $X + Y = 2$. Thus $C(p)$ is the line segment joining $(0,2)$ and $(2,0)$. $F(p) = \{(0,2),(1,1),(2,0)\}$ and $E(p) = \{(0,2),(2,0)\}$.

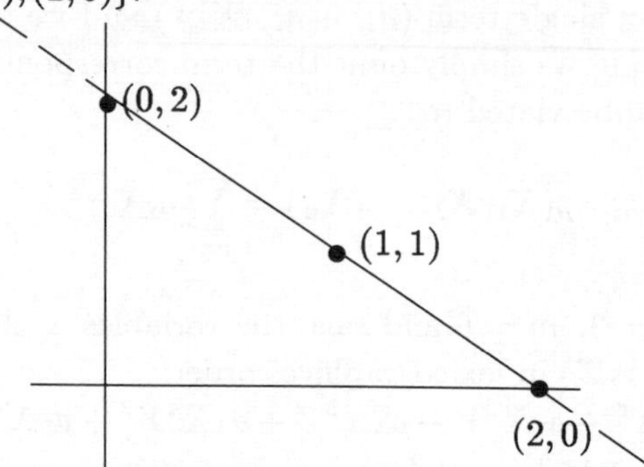

Exercise. Prove that $C(p^2) = 2C(p)$.

So now let $p(X_1, X_2, X_3) = \sum_{i=1}^{28} a_i \underline{X}^{\underline{r}_i}$ be a PSD ternary sextic. We prove a series of necessary conditions for p to be in $\Delta_{3,6}$. Let then $p \in \Delta_{3,6}$.

Lemma 1. *If p is PSD and $\underline{r}_\lambda \in E(p)$, then $a_\lambda > 0$ and $\underline{r}_\lambda$ is an even vector i.e. $r_{\lambda 1}, r_{\lambda 2}, r_{\lambda 3}$ are all even.*

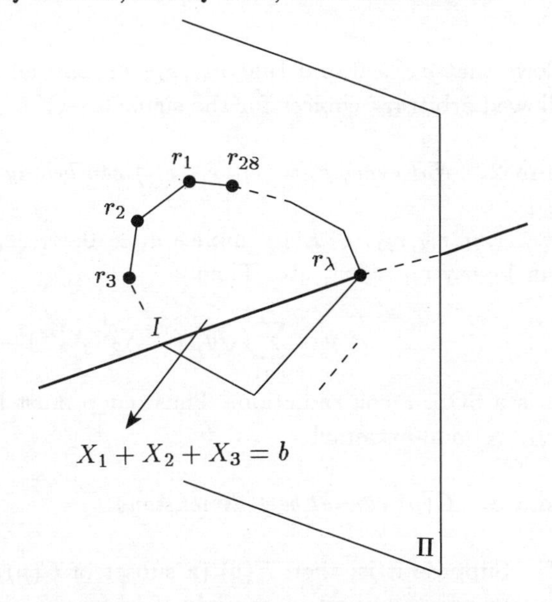

Proof. Choose a plane $\Pi : \underline{bX} = d$ (i.e. $b_1X_1 + b_2X_2 + b_3X_3 = d$) through $\underline{r}_\lambda$ so that $C(p)$ lies on one side of Π (see the figure). Then since Π has been chosen to go through $\underline{r}_\lambda$ we have $\underline{b} \cdot \underline{r}_\lambda = d$ i.e. $b_1 r_{\lambda 1} + b_2 r_{\lambda 2} + b_3 r_{\lambda 3} = d$; and for all other points of $C(p)$, in particular for the other $\underline{r}_i$ $(i \neq \lambda)$, we have $\underline{b} \cdot \underline{r}_i - d < 0$ i.e.

$$\underline{b} \cdot \underline{r}_i < d = \underline{b} \cdot \underline{r}_\lambda \quad (i \neq \lambda) \tag{8.3}$$

Now take

$$X_1 = \pm t^{b_1}$$
$$X_2 = \pm t^{b_2}$$
$$X_3 = \pm t^{b_3}$$

(independent signs). Then for $t > 0$, since p is PSD, we have the following:

$$\begin{aligned} 0 &\leq t^{-d} p(t) \\ &= t^{-d} \sum_{i=1}^{k} a_i (\pm t^{b_1})^{r_{i1}} (\pm t^{b_2})^{r_{i2}} (\pm t^{b_3})^{r_{i3}} \\ &= \sum_{i=1}^{k} a_i (\pm 1)^{r_{i1}} (\pm 1)^{r_{i2}} (\pm 1)^{r_{i3}} t^{b_1 r_{i1} + b_2 r_{i2} b_3 r_{i3} - d}. \end{aligned}$$

Here the power of t is

$$\begin{cases} \underline{b} \cdot \underline{r}_i - d & < \ 0 \text{ if } i \neq \lambda \text{ (by (8.3))}. \\ & = \ 0 \text{ if } i = \lambda. \end{cases}$$

Now let $t \to \infty$ and we see that all the terms in the summation tend to 0 except $i = \lambda$ which becomes $a_\lambda (\pm 1)^{r_{\lambda 1}} (\pm 1)^{r_{\lambda 2}} (\pm 1)^{r_{\lambda 3}}$. Hence

$$0 \leq a_\lambda (\pm 1)^{r_{\lambda 1}} (\pm 1)^{r_{\lambda 2}} (\pm 1)^{r_{\lambda 3}} \tag{8.4}$$

It follows that $a_\lambda > 0$ and that $r_{\lambda 1}, r_{\lambda 2}, r_{\lambda 3}$ are all even since in (8.4) we are allowed arbitrary choices for the signs $+, -$. □

Lemma 2. *Not every* $\underline{r}_i = (r_{i1}, r_{i2}, r_{i3})$ *can belong to* $E(p)$.

Proof. $(r_{i1}, r_{i2}, r_{i3}) \in E(p)$ implies $a_i > 0$, r_{i1}, r_{i2}, r_{i3} are all even (by Lemma 1); say $r_{i1} = 2r'_{i1}$ etc. Then

$$p = \sum_{i=1}^{3} (\sqrt{a_i} X_1^{r'_{i1}} X_2^{r'_{i2}} X_3^{r'_{i3}})^2$$

which is a SOS, a contradiction. Thus our p must have at least one $\underline{r}_i = (r_{i1}, r_{i2}, r_{i3})$ non-extremal. □

Lemma 3. $C(p)$ *cannot be* 1-*dimensional.*

Proof. Suppose it is; then $E(p)$ (a subset of $C(p)$) is also 1-dimensional (i.e. a line segment) and so consists of just two points, $(r_{\lambda 1}, r_{\lambda 2}, r_{\lambda 3})$ and $(r_{\mu 1}, r_{\mu 2}, r_{\mu 3})$, say, and by reindexing if necessary, we may take these points as

$$\underline{r}_1 = (r_{11}, r_{12}, r_{13}),\ \underline{r}_k = (r_{k1}, r_{k2}, r_{k3}).$$

Note λth is called the first and μth the kth; k remember is the number of terms in p i.e. 28; and we want to determine the minimal value of k for which $p \in \Delta_{3,6}$. Since $\underline{r}_1, \underline{r}_k$ are extremal, $a_1, a_k > 0$ and all the $r_{11}, r_{12}, r_{13}, r_{k1}, r_{k2}, r_{k3}$ are even (by Lemma 1). Hence

$$\begin{aligned} \underline{r}_k - \underline{r}_1 &= (r_{k1}, r_{k2}, r_{k3}) - (r_{11}, r_{12}, r_{13}) \\ &= 2d(s_1, s_2, s_3), = 2d\underline{s} \text{ say} \end{aligned}$$

where $2d$ is the gcd of $r_{k1} - r_{11}, r_{k2} - r_{12}, r_{k3} - r_{13}$ and where $\gcd(s_1, s_2, s_3) = 1$, so that at least one s_j is odd. Now $C(p)$ is the convex hull of $E(p)$ i.e. all the (r_{j1}, r_{j2}, r_{j3}) $(j \neq k, 1)$ are a convex combination of $\underline{r}_1, \underline{r}_k$, i.e.

$$\begin{aligned} \underline{r}_j &= \alpha_j \underline{r}_k + (1 - \alpha_j)\underline{r}_1 \quad (0 \le \alpha_j \le 1) \\ &= \alpha_j(\underline{r}_k - \underline{r}_1) + \underline{r}_1 \\ &= \alpha_j \cdot 2d\underline{s} + \underline{r}_1 \\ &= \underline{r}_1 + c_j \underline{s}, \end{aligned}$$

say, where $c_j = 2d\alpha_j \le 2d$. So now

$$\begin{aligned} p(X_1, X_2, X_3) &= \sum_{i=1}^{k} a_i \underline{X}^{\underline{r}_i} \\ &= \sum_{i=1}^{k} a_i \underline{X}^{\underline{r}_1 + c_i \underline{s}} \end{aligned}$$

$$= \underline{X}^{\underline{r}_1} \sum_{i=1}^{k} a_i (\underline{X}^{\underline{s}})^{c_i}$$

$$= \underline{X}^{\underline{r}_1} \sum a_i \underline{Y}^{c_i} \text{ (say)} \tag{8.5}$$

where $\underline{Y} = \underline{X}^{\underline{s}} = X_1^{s_1} X_2^{s_2} X_3^{s_3}$. Here, as we have noted above, at least one s_j is odd.

Let $\underline{X}$ vary over all 3-tuples with $X_i \neq 0$ (i.e. avoiding the three axes); then $\underline{Y}$ ranges over all non-zero reals t (the negative ones being got from the odd s_j on taking the corresponding X_j negative). Then (8.5) becomes

$$p(\underline{X}) = \underline{X}^{\underline{r}_1} \sum a_i t^{c_i} \quad (t \in \mathbf{R}^*).$$

But $\underline{X}^{\underline{r}_1} > 0$ since $\underline{r}_1$ is an even vector, being extremal and $p(\underline{X}) \geq 0$ being PSD. It follows that $\sum a_i t^{c_i} \geq 0$ for all $t \neq 0$ and so for all t, including 0, by continuity (if it were negative at $t = 0$, then there would be a neighborhood $[-\varepsilon, \varepsilon]$ of 0 in which it would be negative). Thus $\sum_{i=1}^{k} a_i t^{c_i}$ is a PSD polynomial in one variable t and so it is a SOS - in fact a sum of two squares $f_1^2(t) + f_2^2(t)$; see Theorem 4.1. Going back from t to Y and on to X_1, X_2, X_3, we see that $p(\underline{X})$ is a sum of two squares, indeed;

$$\begin{aligned} p(\underline{X}) &= X_1^{2r'_{11}} X_2^{2r'_{12}} X_3^{2r'_{13}} (f_i^2(X_1, X_2, X_3) + f_2^2(X_1, X_2, X_3)) \\ &= (X_1^{r'_{11}} X_2^{r'_{12}} X_3^{r'_{13}} f_1)^2 + (X_1^{r'_{11}} X_2^{r'_{12}} X_3^{r'_{13}} f_2)^2 \end{aligned}$$

which is a contradiction, since p is not a SOS.

This proves Lemma 3. □

Proof of Theorem 8.1. Our p must have $C(p)$ of dimension two (Lemma 3) and so at least three extremal $\underline{r}_i$ spanning $C(p)$ and at least one non-extremal $\underline{r}_i$ (Lemma 2). Thus k is at least 4. □

Since both forms

$$S(X, Y, Z) = X^4 Y^2 + Y^4 Z^2 + Z^4 X^2 - 3X^2 Y^2 Z^2 \text{ (Robinson)}$$

and

$$M(X, Y, Z) = Z^6 + X^4 Y^2 + X^2 Y^4 - 3X^2 Y^2 Z^2 \text{ (Motzkin)}$$

belong to $\Delta_{3,6}$ and are made up of exactly four terms, we see that they are indeed the simplest possible ones.

One naturally asks if there are more such four term forms in $\Delta_{3,6}$. Reznick gives necessary and sufficient conditions for a four term form to be PSD but not a SOS; in fact for such forms to be extremal. For more details see [R4].

For quaternary quartics, it may likewise be shown that

$$q(X,Y,Z,W) = W^4 + X^2Y^2 + Y^2Z^2 + Z^2X^2 - 4XYZW$$

is the simplest example in $\Delta_{4,4}$ (see [R4]).

The Choi, Lam and Reznick theorem on zeros of forms in $\Delta_{n,m}$

We have come across forms belonging to $\mathcal{P}_{3,6} - \Sigma_{3,6} = \Delta_{3,6}$ and others belonging to $\Delta_{4,4}$. It so happens that known examples of such forms p have $|\mathfrak{S}(p)| < \infty$. The surprise is that this is the case for all forms in $\Delta_{3,6}$ and in $\Delta_{4,4}$. Indeed we have the following remarkable

Theorem 8.2 (Choi, Lam and Reznick) [C12].

(A) *Let $p \in \mathcal{P}_{3,6}$ and suppose $|\mathfrak{S}(p)| > 10$ then $p \in \Sigma_{3,6}$; in fact p is a sum of three squares of cubics.*

(B) *Let $p \in \mathcal{P}_{4,4}$ and suppose $|\mathfrak{S}(p)| > 11$ then $p \in \Sigma_{4,4}$; in fact p is a sum of six squares of quadratics.*

Further in both cases $|\mathfrak{S}(p)| = \infty$.

Remark. We shall only prove (A). The proof of (B), although similar, is distinctly more difficult. The interested reader may consult [C12].

In striking contrast to this result is the fact that Theorem 8.2 is peculiar to ternary sextics and quaternary quartics. In fact beyond these two cases, it is easy to show that if $\Delta_{n,m} \neq \emptyset$, then forms with infinitely many zeros may be found in plentiful in $\Delta_{n,m}$. Indeed we have the following

Theorem 8.3. *Suppose $(n,m) > (3,6)$ (i.e. either $n > 3$, $m \geq 6$ or $n \geq 3$, $m > 6$) or $(n,m) > (4,4)$; then there exist $p \in \Delta_{n,m}$, with $|\mathfrak{S}(p)| = \infty$.*

Proof. First take the case $(n,m) > (3,6)$. Then clearly the form

$$p(X_1, X_2, \ldots, X_n) = X_1^{m-6}.\ S(X_1, X_2, X_3) \in \Delta_{n,m}.$$

We shall prove that $|\mathfrak{S}(p)| = \infty$. If $m = 6$, then $n > 3$ and so $(1,1,1,X_4,\ldots,X_n) \in \mathfrak{S}(p)$. Since $X_4,\ldots,X_n$ $(n \geq 4)$ can take an infinity of values, it follows that $|\mathfrak{S}(p)| = \infty$ as required.

If $m > 6$, then $(0, X_2, \ldots, X_n) \in \mathfrak{S}(p)$ and so again we have an infinity of zeros of p.

If $(n,m) > (4,4)$, we use instead the form

$$p(X_1, X_2, \ldots, X_n) = X_1^{m-6} Q(X_1, X_2, X_3, X_4)$$

and proceed similarly.

Reminder : $S(X,Y,Z) = X^4Y^2 + Y^4Z^2 + Z^4X^2 - 3X^2Y^2Z^2,$

$$Q(X,Y,Z,W) = W^4 + X^2Y^2 + Y^2Z^2 + Z^2X^2 - 4XYZW.$$

In the case $\Delta_{n,m} = \emptyset$ i.e. when $\mathcal{P}_{n,m} = \Sigma_{n,m}$ we have the following result about the cardinality $|\mathfrak{S}(p)|$ of any PSD form $p \in \mathcal{P}_{n,m}$:

Theorem 8.4.

(i) *Let* $p \in \mathcal{P}_{2,m}$, *then* $|\mathfrak{S}(p)| \leq m/2$

(ii) *Let* $p \in \mathcal{P}_{n,2}$, *then* $|\mathfrak{S}(p)| = \begin{cases} 0 & \text{if rank } p = n \\ 1 & \text{if rank } p = n-1 \\ \infty & \text{otherwise} \end{cases}$

(iii) *Let* $p \in \mathcal{P}_{3,4}$, *then either* $|\mathfrak{S}(p)| = \infty$ *or* $|\mathfrak{S}(p)| \leq 4$ *and* 4 *is the best bound.*

Proof. (i) Converting $p(X,Y)$ to a polynomial, we get $p(X) = a_0 + a_1X + \ldots + a_nX^n$. Factorizing this over $\mathbf{R}$, only linear and quadratic factors appear as irreducible factors. Since p is PSD all linear factors occur to an even multiplicity; and since the quadratic factors contribute no real zeros, the maximum possible number of zeros occur when $p(X)$ factors as

$$p(X) = \{(a_1 + b_1X)\ldots(a_d + b_dX)\}^2 \quad (d = m/2)$$

and then we get one zero from each distinct factor so $|\mathfrak{S}(p)| \leq d = m/2$ as required.

(ii) Making a linear change, we may take

$$p(X_1, \ldots, X_n) = X_1^2 + \ldots + X_r^2$$

where r is the rank of p; $r \leq n$. If $r = n$, then being a SOS, p is never 0 over $\mathbf{R}$ except at $X_1 = 0, \ldots, X_n = 0$ i.e. at $(0,0,\ldots,0)$ which is not a projective point at all, so $\mathfrak{S}(p) = \emptyset$ as required. If $r = n-1$, then

$$p = X_1^2 + \ldots + X_{n-1}^2$$

and the only zero of p over $\mathbf{R}$ is clearly $(0,0,\ldots,0,\rho)$ where ρ is any real number. Thus $|\mathfrak{S}(p)| = 1$. If $r < n-1$, then there are at least two missing X's in p and so $(\underbrace{0,0,\ldots,0}_{n-2},\rho_1,\rho_2)$ is a zero of p for all $\rho_1, \rho_2 \in \mathbf{R}$ giving $|\mathfrak{S}(p)| = \infty$ as required.

(iii) The proof of this will be given later. □

Our aim is to prove that if $p \in \Delta_{3,6}$, then $|\mathfrak{S}(p)| < \infty$. We prove a few lemmas first.

Lemma 4. *Let* p *be an* n*-ary* m*-ic and* $\lambda \neq 0$ *an* n*-ary linear form. If* $\mathfrak{S}(\lambda) \subseteq \mathfrak{S}(p)$, *then* $\lambda | p$. *If moreover* p *is PSD, then* $\lambda^2 | p$; *divisibility in the ring* $\mathbf{R}[X_1, \ldots, X_n]$ *of course.*

Proof. Let $\lambda(X_1,\ldots,X_n) = a_1X_1 + \ldots + a_nX_n$, where say without loss of generality, $a_1 \neq 0$. Using the non-singular linear transformation of variables

$$\begin{aligned} X_1 &\to (X_1 - a_2X_2 - \ldots - a_nX_n)/a_1 \\ X_2 &\to \qquad X_2 \\ &\cdots\cdots\cdots\cdots \\ X_n &\to \qquad X_n \end{aligned}$$

we may suppose that

$$\lambda(\underline{X}) = X_1.$$

The truth of the lemma is now immediate. □

The following lemma is crucial:

Lemma 5. *Let $f(X_1\ldots,X_n)$ be an n-ary form with $n \geq 3$. If $|\mathfrak{S}(f)| < \infty$, then one of $\pm f$ is PSD (i.e. any indefinite n-ary form ($n \geq 3$) has infinitely many real zeros).*

Proof. Suppose $f(\underline{X})$ is indefinite. Then there is a point $\underline{a}$ for which $f(\underline{a}) > 0$ and a point $\underline{b}$ for which $f(\underline{b}) < 0$. By continuity, $f(\underline{c}) < 0$ for all $\underline{c}$ in some neighbourhood N of $\underline{b}$. For any such $\underline{c}$, the line $\underline{a}\ \underline{c}$ meets $f(\underline{X}) = 0$. Clearly by varying $\underline{c}$ we obtain infinitely many zeros of $f(\underline{X})$. □

Lemma 6. *Let $p(X_1,\ldots,X_n)$ be an irreducible polynomial in $\mathbb{R}[X_1,\ldots,X_n]$. Then p becomes reducible in $\mathbb{C}[X_1,\ldots,X_n]$ if and only if one of $\pm p$ is a sum of two squares in $\mathbb{R}[X_1,\ldots,X_n]$.*

Proof. If $\pm p = f_1^2 + f_2^2$, $f_1, f_2 \in \mathbb{R}[X_1,\ldots,X_n]$, then $p = \pm(f_1 + if_2)(f_1 - if_2)$ is reducible in $\mathbb{C}[X_1,\ldots,X_n]$. Conversely suppose p factors non-trivially in $\mathbb{C}[X_1,\ldots,X_n]$ as say

$$p = (r_1 + ir_2)(s_1 + is_2);\ r_1, r_2, s_1, s_2 \in \mathbb{R}[X_1,\ldots,X_n].$$

Taking complex conjugates gives $p = (r_1 - ir_2)(s_1 - is_2)$. Multiplying we get $p^2 = (r_1^2 + r_2^2)(s_1^2 + s_2^2)$. But now $\mathbb{R}[X_1,\ldots,X_n]$ is a UFD, so $p = a(r_1^2 + r_2^2)$ for some $a \in \mathbb{R}$. But one of $\pm a$ is a square in $\mathbb{R}$, if say $a = b^2$ then $p = (br_1)^2 + (br_2)^2$ where as if $-a = b^2$ then $-p = (br_1)^2 + (br_2)^2$ as required. □

Lemma 7. *Let $p(X,Y,Z) \in \mathcal{P}_{3,m}$ be irreducible in $\mathbb{R}[X,Y,Z]$. Then*

$$|\mathfrak{S}(p)| \leq \max\left(\frac{m^2}{4}, (m-1)(m-2)/2\right).$$

Proof. First suppose p is reducible in $\mathbf{C}[X,Y,Z]$; then by Lemma 6, $\pm p = r_1^2 + r_2^2$ (r_1, r_2 forms of degree $m/2$ in $\mathbf{R}[X,Y,Z]$). So $\mathfrak{S}(p) = \mathfrak{S}(r_1) \cap \mathfrak{S}(r_2)$ (since $\mathfrak{S}$ denotes the real zeros only and $p = 0$ if and only if $r_1^2 + r_2^2 = 0$ i.e. if and only if $r = 0$ and $r_2 = 0$). Since p is irreducible in $\mathbf{R}[X,Y,Z]$ so r_1, r_2 are relatively prime and so the plane curves $r_1 = 0$ and $r_2 = 0$ can not have any common components. Hence by Bezout's theorem [W1], these two curves can intersect in at most $\frac{m}{2} \cdot \frac{m}{2}$ points in the complex projective plane. In particular $|\mathfrak{S}(p)| = |\mathfrak{S}(r_1) \cap \mathfrak{S}(r_2)| \leq m^2/4$ as required.

For the remaining case we suppose p is irreducible in $C[X,Y,Z]$ and so p defines an irreducible plane algebraic curve C in the complex plane. Since p is PSD each real zero (α, β, γ) of p is a minimum of p, an extreme value, and so (α, β, γ) is also a zero of $\frac{\partial p}{\partial X}, \frac{\partial p}{\partial Y}, \frac{\partial p}{\partial Z}$ i.e (α, β, γ) is a singular point of C; but C, being irreducible has at most $(m-1)(m-2)/2$ singular points and so $\mathfrak{S}(p) \leq (m-1)(m-2)/2$ as required. □

As a corollary we have the following

Theorem 8.5. *Let $p(X,Y,Z)$ be an irreducible PSD ternary quartic with $|\mathfrak{S}(p)| \geq 4$. Then p is a sum of two squares in $\mathbf{R}[X,Y,Z]$.*

Remark. According to Hilbert's theorem (Theorem 6.1(a)(iv)), since p is PSD, it is a sum of three squares; so under the extra condition $|\mathfrak{S}(p)| \geq 4$, two squares suffice.

Proof. Suppose p is irreducible in $\mathbf{C}[X,Y,Z]$; then it has at most three singular points: $(m-1)(m-2)/2 = (4-1)(4-2)/2 = 3$, and so by the proof of Lemma 7, since a zero of p has got to be a singular point, $|\mathfrak{S}(p)| \leq 3$ contrary to our hypothesis. So p is irreducible in $\mathbf{C}[X,Y,Z]$ and hence by Lemma 6 one of $\pm p$ is a sum of two squares in $\mathbf{R}[X,Y,Z]$. But p is PSD so p (and not $-p$) is a sum of two squares. □

Write $\mu(m) = \max\left(\frac{m^2}{4}, \frac{(m-1)(m-2)}{2}\right)$. Since m is the degree of our ternary form, we are only interested in $\mu(m)$ when m is a positive integer.

It is easy to check that:

(i) $\mu(m) = \begin{cases} m^2/4 \text{ if } m \leq 5 \\ (m-1)(m-2)/2 \text{ if } m \geq 6. \end{cases}$

(ii) $\mu(m)/m$ is monotone increasing.

From these one immediately deduces the following so called "superadditive" property of the μ function:

Lemma 8. $\mu(m_1) + \mu(m_2) \leq \mu(m_1 + m_2)$.

Proof. Since $\mu(m)/m$ is increasing

$$\mu(m_1)/m_1 \leq \mu(m_1 + m_2)/(m_1 + m_2)$$
$$\mu(m_2)/m_2 \leq \mu(m_1 + m_2)/(m_1 + m_2).$$

Hence

$$\mu(m_1) + \mu(m_2) \leq m_1 \frac{\mu(m_1 + m_2)}{m_1 + m_2} + m_2 \frac{\mu(m_1 + m_2)}{m_1 + m_2}$$
$$= \mu(m_1 + m_2) \qquad \square$$

We are now in a position to prove the following main result:

Lemma 9. *Let $p \in \mathcal{P}_{3,m}$. The following conditions are equivalent*

(i) $|\mathfrak{S}(p)| > \mu(m)$

(ii) $|\mathfrak{S}(p)| = \infty$

(iii) *p is divisible by the square of some indefinite form.*

Proof. (ii) implies (i) is trivial. (iii) implies (ii) goes as follows: Suppose $f^2|p$ where f is indefinite; then by Lemma 5, $|\mathfrak{S}(p)| = \infty$; but $\mathfrak{S}(f) \subseteq \mathfrak{S}(p)$, so $|\mathfrak{S}(p)| = \infty$.

It remains to prove that (i) implies (iii). We do this by induction on m. For $m = 2$, $\mu(m) = 1$ and the form p in question is a PSD ternary quadratic with $|\mathfrak{S}(p)| > 1$ i.e with at least two zeros. Take p diagonal without loss of generality:

$$p = aX^2 + bY^2 + cZ^2.$$

We now take the two zeros of p to be the points $(0, 0, \gamma), (0, \beta, 0)$ where $\beta, \gamma \neq 0$. Then $c = b = 0$ giving $p = aX^2$. Since p is PSD, we have $a > 0$ and so $p = (\sqrt{a}X)^2 \sim X^2$ i.e. p may be taken to be X^2 and as asserted in (iii), p is divisible by the square of the indefinite form X. $\square$

In general suppose $|\mathfrak{S}(p)| > \mu(m)$. Then by Lemma 7, p is reducible in $\mathbb{R}[X, Y, Z]$, say

$$p = q_1 q_2 \ldots q_r \quad (r \geq 2),$$

where each $q_i \in \mathbb{R}[X, Y, Z]$ is irreducible. Let $m_i = \deg q_i$.

Case 1. *All q_i are semi-definite (i.e. PSD or NSD).* Adjusting by ± 1, if necessary, we may assume that they are all PSD. Then there must exist an index i such that $|\mathfrak{S}(q_i)| > \mu(m_i)$ for otherwise

$$|\mathfrak{S}(p)| \leq \sum_i |\mathfrak{S}(q_i)| \leq \sum_i \mu(m_i) \leq \mu(\sum_i m_i) \quad \text{(Lemma 8)}$$
$$= \mu(m),$$

a contradiction. Suppose without loss of generality that $|\mathfrak{S}(q_1)| > \mu(m_1)$. Since $m_1 < m$, by the induction hypothesis, the PSD form q_1 is divisible by the square of some indefinite form $\varphi : \varphi^2|q_1$; so $\varphi^2|p$ as required.

Case 2. One of the q_i is indefinite. Hence by Lemma 2, $|\mathfrak{S}(q_i)| = \infty$ and so $|\mathfrak{S}(p)| = \infty$ (so in fact (ii) is proved). Now $p = p(X,Y,Z)$ can not be free of all the variables; say without loss of generality that p involves X, so $\frac{\partial p}{\partial X} \not\equiv 0$. Now consider the two plane curves $\mathcal{C}_1, \mathcal{C}_2$ defined by $p = 0$ and $\frac{\partial p}{\partial X} = 0$. Since p is PSD, the zeros of p are extreme values of p (minima) and so satisfy $\frac{\partial p}{\partial X} = 0$. Thus $\mathfrak{S}(p) \subset \mathfrak{S}\left(\frac{\partial p}{\partial X}\right)$ i.e. $\mathcal{C}_1 \cap \mathcal{C}_2$ contains the infinite set $\mathfrak{S}(p)$ and so $\mathcal{C}_1, \mathcal{C}_2$ have a common component; i.e. p and $\frac{\partial p}{\partial X}$ have an irreducible real common factor; call it h. Thus $h|p$, $h|\frac{\partial p}{\partial X}$. Write $p = hg$. Then $\frac{\partial p}{\partial X} = h\frac{\partial g}{\partial X} + g\frac{\partial h}{\partial X}$. Here we have $h|\frac{\partial p}{\partial X}$; hence

$$h|g\frac{\partial h}{\partial X} \tag{8.6}$$

Subcase 1. h is free of X i.e. $h = h(Y,Z)$. Then $\frac{\partial h}{\partial X} \equiv 0$. If h has a real zero, it splits a linear factor $aY + bZ$ (by the factor theorem). Then $aY + bZ|p$ and so by Lemma 1, $(aX + bY)^2|p$ as required. If h has no real zero, then h is semi-definite (for if not, then it would take both positive and negative values and so would vanish somewhere by continuity i.e. have a real zero), and by a change of sign, we may suppose h is PSD. Since $p = hg$, g is also PSD. Now as a ternary form $h(X,Y)$ has the unique zero $(1,0,0)$ and so since $|\mathfrak{S}(p)| = \infty$, we must have $|\mathfrak{S}(g)| = \infty$ i.e. (ii) holds for g and so (i) holds for g and since $\deg g < \deg p$, by the induction hypothesis, g is divisible by the square of an indefinite form $\varphi : \varphi^2|g$ and so $\varphi^2|p$ since $g|p$ as required.

Subcase 2. $\frac{\partial h}{\partial X} \not\equiv 0$. Here since h was chosen irreducible, we have by (8.6), $h|g$ say $g = hq$. Then $p = hg = h^2q$. If h is indefinite then we are through. If not, without loss of generality, h is PSD. Since h is irreducible, by Lemma 7, $|\mathfrak{S}(h)| < \infty$ so $|\mathfrak{S}(q)| = \infty$. Since q is also PSD and $\deg q < \deg p$, by the induction hypothesis, there exists an indefinite form φ with $\varphi^2|q$ so $\varphi^2|p$ as required. □

Remark 1. The moral of Lemma 9 is that if the number of zeros of p $(\in \mathcal{P}_{3,m})$ is sufficiently large $(> \mu(m))$, then this number is infinite and further that p factorizes as $p = h^2q$, where h is indefinite.

Remark 2. Lemma 9 is peculiar to ternary forms. For four (or more) variables, the form $X^2Y^2 + Z^2W^2$ is irreducible but vanishes at all the points $(\alpha, 0, \gamma, 0)$; an infinity of them.

Lemma 10. *Let $p(X, Y, Z)$ be any ternary form. The following statements are equivalent:*

(i) *p is semi-definite (i.e. one of $\pm p$ is PSD)*

(ii) *$p = h^2 q$, where $|\mathfrak{S}(q)| < \infty$ and h is a product of indefinite forms (this product may be $\emptyset$ in which case we agree that $h = 1$).*

Proof. First suppose (i) holds. By Lemma 5, q cannot be indefinite, so q is semi-definite and h^2 is PSD so $p = h^2 q$ is semi-definite as required.

Conversely suppose p is PSD. Use induction on $\deg p$. If $|\mathfrak{S}(p)| < \infty$, then take $h = 1$ and $p = q$ as required by (ii). If $|\mathfrak{S}(P)| = \infty$, then by Lemma 9, $p = h^2 q$ for some indefinite h. Here $\deg q < \deg p$, so by the induction hypothesis, $q = \varphi^2 q^*$, giving $p = (h\varphi)^2 q^*$, where $h\varphi$ is a product of indefinite forms. □

Remark. The moral of Lemma 10 is that the study of PSD ternary forms can be reduced, in some sense to the study of those ternary forms which have only finitely many zeros: replace p by q.

We now give the

Proof of Theorem 8.2 (A). Let $p \in \mathcal{P}_{3,6}$ and suppose $|\mathfrak{S}(p)| > 10$. We have to prove that $|\mathfrak{S}(p)| = \infty$ and that $p \in \Sigma_{3,6}$, and indeed that p is a sum of three squares (of cubics). Since $\mu(6) = 10$, so by Lemma 6, $|\mathfrak{S}(p)| = \infty$ ((i) $\Rightarrow$ (ii)). The same lemma gives a factorization

$$p = h^2 q \quad (h \text{ indefinite of degree at least } 1, q \text{ PSD}).$$

If deg $h = 1$, then deg $q = 4$ i.e. $q \in \mathcal{P}_{3,4}$. Hence by Hilbert's theorem q is a sum of three squares of quadratics:

$$q = f_1^2 + f_2^2 + f_3^2$$

giving $p = (hf_1)^2 + (hf_2)^2 + (hf_3)^2$ as required. If deg $h \geq 2$, then deg $q \leq 2$ i.e. q is a ternary quadratic and so by completing squares

$$\begin{aligned} q &= \text{ a sum of three squares of linear forms} \\ &= f_1^2 + f_2^2 + f_3^2 \\ p &= (hf_1)^2 + (hf_2)^2 + (hf_3)^2, \end{aligned}$$

as required. □

We finally give the left over

Proof of Theorem 8.4, (iii). Let $p \in \mathcal{P}_{3,4}$. Suppose $|\mathfrak{S}(p)| < \infty$, then by Lemma 9

$$|\mathfrak{S}(p)| \leq \mu(4) = \max(4,3) = 4$$

as required.

To show that 4 is the best bound, we consider the ternary quartic

$$p(X,Y,Z) = (X-Z)^2(X-2Z)^2 + (Y-Z)^2(Y-2Z)^2.$$

This has the following four zeros as required:

$$(1,1,1), (1,2,1), (2,1,1), (2,2,1).$$

□

Exercises

Let

$$p(X_1,\ldots,X_n) = X^2 f(X_1,\ldots,X_n) + 2Xg(X_1,\ldots,X_n) + h(X_1,\ldots,X_n)$$

be a quadratic polynomial in X with coefficients f, g, h which are polynomials in $X_1,\ldots,X_n$. Let $D = fh - g^2 \in \mathbf{R}[X_1,\ldots,X_n]$ be the discriminant of p.

1. Show that as a polynomial in X, p is ≥ 0 if and only if $f \geq 0$ and $D \geq 0$ as polynomials in $X_1,\ldots,X_n$.

2. Suppose $\lambda(X_1,\ldots,X_n) \in \mathbf{R}[X_1,\ldots,X_n]$ and let

$$q(Y; X_1,\ldots,X_n) = p(Y + \lambda(X_1,\ldots,X_n); X_1,\ldots,X_n)$$

Show that the discriminant of q with respect to Y is the same as the discriminant of p with respect to X.

3. For a p as above, show that if p is SOS of polynomials then so is D. (If $p = \Sigma(u_iX + v_i)^2, u_i, v_i \in \mathbf{R}[X_1,\ldots,X_n]$, then $f = \Sigma u_i^2, h = \Sigma v_i^2, g = \Sigma u_i v_i$; so $D = (\Sigma u_i^2)(\Sigma v_i^2) - (\Sigma u_i v_i)^2 = \Sigma(u_i v_j - u_j v_i)^2$).

4. Show that the converse of Exercise 3 is false (take $p = W^4 + X^2Y^2 + Y^2Z^2 + Z^2X^2 - 4XYZW$).

5. Let $p \in \mathcal{P}_{3,6}$ and suppose there exists a ternary cubic h such that $p - h^2 \leq 0$. Show, along the following lines that $p \in \Sigma_{3,6}$.
(i) Show that $\mathfrak{S}(h) \subset \mathfrak{S}(p)$ (use $0 \leq p \leq h^2$)
(ii) Show that h is indefinite (use that h is a cubic)
(iii) Show that $|\mathfrak{S}(h)| = \infty$ (use Lemma 5)
(iv) Now use Theorem 8.2 to show that $p \in \Sigma_{3,6}$ Deduce that if $p \in \mathcal{P}_{n,m}$ and $p - h^2 \leq 0$ for some n-ary $\frac{m}{2}$-ic h, then $p \in \Sigma_{n,m}$ if

(a) $n = 2$, all m;

(b) all $n, m = 2$;
(c) $n = 3, m = 4$;
(d) $n = 3, m = 6$.
(In all cases except the last $p \in \Sigma_{n,m}$ anyhow.)

6. Prove the converse of Exercise. 5, viz. that if $p \in \mathcal{P}_{n,m}$ and $p - h^2 \leq 0$ for some n-ary $\frac{m}{2}$-ic h, then $p \in \Sigma_{n,m}$ only in the four cases listed above, by constructing counterexamples in all the other cases on the following lines:
(i) First let $n, m \geq 4$. For the form

$$Q = W^4 + X^2Y^2 + Y^2Z^2 + Z^2X^2 - 4XYZW$$

let $N = \max Q$ on the sphere $X^2 + Y^2 + Z^2 + W^2 = 1$. Show that $Q \leq N(X^2 + Y^2 + Z^2 + W^2)$ for all X, Y, Z, W i.e. that

$$X^{m-4}Q \leq [\sqrt{N}X^{\frac{m}{2}-2}(X^2 + Y^2 + Z^2 + W^2)]^2$$

Since $X^{m-4}Q \notin \Sigma_{n,m}$, we get an example of a form p for each $n, m \geq 4$ for which $p \in \mathcal{P}_{n,m}, p - h^2 \leq 0$ for some h, but still $p \notin \Sigma_{n,m}$ as required.
(ii) Now let $n = 3, m \geq 8$. For the form

$$S = X^4Y^2 + Y^4Z^2 + Z^4X^2 - 3X^2Y^2Z^2$$

prove using an argument as above that

$$S(X, Y, Z) \leq M(X^2 + Y^2 + Z^2)^2 \quad \text{for some } M > 0.$$

Then

$$\begin{aligned} X^{m-6}S &\leq X^{m-8}(X^2 + Y^2 + Z^2)S \\ &\leq [\sqrt{M}X^{\frac{m}{2}-4}(X^2 + Y^2 + Z^2)]^2; \end{aligned}$$

but $X^{m-6}S \notin \Sigma_{3,m}$. This covers all cases.

The next few exercises are on Reznick's result (Theorem 8.1). Let $p = a_1X^{\mathbf{r}_1} + a_2X^{\mathbf{r}_2} + a_3X^{\mathbf{r}_3} + a_4X^{\mathbf{r}_4} \in \Delta_{3,6}$ have its minimal number of 4 terms, where by Lemma 1, (i) $a_1, a_2, a_3 > 0$, (ii) $\mathbf{r}_1, \mathbf{r}_2, \mathbf{r}_3$ are even vectors. Furthermore, (iii) $E(p) = \{\mathbf{r}_1, \mathbf{r}_2, \mathbf{r}_3\}$ and (iv) $\mathbf{r}_4 \notin E(p)$.

7. Prove, by following the steps given below, that $\mathbf{r}_4$ lies strictly in the interior of $\Delta = T(\mathbf{r}_1, \mathbf{r}_2, \mathbf{r}_3)$.

Suppose without loss of generality $\mathbf{r}_4 \in$ the edge $\mathbf{r}_1, \mathbf{r}_2$.
(i) Show that we can find a vector $\mathbf{b} = (b_1, b_2, b_3)$ such that $\mathbf{b} \cdot \mathbf{r}_1 = \mathbf{b} \cdot \mathbf{r}_2 = \mathbf{b} \cdot \mathbf{r}_4 = d > \mathbf{b} \cdot \mathbf{r}_3$.
(ii) Put $X_i = c_i t^{b_i}$ $(i = 1, 2, 3)$ and show that $0 \leq t^{-d} \cdot p(t) = \Sigma_1^4 a_i c_1^{r_{i1}} c_2^{r_{i2}} c_3^{r_{i3}} t^{b_1 r_{i1} + b_2 r_{i2} + b_3 r_{i3} - d}$
(iii) Using (i), show by letting $t \to \infty$ that $0 \leq a_1 c^{\mathbf{r}_1} + a_2 c^{\mathbf{r}_2} + a_4 c^{\mathbf{r}_4} = q(c)$ say. Thus $q(X)$ is PSD.
(iv) Show that $c(q)$, the cage of q, is 1-dimensional and hence by Lemma 3, $q(X)$ is SOS.

(v) Using Lemma 1, show that $a_3 > 0, \mathbf{r}_3$ is even ($= 2\mathbf{r}'_3$ say) giving $a_3 X^{\mathbf{r}_3} = (\sqrt{a_3} X^{\mathbf{r}'_3})^2$. Deduce that $p(X) = q(X) + a_3 X^{\mathbf{r}_3}$ is a SOS giving a contradiction; hence $\mathbf{r}_4$ is not contained in any edge.

8. Show step by step, that by a suitable transformation of the type $X_i \to \pm X_i$ $(i = 1, 2, 3)$ $(*)$, we can make $a_4 < 0$ (and of course leave $a_1, a_2, a_3 > 0$). Suppose $a_4 > 0$.
(i) Prove that $\mathbf{r}_4$ can not be even (otherwise all $a_j X^{\mathbf{r}_j}$ $(j = 1, 2, 3, 4)$ are squares, so p is a SOS; a contradiction).
(ii) Show that by suitably choosing signs in $(*)$, we can get what is required.

Now by Exercise 7, $\mathbf{r}_4$ is in the interior of Δ and so it is a convex combination of $\mathbf{r}_1, \mathbf{r}_2, \mathbf{r}_3 : \mathbf{r}_4 = \sum_{j=1}^{3} \lambda_j \mathbf{r}_j, \Sigma\lambda_j = 1, \quad 0 \le \lambda_j \le 1$.

9. Show that in the above, $0 < \lambda_j < 1$ (use Exercise 7 and $\Sigma\lambda_j = 1$).

10. Show that by a suitable substitution of the type $X_j \to v_j X_j$ $(j = 1, 2, 3)$, $p(X)$ can be got in the form

$$p(X) = \lambda_1 X^{\mathbf{r}_1} + \lambda_2 X^{\mathbf{r}_2} + \lambda_3 X^{\mathbf{r}_3} - \alpha X^{\mathbf{r}_4},$$

where $\mathbf{r}_4 = \Sigma_1^3 \lambda_j \mathbf{r}_j, \Sigma\lambda_j = 1, 0 < \lambda_j < 1, \alpha = a_4 v^{\mathbf{r}_4} (= a_4 v_1^{r_{41}} v_2^{r_{42}} v_3^{r_{43}})$:
(i) Choose $v = v_1 v_2 v_3$ to satisfy $v^{\mathbf{r}_i} = \lambda_i / a_i$ $(i = 1, 2, 3)$. Show that the three equations got from these by taking logs have a solution v_1, v_2, v_3 (determinant of non-zero coefficients).
(ii) Show that the transformation $X_j \to v_j X_j$ takes $p(X_1, X_2, X_3)$ to the required form with $\alpha = a_4 v^{\mathbf{r}_4}$.

Remark. If in addition to the requirement $p \in \Delta_{3,6}$, we also demand that p is extremal, then we can fully fix the choice of α. Indeed up to a change of variables $X_j \to v_j X_j$ $(j = 1, 2, 3)$,

$$p = \lambda_1 X^{\mathbf{r}_1} + \lambda_2 X^{\mathbf{r}_2} + \lambda_3 X^{\mathbf{r}_3} - X^{\mathbf{r}_4},$$

where $\mathbf{r}_4 = \sum_1^3 \lambda_j \mathbf{r}_j, \sum_1^3 \lambda_j = 1, \lambda_j > 0$ (for details see [R4], Theorem 2).

9

Theorems of Choi, Calderon and of Robinson [C7], [C1], [R8]

A rather special class of quartics is the class of biquadratics

$$F(X_1, X_2, \ldots, X_m;\ Y_1, Y_2, \ldots, Y_n) = \sum_{j\le k,\ p\le q} \alpha_{jkpq} X_j X_k Y_p Y_q$$

where exactly two X's and exactly two Y's occur in each term. We may ask Hilbert's question of the class $\mathcal{B}(m,n)$ of all PSD biquadratic forms:

If F is PSD, must there exist bilinear forms

$$f_i(X_1, \ldots, X_m, Y_1, \ldots, Y_n) = \Sigma \beta_{jp}^{(i)} X_j Y_p$$

such that $F = \Sigma f_i^2$?

It was thought for some time (see Koga's paper [K3]) that the answer to this question is in the affirmative and indeed for $m = 2$ (or $n = 2$) Calderon [C1] proved the result true. We shall give a proof of Calderon's result; but first let us ask what happens if $m \geq 3$, $n \geq 3$. It is enough to consider the simplest case $m = n = 3$ (why?) and prove the following

Theorem 9.1 (Choi). *There is a PSD biquadratic form that is not a SOS of bilinear forms.*

Proof. The following simple example was given by Choi [C7]:

$$F(X_1, X_2, X_3, Y_1, Y_2, Y_3) = X_1^2Y_1^2 + X_2^2Y_2^2 + X_3^2Y_3^2$$
$$- 2(X_1X_2Y_1Y_2 + X_2X_3Y_2Y_3 + X_3X_1Y_3Y_1) + 2(X_1^2Y_2^2 + X_2^2Y_3^2 + X_3^2Y_1^2).$$

We shall show that (i) F is PSD (ii) F is not a SOS.

To prove (i), note that F is invariant under the cyclic permutation of the subscripts $1, 2, 3$. Now since one of $|X_1| \le |X_2|$, $|X_2| \le |X_3|$, $|X_3| \le |X_1|$ must hold, it suffices to show that $F \ge 0$ whenever $|X_1| \le |X_2|$. This can be seen as follows: We have

$$\begin{aligned}F(X_1, X_2, X_3; Y_1, Y_2, Y_3) = (X_1Y_1 - X_2Y_2 + X_3Y_3)^2 + 2X_1^2Y_2^2 \\ + 2(X_2^2Y_3^2 + X_3^2Y_1^2 - 2X_3X_1Y_3Y_1).\end{aligned}$$

Here the first two terms are squares, so non-negative, while, using $X_2^2 \ge X_1^2$, since $|X_2| \ge |X_1|$, the third is not less than

$$X_1^2Y_3^2 + X_3^2Y_1^2 - 2X_3X_1Y_3Y_1 = (X_1Y_3 - X_3Y_1)^2$$

as required.

To show that F is not a SOS, suppose to the contrary that $F = \Sigma f_i^2$, with f_i bilinear in X's, Y's. Since the terms $X_1^2Y_3^2, X_2^2Y_1^2, X_3^2Y_2^2$ are absent in F, they are absent in each f_i^2 i.e. X_1Y_3, X_2Y_1, X_3Y_2 do not appear in f_i so $f_i(X_1, X_2, X_3; Y_1, Y_2, Y_3)$ involves the following terms, the underlined ones being absent:

$$X_1Y_1, X_1Y_2, \underline{X_1Y_3}; \underline{X_2Y_1}, X_2Y_2, X_2Y_3; X_3Y_1, \underline{X_3Y_2}, X_3Y_3.$$

Write $f_i = g_i + h_i$ where g_i involves X_1Y_1, X_2Y_2, X_3Y_3 only and h_i involves the remaining terms. Then

$$F = \Sigma g_i^2 + 2\Sigma g_ih_i + \Sigma h_i^2$$

which gives

$$\begin{aligned}X_1^2Y_1^2 + X_2^2Y_2^2 + X_3^2Y_3^2 + 2(X_1^2Y_2^2 + X_2^2Y_3^2 + X_3^2Y_1^2) \\ - 2(X_1X_2Y_1Y_2 + X_2X_3Y_2Y_3 + X_3X_1Y_3Y_1 \\ = \sum(a_i^{(1)}X_1Y_1 + a_i^{(2)}X_2Y_2 + a_i^{(3)}X_3Y_3)^2 \\ + 2\sum(a_i^{(1)}X_1Y_1 + a_i^{(2)}X_2Y_2 + a_i^{(3)}X_3Y_3)\times \\ + (b_i^{(1)}X_1Y_2 + b_i^{(2)}X_2Y_3 + b_i^{(3)}X_3Y_1) \\ + \sum(b_i^{(1)}X_1Y_2 + b_i^{(2)}X_2Y_3 + b_i^{(3)}X_3Y_1)^2.\end{aligned}$$

Consider the middle term on the right hand side; no cross product in it occurs on the left hand side, nor in the other two terms of the right hand side, to cancel out. Hence $\Sigma 2g_ih_i = 0$. The first sum now gives the identity

$$\sum g_i^2 \equiv X_1^2Y_1^2 + X_2^2Y_2^2 + X_3^2Y_3^2 - 2(X_1X_2Y_1Y_2 + X_2X_3Y_2Y_3 + X_3X_1Y_3Y_1)$$

since no term in the third sum occurs here. In this put $X_j = 1 = Y_p$ (for all j, p) to give

$$\sum(a_i^{(1)} + a_i^{(2)} + a_i^{(3)})^2 = 3 - 2 \cdot 3 = -3$$

which is impossible. □

Having done the simplest case $n = m = 3$, it is now an easy matter to cover all higher cases; we simply regard $F(X_1, X_2, X_3; Y_1, Y_2, Y_3)$ as a form in $\mathcal{B}(m, n)$ for any $m \geq 3$, $n \geq 3$ and show that it is not a SOS of bilinear forms involving all the variables. For example let us show that $f \neq$ a SOS of $f_i(X_1, X_2, X_3, X_4; Y_1, Y_2, Y_3)$. Suppose to the contrary that $F = \Sigma f_i^2$ where

$$f_i = a_i X_4 Y_1 + b_i X_4 Y_2 + c_i X_4 Y_3 + g_i(X_1, X_2, X_3; Y_1, Y_2, Y_3) \qquad (9.1)$$

Then

$$F = X_4^2 \Sigma(a_i Y_1 + b_i Y_2 + c_i Y_3)^2 + \Sigma g_i^2$$
$$+ 2X_4 \Sigma(a_i Y_1 + b_i Y_2 + c_i Y_3) \cdot g_i(X_1, X_2, X_3; Y_1, Y_2, Y_3).$$

Since X_4 does not appear on the left hand side, we see that, in particular $\Sigma(a_i Y_1 + b_i Y_2 + c_i Y_3)^2 = 0$. It follows that $(a_i Y_1 + b_i Y_2 + c_i Y_3) = 0$ for all i and all $Y_1, Y_2, Y_3 \in \mathbf{R}$. Taking $Y_2 = Y_3 = 0$, $Y_1 < 0$, we get $a_i = 0$, and similarly $b_i = c_i = 0$ and so $f_i = g_i$ (by (9.1)) i.e. $F = \Sigma g_i^2$ giving a contradiction to the case $m = n = 3$. □

Remark 1. The form F of Theorem 9.1 serves also as a 6-variable quartic (6-ary, 4-ic) contained in $\mathcal{P}_{6,4} - \Sigma_{6,4}$.

Remark 2. Putting $X_1 = Y_1 = r_1$, $X_2 = Y_2 = r_2$, $X_3 = s_1$, $Y_3 = s_2$ in the F of Theorem 9.1, we get a PSD quaternary quartic

$$F_1(r_1, r_2; s_1, s_2) = r_1^4 + r_2^4 - 2(r_1^2 + r_2^2)s_1 s_2$$
$$+ s_1^2 s_2^2 + 2(r_1^2 s_1^2 + r_2^2 s_2^2).$$

It is easy to verify directly that F_1 is not a SOS: Suppose

$$F_1 = \Sigma f_i^2 \quad (f_i(r_1, r_2, s_1, s_2)$$

are quadratic forms. As in the proof of Theorem 9.2, write $f_i = g_i + h_i$ where g_i involves only $r_1^2, r_2^2, s_1 s_2$ and h_i involves only $r_1 r_2, r_1 s_2, r_2 s_2$. Equating coefficients gives

$$\Sigma g_i^2 = r_1^4 + r_2^4 - 2(r_1^2 + r_2^2)s_1 s_2 + s_1^2 s_2^2 +$$

a term in $r_1^2 r_2^2$ with a coefficient which is at most zero.

and this is negative when $r_1 = r_2 = s_1 = s_2 = 1$ giving a contradiction. So $F_1 \in \mathcal{P}_{4,4} - \Sigma_{4,4}$.

Remark 3. Put $Y_1 = X_1 X_2$, $Y_2 = X_2 X_3$, $Y_3 = X_3 X_1$ in the F of Theorem 9.1 and we get a PSD ternary sextic

$$F_2(X_1, X_2, X_3) = X_1^4 X_2^2 + X_2^4 X_3^2 + X_3^4 X_1^2 + 6X_1^2 X_2^2 X_3^2$$
$$- 2X_1 X_2 X_3 (X_1 X_2^2 + X_2 X_3^2 + X_3 X_1^2).$$

Again it is easy to verify that F_2 is not a SOS: suppose $F_2 = \Sigma f_i^2$. As above write $f_i = g_i + h_i$ where g_i involves only $X_1^2 X_2, X_2^2 X_3, X_3^2 X_1$ and h_i involves only $X_1 X_2 X_3$. Equating coefficients gives

$$\Sigma g_i^2 = F_2 - 6(X_1 X_2 X_3)^2,$$

which is negative when $X_1 = X_2 = X_3 = 1$, giving a contradiction. Thus $F_2 \in \mathcal{P}_{3,6} - \Sigma_{3,6}$.

We shall now prove Calderon's theorem. In our notation m is the number of X's and n the number of Y's. The degree is 4. Since $m = 2$ or $n = 2$, we may suppose without loss of generality, the latter and call the two Y's s and t and write $\underline{X} = (X_1, \ldots, X_m)$. Let the biquadratic form be then

$$Q(X_1, \ldots, X_m; s, t) = a(\underline{X}, \underline{X})s^2 + b(\underline{X}, \underline{X})st + c(\underline{X}, \underline{X})t^2,$$

where $a(\underline{X}, \underline{X}), b(\underline{X}, \underline{X}), c(\underline{X}, \underline{X})$ are quadratic forms in $\underline{X}$; in fact $a(\underline{X}, \underline{X}')$ is the bilinear form associated with the quadratic form $a(\underline{X}, \underline{X})$ so that

$$a(\underline{v}, \underline{w}) = \frac{1}{2}\left(a(\underline{v} + \underline{w}, \underline{v} + \underline{w}) - a(\underline{v}, \underline{v}) - a(\underline{w}, \underline{w})\right) \tag{9.2}$$

We have the following

Theorem 9.2 (Calderon). *Suppose Q is PSD; then there exist $3m(m+1)/2$ linear forms $u_i(\underline{X})$, $v_i(\underline{X})$, $1 \le i \le 3m(m+1)/2$, such that*

$$Q = \sum_{i=1}^{3m(m+1)/2} \{u_i(\underline{X})s + v_i(\underline{X})t\}^2 .$$

First note that the dimension of the linear space S of forms $\{u(\underline{X})s + v(\underline{X})t\}^2$, where $u(\underline{X}), v(\underline{X})$, linear forms in $\underline{X}$ is $3m(m+1)/2$: verify this for say $m = 1, 2$ and then use induction. Thus for $m = 1$ we have

$$\begin{aligned}(u(\underline{X})s + v(\underline{X})t)^2 &= (\alpha X_1 s + \beta X_1 t)^2 \\ &= \alpha^2 X_1^2 s^2 + \beta^2 X_1^2 t^2 + 2\alpha\beta X_1^2 st,\end{aligned}$$

so that the required dimension is 3, a basis being $X_1^2 s^2$, $X_1^2 t^2$, $X_1^2 st$ and also $3m(m+1)/2 = 3$ for $m = 1$.

For $m = 2$, the typical element of S is

$$((\alpha_1 X_1 + \alpha_2 X_2)s + (\beta_1 X_1 + \beta_2 X_2)t)^2$$

and expanding, we see that a basis for the space is

$$X_1X_2st, X_1^2s^2, X_2^2s^2, X_1^2t^2, X_2^2t^2, X_1X_2s^2, X_1X_2t^2, X_1^2st, X_2^2st$$

so that the dimension is 9 as also is $3m(m+1)/2$ for $m = 2$.

Let Γ be the space (closed cone) of all PSD biquadratic forms. First we shall show that the

$$\text{extreme elements (rays) } \mathcal{E}(\Gamma) \text{ of } \Gamma \text{ are all in } S \tag{9.3}$$

$$\text{i.e. } \mathcal{E}(\Gamma) \subset S \subset \Gamma.$$

Then Γ is a convex combination of elements of $\mathcal{E}(\Gamma)$ and each element of $\mathcal{E}(\Gamma)$ (being in S by (9.3)) is a linear combination of a basis of S as required.

To prove (9.3), we need the following

Lemma. *Let $Q(\not\equiv 0) \in \Gamma$. Then there is a $Q_1(\not\equiv 0) \in S$ such that $Q \ge Q_1$ (i.e. $Q - Q_1$ is PSD).*

Proof. We use induction on m. For $m = 1, \underline{X} = X_1 = X$ say and

$$Q = Q(\underline{X}; s, t) = \alpha X^2 s^2 + \beta X^2 st + \gamma X^2 t^2$$
$$= \alpha X^2 \left(\left(s + \frac{\beta}{2\alpha} t \right)^2 + \frac{4\alpha\gamma - \beta^2}{4\alpha^2} t^2 \right).$$

Now being PSD, $\alpha > 0$, $0 \leq 4\alpha\gamma - \beta^2$ $(= \delta$ say), so the above equals $\left(\sqrt{\alpha} X\beta + \frac{\beta}{2\sqrt{\alpha}} Xt\right)^2 + \left(\frac{\sqrt{\alpha\delta}}{2\alpha} Xt\right)^2$. So suppose the result is proved when the number of variables is at most $m - 1$ and let $Q = as^2 + bst + ct^2$ with a, b, c having m variables. Being PSD, we see, on completing squares that

$$a(\underline{X}, \underline{X}) \geq 0, c(\underline{X}, \underline{X}) \geq 0, 4a(\underline{X}, \underline{X}) \cdot c(\underline{X}, \underline{X}) \geq b^2(\underline{X}, \underline{X}) \tag{9.4}$$

Case 1: $\mathfrak{S}(Q) = \emptyset$ i.e. $Q = 0$ if and only if $\underline{X} = 0, s = t = 0$. Then for $X_1, \ldots, X_m, s, t$ satisfying $X_1^2 + \ldots + X_m^2 = 1$; $s^2 + t^2 = 1$, the continuous function $Q(\underline{X}, s, t)/(u(\underline{X}))^2 t^2$ is bounded below by a positive number μ say (as in the first proof of Hilbert's Theorem 6.1, main part) where $u(\underline{X})$ is any given linear functional in $\underline{X}$, i.e.

$$Q \geq \mu u^2(\underline{X}) t^2 \quad \text{for } \underline{X}, s, t \text{ satisfying}$$
$$X_1^2 + \ldots + X_m^2 = 1, \; s^2 + t^2 = 1 \tag{9.5}$$

and so this inequality holds for *all* $\underline{X}, s, t$, for let $\underline{X} = (X_1, \ldots, X_m), s, t$ be any $\underline{X}, s, t$; then the coordinates

$$\left(\frac{X_1}{\sqrt{\Sigma X_j^2}}, \ldots, \frac{X_m}{\sqrt{\Sigma X_j^2}} \right), \frac{s}{\sqrt{s^2 + t^2}}, \frac{t}{\sqrt{s^2 + t^2}} \tag{9.6}$$

satisfy (9.5) and so for the coordinates (9.6)

$$Q \geq \mu u^2(\underline{X}) t^2.$$

Substituting (9.6) in this, we see that the denominators disappear and the result holds for all $\underline{X}, s, t$ as required.

Case 2: $|\mathfrak{S}(Q)| \geq 1$ i.e. Q has at least one non-trivial zero. By linear substitution of the variables s, t we may assume that

$$Q(\xi_1, \xi_2, \ldots, \xi_m, 1, 0) = 0.$$

Let R^m be the space of points $\underline{X} = (X_1, \ldots, X_m)$ and $\mathfrak{A}$ be the one–dimensional space spanned by $\underline{e} = (\xi_1, \ldots, \xi_m)$ so that $\mathfrak{A} \subset \mathbf{R}^m$. Let $\mathfrak{B}$ be a complement of $\mathfrak{A}$ such that

$$c(\underline{e}, \underline{Y}) = 0 \text{ for all } \underline{Y} \in \mathfrak{B}; \tag{9.7}$$

this is *like* choosing the orthogonal complement; instead of the condition of orthogonality, we have (9.7). Then $a(\underline{X}, \underline{X}) = \underline{e}$ together with $\mathfrak{B}$ span $\mathbf{R}^m$;

i.e. each $\underline{X}$ in $\mathbf{R}^m$ may be written

$$\underline{X} = \lambda\underline{e} + \underline{Y} \quad (\underline{Y} \in B,\ \lambda \text{ a scalar}).$$

Then, by (9.2),

$$\begin{aligned} a(\underline{X},\underline{X}) &= a(\lambda\underline{e},\lambda\underline{e}) + 2a(\lambda\underline{e},Y) + a(\underline{Y},\underline{Y}) \\ &= \lambda^2 a(\underline{e},\underline{e}) + 2\lambda a(\underline{e},\underline{Y}) + a(\underline{Y},\underline{Y}). \end{aligned}$$

But $Q(\underline{e},1,0) = 0$ by the choice of our zero of Q, i.e.

$$a(\underline{e},\underline{e})\cdot 1^2 + b(\underline{e},\underline{e})\cdot 1\cdot 0 + c(\underline{e},\underline{e})\cdot 0^2 = 0.$$

Hence $a(\underline{e},\underline{e}) = 0$ and so $a(\underline{X},\underline{X}) = 2\lambda a(\underline{e},\underline{Y}) + a(\underline{Y},\underline{Y})$. But a is PSD (by (7.9)) and unless $a(\underline{e},\underline{Y}) = 0$, this can be made negative by choosing λ suitably. Hence $a(\underline{e},\underline{Y}) = 0$ and so

$$a(\underline{X},\underline{X}) = a(\underline{Y},\underline{Y}) \tag{9.8}$$

Similarly we have

$$b(\underline{X},\underline{X}) = b(\underline{Y},\underline{Y}) + 2b(\underline{e},\underline{Y})\lambda + b(\underline{e},\underline{e})\lambda^2.$$

Here $b(\underline{e},\underline{e}) = 0$ for taking $\underline{Y} = \underline{0}$ (i.e. $\underline{X} = \lambda\underline{e}$) in (9.4) gives

$$4a(\underline{e},\underline{e})\lambda^2\cdot c(\underline{e},\underline{e})\lambda^2 \geq b^2(\underline{e},\underline{e})\lambda^4$$

i.e. $0 \geq b^2(\underline{e},\underline{e})$ so $b(\underline{e},\underline{e}) = 0$. Thus

$$b(\underline{X},\underline{X}) = b(\underline{Y},\underline{Y}) + 2b(\underline{e},\underline{Y})\lambda. \tag{9.9}$$

Also, using (9.7),

$$c(\underline{X},\underline{X}) = c(\underline{Y},\underline{Y}) + c(\underline{e},\underline{e})\lambda^2. \tag{9.10}$$

Now use (9.8), (9.9), (9.10) to get:

Case 1: $c(\underline{e},\underline{e}) = 0$; then

$$\begin{aligned} Q(\underline{X},s,t) &= a(\underline{X},\underline{X})s^2 + b(\underline{X},\underline{X})st + c(\underline{X},\underline{X})t^2 \\ &= a(\underline{Y},\underline{Y})s^2 + (b(\underline{Y},\underline{Y}) + 2b(\underline{e},\underline{Y})\lambda)st + c(\underline{Y},\underline{Y})t^2 \end{aligned}$$

which is linear in λ and so can be made negative by a suitable choice of λ - a contradiction since Q is PSD. Thus in this case $b(\underline{e},\underline{Y}) = 0$ giving

$$Q(\underline{X},s,t) = a(\underline{Y},\underline{Y})s^2 + b(\underline{Y},\underline{Y})st + c(\underline{Y},\underline{Y})t^2$$

where $\underline{Y} \in \mathfrak{B}$ has $n-1$ variables and so by the induction hypothesis $Q \geq Q_1$.

Case 2: $c(\underline{e},\underline{e}) \neq 0$. First we verify that if $p(\lambda) = A\lambda^2 + B\lambda + C$ is a quadratic polynomial which is PSD, then

$$p(\lambda) \geq \frac{1}{4A}\left(\frac{\partial p}{\partial \lambda}\right)^2 \tag{9.11}$$

Now $p(\lambda) \geq 0$ gives as usual that $A \geq 0,\ 4AC \geq B^2$. Then

$$\begin{aligned} p(\lambda) &= A\left[\left(\lambda + \frac{B}{2A}\right)^2 + (4AC - B^2)/4A^2\right] \\ &= \frac{A}{4A^2}\left[(2A\lambda - B)^2 + (4AC - B^2)\right] \\ &= \frac{1}{4A}\left[\left(\frac{\partial p}{\partial \lambda}\right)^2 + 4AC - B^2\right] \geq \frac{1}{4A}\left(\frac{\partial p}{\partial \lambda}\right)^2 \end{aligned}$$

as required. So now

$$\begin{aligned} Q(\underline{X}, s, t) &= a(\underline{Y}, \underline{Y})s^2 + (b(\underline{Y}, \underline{Y}) + 2b(\underline{e}, \underline{Y})\lambda)st \\ &\qquad + (c(\underline{Y}, \underline{Y}) + c(\underline{e}, \underline{e})\lambda^2)t^2 \\ &= (c(\underline{e}, \underline{e})t^2)\lambda^2 + (2b(\underline{e}, \underline{Y})st)\lambda \\ &\qquad + (a(\underline{Y}, \underline{Y})s^2 + b(\underline{Y}, \underline{Y})st + c(\underline{Y}, \underline{Y})t^2), \end{aligned}$$

a quadratic polynomial in λ which is PSD. Applying (9.11) we get

$$\begin{aligned} Q = p(\lambda) &\geq \frac{1}{4c(\underline{e}, \underline{e})t^2}(2c(\underline{e}, \underline{e})t^2\lambda + 2b(\underline{e}, \underline{Y})st)^2 \\ &= \left(\frac{1}{\sqrt{c(\underline{e}, \underline{e})}}(b(\underline{e}, \underline{Y})s + c(\underline{e}, \underline{e})\lambda t)\right)^2 \\ &= \left(\frac{b(\underline{e}, \underline{Y})}{\sqrt{c(\underline{e}, \underline{e})}}s + \sqrt{c(\underline{e}, \underline{e})}\lambda t\right)^2 \\ &= Q_1^2 \end{aligned}$$

as required by the lemma.

Proof of Theorem 9.2. Suppose $Q \notin S$, so Q is not a multiple of Q_1 (of the lemma) and

$$Q = \frac{1}{2}(Q - Q_1) + \frac{1}{2}(Q + Q_1).$$

Here both $\frac{1}{2}(Q - Q_1)$ and $\frac{1}{2}(Q + Q_1)$ are PSD since $Q \geq Q_1$ and neither are multiples of Q, otherwise Q would become a multiple of Q_1. Thus Q is a convex combination of elements of Γ which are not multiples of Q, so Q is not an extreme element of Γ. □

Another very interesting result due to Robinson [R8] is the following

Theorem 9.3.

(i) *Let $F(X_1, \ldots, X_n)$ be any real form of degree $m = 2d$; then*

$$F(X_1, \ldots, X_n) + \beta(X_1^{2d} + \ldots + X_n^{2d})$$

is a SOS of real forms of degree d, when β is sufficiently large. Furthermore, these forms may be chosen to be monomials or binomials (i.e. one- or two-term forms).

In striking contrast to this is the next part:
(ii) *Let $F(X_1,\dots,X_n)$ be a form as in (i) which is PSD but not a SOS; then for β sufficiently small, the strictly positive definite form*

$$F(X_1,\dots,X_n)+\beta(X_1^{2d}+\dots+X_n^{2d})$$

is still not a SOS.

Proof. (i) We shall successively subtract squares of real two-term forms of degree d from F and show that, done suitably, there will remain only terms of the type $\alpha_1X_1^{2d}+\dots+\alpha_nX_n^{2d}$. Thus say if

$$F-(f_1^2+f_2^2+\dots)=\alpha_1X_1^{2d}+\dots+\alpha_nX_n^{2d}$$

(f_j monomials or binomials), then for sufficiently large β ($\geq \max|\alpha_j|$) we shall have

$$F+\beta(X_1^{2d}+\dots+X_n^{2d})=f_1^2+\dots+f_1^2+ \\ (\beta+\alpha_1)X_1^{2d}+\dots+(\beta+\alpha_n)X_n^{2d}$$

and here each $(\beta+\alpha_j)X_j^{2d}=(\sqrt{\beta+\alpha_j}X_j^d)^2$ as required. To achieve this subtracting process we proceed as follows:

First eliminate all terms in which any variable occurs to an odd power. There will be an even number of such variables in any term (since the degree of the form, being $2d$, is even). If the term has the form $cX_1X_2X^2$ (X a product of powers of $X_1,\dots,X_n$) then we may subtract

$$(X_1+\frac{cX_2}{2})^2X^2,$$

so that $cX_1X_2X^2$ and $X_1^2X^2$ and $X_2^2X^2$ are added on.

If the term has the form $cX_1X_2X_3X_4X^2$, then we subtract

$$(X_1X_2+\frac{cX_3X_4}{2})^2X^2;$$

and so on.

We are left with a form Φ in which only even powers of the variables occur. For such forms we shall prove by induction on n, that all terms involving more than one variable can be removed by successively subtracting squares of real binomials.

The case $n=1$ is trivial: the form is αX_1^{2d} and for β sufficiently large ($\beta+\alpha$ should be positive) $\alpha X_1^{2d}+\beta X_1^{2d}=(\sqrt{\beta+\alpha}X_1^d)^2$ as required.

So let $n=2$ (this, we shall see, is the main case to consider; the others follow easily from it). We use the identities

$$(X_1^2 - X_2^2)(X_1^2 - X_2^2) = X_1^4 - 2X_1^2X_2^2 + X_2^4$$
$$(X_1^2 - X_2^2)(X_1^2 - X_2^4) = X_1^6 - X_1^4X_2^2 - X_1^2X_2^4 + X_2^6$$
$$(X_1^2 - X_2^2)(X_1^4 - X_2^4) = X_1^8 - X_1^6X_2^2 - X_1^2X_2^6 + X_2^8$$
$$\cdots\cdots\cdots\cdots\cdots\cdots\cdots\cdots\cdots\cdots$$

and so forth.

Here, on the left side, there is a second $X_1^2 - X_2^2$ factoring out of the other bracket so that each term (on the left) is $(X_1^2 - X_2^2)^2$ times a sum of squares of monomials: Thus the polynomials on the right side are SOS of binomials; e.g. the right side of the third equation equals:

$$(X_1^2 - X_2^2)^2(X_1^4 + X_1^2X_2^2 + X_2^4)$$
$$= (X_1^4 - X_1^2X_2^2)^2 + (X_1^3X_2 - X_1X_2^3)^2 + (X_2^2X_1^2 - X_2^4)^2$$

which is a SOS of two-term forms as desired.

We now multiply these right hand polynomials by large positive multiples of powers of $X_1^2X_2^2$ and subtract this from the form Φ. We see that all cross products $X_1^{2r}X_2^{2d-2r}$ $(0 < r < d)$ i.e. all except X_1^{2d}, X_2^{2d} in Φ with negative coefficients will gradually move towards positive coefficients. An example will make this clearer: Say

$$\Phi = 3X_1^8 - 7X_1^6X_2^2 - 12X_1^4X_2^4 - 4X_1^2X_2^6 - 5X_2^8.$$

Then we subtract 19 times the third equation and 12 $X_1^2X_2^2$ times the first equation from Φ to get :

$$\Phi - 19((X_1^4 - X_1^2X_2^2)^2 + (X_1^3X_2 - X_1X_2^3)^2 + (X_2^2X_1^2 - X_2^4)^2)$$
$$- 12X_1^2X_2^2((X_1^2 - X_2^2)^2)$$
$$= -16X_1^8 + 3X_1^2X_2^6 - 24X_2^8$$

(on the left side we use the left side of the corresponding equation and on the right, the right side). In doing this remember that if d is even (i.e. $4|m$) then we use the first, third, fifth, ... identities in turn, while if d is odd, use the second, fourth, sixth, ... identities. Eventually all coefficients will be made positive except for those of X_1^{2d}, X_2^{2d} (in the above example we did this in a single step). These positive terms may now be removed by subtracting squares of monomials, so that only multiples of X_1^{2d} and X_2^{2d} remain. Thus in the above example we get: $\Phi - 19$ (the left side of third identity) $-12X_1^2X_2^2$ (the left side of first identity) $-(\sqrt{3}X_1X_2^3)^2 = -16X_1^8 - 24X_2^8$. Then adding a large multiple, say β, of $X_1^{2d} + X_2^{2d}$ to Φ and taking the rest to the right side will do what is required. In the above example $\beta = 24$ gives $\Phi + \beta(X_1^8 + X_2^8)$ to be a sum of four binomials and two monomials.

Now suppose $n > 2$ and that the possibility of removing all terms involving more than one variable by subtracting squares of binomials and monomials is known for any smaller number of variables. Consider the terms of $\Phi(X_1, \ldots, X_n)$ involving various powers of

$$X_n : X_n^2, X_n^4, \ldots, X_n^{2k}, \ldots, X_n^{2d}.$$

Take a general one X_n^{2k}. Applying the inductive hypothesis for $n-1$ variables we see that all of these terms can be eliminated except for

$$X_1^{2d-2k}X_n^{2k}, \ldots, X_{n-1}^{2d-2k}X_n^{2k} \quad (k = 0, 1, \ldots, d).$$

Do this for all $k = 0, 1, \ldots, d$ and then apply the 2 variable process to eliminate terms involving both X_l and X_n $(l = 1, 2, \ldots, n-1)$. The only remaining terms will now be multiples of $X_1^{2d}, \ldots, X_n^{2d}$ as required. This completes the proof of (i). □

(ii) In the notation of Chapter 6, we see that polynomials which are not SOS of real polynomials will form an open set in V - being the complement of $\mathcal{C}_2$ which is a closed set. This means that if $F(X_1, \ldots, X_n)$ is not a SOS, then the same will be true if we modify the coefficients of F slightly.

Exactly similar reasoning works for forms, except that the coefficient space V for forms of degree $m = 2d$ in n variables is $\binom{2d+n-1}{n-1}$ - dimensional.

Thus if $F(X_1, \ldots, X_n)$ is a form of degree $m = 2d$, which is PSD but not SOS, then for β sufficiently small, the strictly positive definite form $F(X_1, \ldots, X_n) + \beta(X_1^{2d} + \ldots, X_n^{2d})$ will still not be a SOS of real forms.

□

Some very interesting questions arise out of this theorem. Let us be explicit and consider the following example:

Let $q_\alpha(X, Y, Z, W) = q(X, Y, Z, W) + \alpha X^2W^2 \quad (\alpha \geq 0)$ where q is the quaternary quartic of our basic examples of Chapter 6. Since $q \notin \Sigma_{4,4}$ we see that $q_\alpha \notin \Sigma_{4,4}$ for sufficiently small α. On the other hand we have

$$q_\alpha = w^4 + X^2Y^2 + Z^2X^2 + (YZ - 2XW)^2 + (\alpha - 4)X^2W^2,$$

so $q_\alpha \in \Sigma_{4,4}$ for $\alpha \geq 4$. Allowing α to vary from 0 to ∞, we see that there must exist a minimum value α_0 of α, $\alpha_0 \leq 4$, for which $q_{\alpha_0} \in \Sigma_{4,4}$. What is the precise value of this α_0?

Extending the method of proof of Chapter 6 we can in fact show that $\alpha_0 = 4$. For details, see [C9].

Questions of this kind can be posed for any of the forms we have come across before. It can, at times, become quite difficult to calculate the exact place at which the turning point appears; for example let

$$\begin{aligned} S(X,Y,Z) = &X^6 + Y^6 + Z^6 - (X^4Y^2 + X^4Z^2 + Y^4X^2 + \\ &Y^4Z^2 + Z^4X^2 + Z^4Y^2) + 3X^2Y^2Z^2. \end{aligned}$$

Then we know (see later in this chapter) that S is PSD but not a SOS. By

Robinson's theorem, there exists a β_0 such that $S+\beta(X^6+Y^6+Z^6)$ is a SOS if and only if $\beta \geq \beta_0$. What is the exact value of β_0? In [R8], Robinson gave the estimate

$$.05 \approx 2.5-\sqrt{6} \leq \beta_0 \leq 1.$$

In [C11], Choi, Lam and Reznick finally prove that in fact $\beta_0 = 1/8$.

One can also consider natural extensions of Hilbert's question (6.1) to various special classes of forms as for example the biquadratics considered at the beginning of this chapter: As other obvious examples one could consider the following classes of forms:

1. $\mathcal{S} =$ {symmetric forms};
2. $\mathcal{E} =$ {even forms}, i.e. in which all the intervening exponents are even;
3. $\mathcal{K}^e =$ {forms in which all the intervening exponents are $\leq$ a fixed integer e};
4. $\mathcal{I} = \{\text{forms } f \mid |\mathfrak{S}(f)| = \infty\}$.

Theorem 8.2 deals with class 4 above. If $\mathcal{J}$ denotes any one of the above four families, we can replace Hilbert's question (6.1) by the following more specialized one

$$Q(\mathcal{J}) : \textit{For what pairs } (n,m) \textit{ will } \mathcal{J} \cap \mathcal{P}_{n,m} \subseteq \Sigma_{n,m}? \qquad (9.12)$$

By Hilbert's theorem, the answer is always "yes" if either $m=2$ or $n=2$, so in future let $n \geq 3$, $m \geq 4$. Theorem 8.2 says the answer is yes if $\mathcal{J} = \mathcal{I}$. We shall briefly discuss the solution to $Q(\mathcal{S})$. Actually $\mathcal{S}$ is the first obvious example that comes to ones mind. This is an account of the fact that although the examples of forms given to prove (iv) of Theorem 6.1 have certain symmetries in their structure, they are not fully symmetric and the proofs of their not being in $\Sigma_{n,m}$ seem to have depended, to some extent, on their lack of symmetry. So it may be that a symmetric form in $\mathcal{P}_{3,6}$ (or in $\mathcal{P}_{4,4}$) will always lie in $\Sigma_{3,6}$ (respectively $\Sigma_{4,4}$). We have the following result of which we shall only sketch a proof. For details of references the reader is referred to [C8].

Theorem 9.4. *$\mathcal{S} \cap \mathcal{P}_{n,m} \subseteq \Sigma_{n,m}$ iff $n=2$, all m or $m=2$, all n or $(n,m)=(3,4)$ i.e. the answer to (9.12) for $\mathcal{J}=\mathcal{S}$ is given by the same chart as that given in the remarks after the statement of Theorem 6.1.*

To prove this we must find symmetric forms $\in \mathcal{P}_{n,m} - \Sigma_{n,m}$ for all pairs $(n,4)$ for $n \geq 4$ and for the pair $(3,6)$, for once such F's are found, we can construct symmetric forms of higher degree by taking $(X_1+\ldots+X_n)^{2i}F$, which is easily seen to be in $\mathcal{P}_{n,m+2i} - \Sigma_{n,m+2i}$. Unfortunately to construct a symmetric form $\in \mathcal{P}_{n,m} - \Sigma_{n,m}$, the construction of a symmetric form

$F_{4,4} \in \mathcal{P}_{4,4} - \Sigma_{4,4}$ will not do, for, this form, although in $\mathcal{P}_{n,4}$ for $n > 4$ is *not* symmetric as a form of $\mathcal{P}_{n,4}(n > 4)$. Thus a different $F_{n,4} \in (\mathcal{P}_{n,4} \cap \mathcal{S}) - \Sigma_{n,4}(n > 4)$ is required for each n and it needs a considerable effort to get it. We shall therefore only record the special form $F_{4,4} \in (\mathcal{P}_{4,4} \cap \mathcal{S}) - \Sigma_{4,4}$ viz.

$$\begin{aligned} & X^2Y^2 + X^2Z^2 + X^2W^2 + Y^2Z^2 + Y^2W^2 + Z^2W^2 - 2XYZW \\ & \quad + X^2YZ + X^2YW + X^2ZW + Y^2ZX + Y^2XW + Y^2WZ \\ & \quad\quad + Z^2XY + Z^2YW + Z^2WX + W^2XY + W^2YZ + W^2ZX \\ & = \Sigma X^2Y^2 + \Sigma X^2YZ - 2XYZW \quad \text{(full symmetric sums).} \end{aligned}$$

As a form $F_{3,6} \in (\mathcal{P}_{3,6} \cap \mathcal{S}) - \Sigma_{3,6}$ we have the beautiful even symmetric ternary sextic constructed by Robinson:

$$\begin{aligned} R(X,Y,Z) = {} & X^6 + Y^6 + Z^6 + 3X^2Y^2Z^2 \\ & - (X^4Y^2 + Y^4X^2 + Y^4Z^2 + Z^4Y^2 + Z^4X^2 + X^4Z^2) \end{aligned}$$

Proof. We first note that R is PSD. To see this write

$$x = X^2, \ y = Y^2, \ z = Z^2.$$

Then

$$\begin{aligned} R &= x^3 + y^3 + z^3 - (x^2y + y^2x + y^2z + z^2y \\ &\quad + z^2x + x^2z) + 3xyz \\ &= x(y - x)(z - x) + (z + y - x)(z - y)^2 \end{aligned}$$

which is non-negative if $0 \le x \le y \le z$ and so non-negative for $0 \le x, 0 \le y, 0 \le z$ by symmetry. But x, y, z are all indeed non-negative, being equal to X^2, Y^2, Z^2.

To see that $R \notin \Sigma_{3,6}$, suppose to the contrary that

$$R = \Sigma q_j^2 \quad (\deg q_j = 3)$$

say

$$\begin{aligned} q_j(X,Y,Z) &= a_jX^3 + b_jX^2Y + c_jX^2Z + d_jXY^2 + e_jXZ^2 \\ &\qquad + \gamma_jXYZ + \rho_j(Y,Z) \\ &= X(a_jX^2 + d_jY^2 + e_jZ^2 + \gamma_jYZ) \\ &\qquad + b_jX^2Y + e_jX^2Z + \rho_j(Y,Z) \\ &= Xf_j + g_j, \end{aligned}$$

say, where f_j, g_j involve only powers of X that are 0 or 2. Then

$$R \equiv \Sigma(X^2f_j^2 + 2Xf_jg_j + g_j^2). \tag{9.13}$$

Comparing coefficients of X^6 we get

$$1 = \Sigma a_j^2 = \Sigma f_j^2(1,0,0) \tag{9.14}$$

Further, the left side of (9.13) is even in X, hence the coefficients of the odd powers of X on the right side must be zero i.e. $\Sigma 2Xf_jg_j = 0$; the remaining terms being squares give

$$R(X,Y,Z) \geq X^2 f_j^2 \quad \text{(for each } j\text{)} \tag{9.15}$$

Plugging in $Y = \pm X$ in this gives

$$\begin{aligned} R(X,\pm X,Z) &= Z^2(Z^2 - X^2)^2 \\ &\geq X^2 f_j^2(X,\pm X,Z) \\ &= X^2((a_j + d_j)X^2 \pm \gamma_j XZ + e_j Z^2)^2. \end{aligned}$$

Thus $Z^2(Z^2 - X^2)^2 \geq X^2((a_j + d_j)X^2 \pm \gamma_j XZ + e_j Z^2)^2$ and this is impossible: take Z small and the left side can be made arbitrarily small (on keeping X constant but large enough to ensure that the right side is, say, at least 1). The only way out is $f_j(X,\pm X,Z) = 0$, i.e. $f_j(X,Y,Z)$ vanishes when $Y = \pm X$. It follows that $X+Y$ and $X-Y$ are both factors of f_j and so

$$X^2 - Y^2 | f_j.$$

Similarly plugging in $Z = \pm X$ (or by symmetry) we see that

$$X^2 - Z^2 | f_j.$$

Thus $(X^2 - Y^2)(X^2 - Z^2)|f_j$. But f_j is quadratic, so this is only possible if $f_j \equiv 0$ for all j; so $f_j(1,0,0) = 0$ for all j giving a contradiction to (9.14). □

Remark. One can prove likewise that the symmetric form

$$\begin{aligned} H_\mu(X,Y,Z) &= X^{2\mu}(X^2 - Y^2)(X^2 - Z^2) + Y^{2\mu}(Y^2 - Z^2)(Y^2 - X^2) \\ &\quad + Z^{2\mu}(Z^2 - X^2)(Z^2 - Y^2) \\ &= \sum_3 X^{2\mu+4} - \sum_6 X^{2\mu+2}Y^2 + \sum_3 X^{2\mu}Y^2Z^2 \\ &\in \mathcal{P}_{3,2\mu+4} - \Sigma_{3,2\mu+4}; \end{aligned}$$

see [C9], Proposition 2.7. So much for symmetric forms.

There are so many other interesting topics we have not even touched on; for example we have said nothing concrete about extremal forms. In general, to determine the set $\mathcal{E}(\mathcal{P}_{n,m})$ of all the extremal PSD forms is difficult. Even to tell, for a given form $f \in \mathcal{P}_{n,m}$, whether or not f belongs to $\mathcal{E}(\mathcal{P}_{n,m})$ is, in general difficult. In [C9], Choi, Lam and Reznick give excellent illustrations of this problem. In [C11], these three authors deal with all the even symmetric sextics. The paper provides a very good concrete example of a typical situation. A lot more work is being done on this topic and for further details, the reader may consult the bibliography given at the end viz. [C10], [C13], [C14], [C16]. For those who desire a basic knowledge of algebra behind all this, the best reference is [L3].

Exercises

Let $B_\mu(X_1, X_2, X_3, Y_1, Y_2, Y_3) = B_\mu(\mathbf{X}, \mathbf{Y}) = X_1^2Y_1^2 + X_2^2Y_2^2 + X_3^2Y_3^2 + \mu(X_1^2Y_2^2 + X_2^2Y_3^2 + X_3^2Y_1^2) - 2(X_1X_2Y_1Y_2 + X_2X_3Y_2Y_3 + X_3X_1Y_3Y_1)$, so that for $\mu = 2$ we get the counterexample of Choi, discussed in Theorem 9.1: $B_2 \in \mathcal{B}_{6,4}$ but is not a SOS of bilinear forms. We wish to show that B_1 is PSD and indeed extremal among all PSD biquadratic forms. This is done in Exercises 1– 5 below.

1. Let $a, b, c \in \mathbf{R}$. Prove that $a^2 + b^2 + c^2 + 3(abc)^{2/3} \geq 2(ab + bc + ca)$. Suffice to prove this for $a, b, c \geq 0$ and by symmetry for $a, b \geq c$. Apply the arithmetic-geometric mean inequality to $c^2, (abc)^{2/3}, (abc)^{2/3}, (abc)^{2/3}$ to get $c^2+3(abc)^{2/3} \geq 4c\sqrt{ab}$. So $a^2+b^2+c^2+3(abc)^{2/3}-2(ab+bc+ca) \geq (a-b)^2+4c\sqrt{ab}-2(a+b)c = (a-b)^2-2c(\sqrt{a}-\sqrt{b})^2 = (\sqrt{a}-\sqrt{b})^2[(\sqrt{a}+\sqrt{b})^2-2c] \geq 0$.

2. Show that B_1 is PSD. (Apply the arithmetic-geometric mean inequality to $X_1^2Y_2^2, X_2^2Y_3^2, X_3^2Y_1^2$ and Exercise 1 to $a = X_1Y_1, b = X_2Y_2, c = X_3Y_3$ then add).

3. Prove that the substitutions
(i) $X_1, X_2, X_3; Y_1, Y_2, Y_3 \to YZ, ZX, XY, X, Y, Z$
(ii) $X_1, X_2, X_3; Y_1, Y_2, Y_3 \to X, W, Z, Y, Z, W$
take B_1 to $S(X, Y, Z)$ and $Q(X, Y, Z, W)$ respectively (see Chapter 6). Prove also that $B_1(Z, X, W, W, Y, Z) = B_1(W, Z, X, Z, W, Y) = Q(X, Y, Z, W)$.

4. Let $f(X_1, X_2, X_3)$ be a quadratic form such that (i) $f(X_1, \beta_2, \gamma_3) \equiv 0$, (ii) $f(\alpha_1, X_2, \gamma_3') \equiv 0$, (iii) $f(\alpha_1', \beta_2', X_3) \equiv 0$. Show that $f = 0$. (Hint: (i)$\Rightarrow$ $\gamma_3X_2 - \beta_2X_3 | f$, (ii)$\Rightarrow$ $\gamma_3'X_1 - \alpha_1X_3 | f$, (iii)$\Rightarrow$ $\beta_2'X_1 - \alpha_1'X_2 | f$. Since f is of degree 2, $f(X_1, Y_3, Y_2) = 0$.

5. Prove by the following procedure that B_1 is extremal as a PSD biquadratic:
(i) If $B_1(\mathbf{X}, \mathbf{Y}) \geq F(\mathbf{X}, \mathbf{Y}) \geq 0$, then $Q \geq B_1(X, W, Z, Y, Z, W) \geq F(X, W, Z, Y, Z, W) \geq 0$; so since Q is extremal we see that

$$F(X, W, Z, Y, Z, W) = \lambda_1 Q$$

and similarly

$$F(Z, X, W, W, Y, Z) = \lambda_2 Q \text{ and } F(W, Z, X, Z, W, Y) = \lambda_3 Q.$$

(ii) Comparing coefficients of X^2Y^2, Y^2Z^2, Z^2W^2, show that $\lambda_1 = \lambda_2 = \lambda_3 = \lambda$ say.
(iii) Now take Y_1, Y_2, Y_3 to be fixed reals and let

$$f(X_1, X_2, X_3) = (F - \lambda B_1)(X_1, X_2, X_3; Y_1, Y_2, Y_3).$$

Show, using Exercise 3 that

$$f(X_1, X_3, Y_3, Y_2) \equiv 0, f(Y_3, X_2, Y_1) \equiv 0, f(Y_2, Y_1, X_3) \equiv 0$$

(iv) Using Exercise 4, show that $f \equiv 0$

(v) Conclude, by continuity, that $F = \lambda B_1$ i.e. B_1 is extremal among biquadratic forms.

The next four exercises are from Robinson's comments on Motzkin's ternary sextic [R8].

6. Let $P(X_1, \dots, X_n)$ be a polynomial of degree less than $2n$ and let

$$F(X_1, \dots, X_n) = X_1^2 X_2^2 \cdots X_n^2 \; P(X_1, \dots, X_n) + 1.$$

Suppose F is a SOS of m polynomials:

$$F = \Sigma_1^m f_j^2(X_1, \dots, X_n) \qquad (*)$$

Show step by step, that P is then a SOS of m polynomials.

(i) Prove that $f_j = X_1 X_2 \cdots X_n \; h_j(X_1, \dots, X_n) + c_j$ ($c_j \in \mathbf{R}$) (put $X_1 = 0$ in $(*)$ to get $1 = \Sigma f_j^2(1, X_2, \dots, X_n)$, giving a contradiction to the formal reality of $\mathbf{R}$, unless $f_j =$ constant. Thus each term of f_j, except the constant term, is divisible by X_1 and similarly by $X_2, \dots, X_n$).

Note that $\deg P, \deg f_j < 2n, \deg F < 4n, \deg h_j < n$.

(ii) Substituting (i) in $(*)$ show that

$$F = X_1^2 \dots X_n^2 \Sigma_1^m h_j^2 + 2X_1 \dots X_n \Sigma_1^m c_j h_j + \Sigma_1^m c_j^2.$$

Show that the middle term on the right is identically 0 (its degree is less than n so it can contribute terms absent in F).

Deduce that $P(X_1, \dots, X_n)$ is a SOS of m real polynomials

7. Show that Exercise 6 is false if $\deg P = 2n$ (take $P = X_1^2 \dots X_n^2 - 2$).

8. Let $n \geq 2$ and take $P = X_1^2 \dots X_n^2 - \alpha$. Prove that the corresponding F of Exercise 6 is PSD if $\alpha = n + 1$ and so also for all $\alpha \leq n + 1$ (apply the arithmetic-geometric mean inequality to the $n + 1$ quantites $X_1^2, \dots, X_n^2, 1/X_1^2 \dots X_n^2$, giving

$$\frac{1}{n+1}\left(X_1^2 + \cdots + X_n^2 + \frac{1}{X_1^2 \cdots X_n^2}\right) \geq 1,$$

and this just says that F is PSD for $\alpha = n + 1$.

Remark. Here P is indefinite, so not a SOS, but F is PSD. If F is a SOS, then by Exercise 6, so is P, hence F is not a SOS but is PSD.

10

The (r,s,n)-identities and the theorem of Hurwitz-Radon (1922-3)

The aim of this chapter is to look at the second of the three generalizations of the identity (1.4):

$$(X_1^2+\ldots+X_n^2)(Y_1^2+\ldots,Y_n^2)=Z_1^2+\ldots,Z_n^2.$$

We have already dealt with the first generalization viz. in which we allowed the Z_k to belong to the field $K(X_1,\ldots,X_n,Y_1,\ldots,X_n)$ and realized that for each power $n=2^m$ of 2, we have such an identity with $Z_k\in K(X_1,\ldots,X_n,Y_1,\ldots,Y_n)$ and indeed that the Z_k may be chosen linear in the Y_j with coefficients in $K(X_1,\ldots,X_n)$.

Thus we consider now the identity

$$(X_1^2+\ldots+X_r^2)(Y_1^2+\ldots+Y_s^2)=Z_1^2+\ldots+Z_n^2 \tag{10.1}$$

where we of course restrict the Z_k to be bilinear in the X_i and the Y_j with coefficients in K (we suppose char $K\neq 2$).

Definition 10.1. Call the triple (r,s,n) *admissible* over K if (10.1) holds with the Z_k bilinear functions of the X_i and the Y_j with coefficients in K.

Our main problem is to determine what triples (r,s,n) are admissible. Obviously (r,s,rs) is trivially admissible so that what we really want is the most economical n for which (r,s,n) is admissible for the given pair r,s. With this in view we have the

Definition 10.2. Denote by r_*s (or rather $r*_K s$) the least n for which (r,s,n) is admissible over K.

We have the trivial bounds

$$\max(r,s)\le r_*s\le r\cdot s.$$

It is not easy to determine $r_* s$, even for small values of r, s and indeed the main problem of determining exactly what triples (r, s, n) are admissible is far from being solved.

Alternatively we could ask, for given s, n the maximum value of r for which (r, s, n) is admissible; and this is the approach we adapt here. Our aim is to describe the special case $s = n$ solved by Radon [R2] in 1922 for the field $\mathbf{R}$ of real numbers and simultaneously by Hurwitz [H6], published posthumously in 1923, for the field $\mathbf{C}$ of complex numbers. Various authors have since dealt with other fields (see the references). Actually Radon's proof would work for all real closed fields ($\mathbf{R}$ being such a field) where as Hurwitz's proof would work for all algebraically closed fields ($\mathbf{C}$ being such a field). It would not be worth our while to give all the details here. Basically it is a game played with matrices and we give here, as a sample, the typical case of the real number field treated by Radon. We have been, no doubt, rather partial towards real numbers; the presence of Chapters 4–9 testifies to this. After all we do live in a real world. Some examples are now called for before we proceed with Radon's solution to Hurwitz's problem.

Examples.

(i) (n, n, n) is admissible over $\mathbf{R}$, indeed over any field K, char $K \neq 2$, iff $n = 1, 2, 4, 8$. Thus is Hurwitz's theorem (Theorem 1.1).

(ii) $(1, n, n)$, and indeed (r, s, rs), is admissible for all n, r, s over any field K.

(iii) If char $K = 2$, then $r *_K s = 1$ for all r, s for then $a^2 + b^2 = (a + b)^2$.

(iv) $8 * 8 = 8$ for $\max(8, 8) \leq 8 * 8 \leq 8$. Similarly $4 * 4 = 4$ and $2 * 2 = 2$.

(v) *The 16-square problem:* Before Hurwitz, studies about the (r, s, n)-identites (10.1) were exclusively restricted to the polynomial ring $\mathbf{Z}[X_1, \ldots, X_r, Y_1, \ldots, Y_s]$ over $\mathbf{Z}$. One then speaks of the $(r, s, n)_{\mathbf{Z}}$-identities. It has recently been confirmed that $16 *_{\mathbf{Z}} 16 = 32$, thereby completing the solution of the so-called 16-square problem in the integer coefficient case (see [Y2]). However, the integer $\nu = 16 *_{\mathbf{R}} 16$ is not known to date. Various methods developed by K. Y. Lam and J. Adem narrow down the range of ν to $23 \leq \nu \leq 32$. The values $23, 24$ were subsequently ruled out by Lam and Yuzvinski. By going more deeply into the geometry of sums of square formulae and using sophisticated algebraic topology, it has now been established by Lam and Yiu that $29 \leq \nu \leq 32$.

It is trivial to see that $\nu \leq 32$; indeed

$$\left(\sum_1^{16} X_j^2\right)\left(\sum_1^{16} Y_j^2\right) = \left(\sum_1^{8} X_j^2 + \sum_9^{16} X_j^2\right)\left(\sum_1^{8} Y_j^2 + \sum_9^{16} Y_j^2\right)$$

which is a sum of 32 squares, using the 8-square identity four times.

(vi) Amongst small values of r, s, n, even (10,11,25) is not known to be admissible or otherwise.

Definition 10.3. For any positive integer n, define the so-called *Radon function* $\rho(n)$ as follows:

Write $n = 2^m \cdot u$ (u odd); then

$$\rho(n) = \begin{cases} 2m+1 \\ 2m \\ 2m \\ 2m+2 \end{cases} \quad \text{according as } m \equiv \begin{cases} 0 \\ 1 \\ 2 \\ 3 \end{cases} \quad \text{(modulo 4)}.$$

Equivalently write $m = 4a + b$, $0 \le b \le 3$; then

$$\rho(n) = 8a + 2^b.$$

We now have the following

Theorem 10.1. (Radon, Hurwitz - 1922, 1923) *The triple (r,n,n) is admissible over the field of real numbers (indeed over any field K, char $K \neq 2$) iff $r \le \rho(n)$.*

For the proof we convert the general identity (10.1) into a system of matrix equations (cf. Chapter 1). Write

$$\mathbf{X} = \begin{pmatrix} X_1 \\ \vdots \\ X_r \end{pmatrix}, \quad \mathbf{Y} = \begin{pmatrix} Y_1 \\ \vdots \\ Y_s \end{pmatrix}, \quad \mathbf{Z} = \begin{pmatrix} Z_1 \\ \vdots \\ Z_n \end{pmatrix};$$

then $\mathbf{Z} = A\mathbf{Y}$, where A is an $n \times s$ matrix whose entries are bilinear forms in the X_i with coefficients in K. Then (10.1) becomes:

$$(X_1^2 + \ldots + X_r^2)(Y_1, \ldots, Y_s) \begin{pmatrix} Y_1 \\ \vdots \\ Y_s \end{pmatrix} = (Z_1, \ldots, Z_n) \begin{pmatrix} Z_1 \\ \vdots \\ Z_n \end{pmatrix} = \mathbf{Y}'A'A\mathbf{Y}$$

i.e.

$$\mathbf{Y}' \left(\left\{ \sum_{i=1}^{r} X_i^2 \right\} I_s - A'A \right) \mathbf{Y} = 0.$$

It follows that

$$A'A = \left(\sum_{i=1}^{r} X_i^2 \right) I_s.$$

Now write $A = A_1X_1 + \ldots + A_rX_r$ where each A_i is an $n \times s$ matrix over K. This then gives the following system of equations called the Hurwitz matrix equations:

$$\left.\begin{aligned} A_i'A_i &= I_s \quad (1 \le i \le r) \\ A_i'A_j + A_j'A_i &= 0 \quad (1 \le i,j \le r,\ i \neq j) \end{aligned}\right\} \qquad (10.2)$$

Conversely if there are r matrices $A_1, \ldots, A_r$ of type $n \times s$ satisfying (10.2), then (r, s, n) is admissible over K.

Now let $s = n$ and $K = \mathbf{R}$ (the Radon case) so that the A_i become $n \times n$ real matrices and the system (10.2) becomes

$$A_i' A_i = I_n \quad (1 \le i \le r).$$

$$A_i' A_j + A_j' A_i = 0 \quad (i \neq j,\ 1 \le i,\ j \le r).$$

Notice that we may suppose that $r > 1$ since for $r = 1$ the second equation above is vacuous and as we know $(1, n, n)$ is trivially admissible.

Step 1. Let $B_i = A_r' A_i$ $(i = 1, 2, \ldots, r-1)$. The B's are easily checked to satisfy (see Chapter 1)

$$\left.\begin{array}{ll} (1) & B_i + B_i' = 0 \\ (2) & B_i^2 = -I_n \\ (3) & B_i B_j + B_j B_i = 0 \ (i \neq j) \end{array}\right\} \quad 1 \le i,\ j \le r-1. \tag{10.3}$$

(1) implies $|B_i| = -|B_i'| = (-1)^n |B_i|$ and since $|B_i| \neq 0$ (by (2)) it follows that n is even. Thus if n is odd, the set of matrices B_i is empty i.e. $r - 1 = 0$ i.e. $r = 1$ i.e. the largest r for which the triple (r, n, n) (n odd) is admissible, is 1.

So we shall suppose in future that n is even.

Step 2. By the corollary to Proposition 2 in Appendix 1, find a real orthogonal matrix O such that $B_{r-1} = O C_{r-1} O'$ where

$$C_{r-1} = \begin{pmatrix} 0 & I_{n/2} \\ -I_{n/2} & 0 \end{pmatrix}.$$

Put $C_i = O B_i O'$ $(i = 1, 2, \ldots, r-2)$. The $C_1, C_2, \ldots, C_{r-2}, C_{r-1}$ satisfy the following system of equations got directly from (10.3):

$$\left.\begin{array}{ll} (1) & C_i + C_i' = 0 \\ (2) & C_i^2 = -I_n \\ (3) & C_i C_j + C_j C_i = 0 \ (i \neq j) \end{array}\right\} \quad (1 \le i,\ j \le r-1) \tag{10.4}$$

Write, for $1 \le i \le r-2$, $C_i = \begin{pmatrix} R_i & S_i \\ -S_i' & W_i \end{pmatrix}$, where R_i, S_i, W_i are $n/2 \times n/2$ matrices; note also that since the C_i are skew-symmetric, so are R_i, W_i. Put $j = r-1$ in (3) of (10.4) and let $1 \le i \le r-2$. We get

$$\begin{pmatrix} R_i & S_i \\ -S_i' & W_i \end{pmatrix} \begin{pmatrix} 0 & I_{n/2} \\ -I_{n/2} & 0 \end{pmatrix} + \begin{pmatrix} 0 & I_{n/2} \\ -I_{n/2} & 0 \end{pmatrix} \begin{pmatrix} R_i & S_i \\ -S_i' & W_i \end{pmatrix} = 0$$

or

$$\begin{pmatrix} -S_i & R_i \\ -W_i & -S_i' \end{pmatrix} + \begin{pmatrix} -S_i' & W_i \\ -R_i & -S_i \end{pmatrix} = 0.$$

In other words, $W_i + R_i = 0$ and $S_i + S_i' = 0$ (i.e. S_i is skew-symmetric).

Now put $D_i = R_i + \imath S_i$ $(i = 1, 2, \ldots, r-2, \imath = \sqrt{-1})$. The D_i are skew-symmetric since the R_i and S_i are. Then

$$\begin{aligned} C_iC_j &= \begin{pmatrix} R_i & S_i \\ S_i & -R_i \end{pmatrix}\begin{pmatrix} R_j & S_j \\ S_j & -R_j \end{pmatrix} \\ &= \begin{pmatrix} R_iR_j + S_iS_j & R_iS_j - S_iR_j \\ S_iR_j - R_iS_j & S_iS_j + R_iR_j \end{pmatrix} \\ &= \begin{pmatrix} \mathcal{R}(D_i\overline{D}_j) & -\mathcal{I}(D_i\overline{D}_j) \\ \mathcal{I}(D_i\overline{D}_j) & \mathcal{R}(D_i\overline{D}_j) \end{pmatrix} \end{aligned}$$

since $D_i\overline{D}_j = R_iR_j + S_iS_j + \imath(R_iS_j - S_iR_j) = \mathcal{R}(D_i\overline{D}_j) + \imath\mathcal{I}(D_i\overline{D}_j)$. Then (10.4) becomes

$$\left.\begin{array}{ll} (1) & D_i + D_i' = 0 \\ (2) & D_i\overline{D}_i = -I_{n/2} \\ (3) & D_i\overline{D}_j + D_j\overline{D}_i = 0\ (i \neq j) \end{array}\right\} \quad 1 \le i,\ j \le r-2 \qquad (10.5)$$

By (2) of (10.5) we get $|D_i|\ |\overline{D}_i| = (-1)^{n/2}$ i.e.

$$|\det\,(D_i)|^2 = (-1)^{n/2}.$$

Here since $|\det\ D_i|$ is a real number, $|\det\ D_i|^2 > 0$. It follows that $n/2$ is even. Thus for $n/2$ odd (i.e. for $2 \parallel n$), the set of matrices $D_i (1 \le i \le r-2)$ is empty i.e. $r = 2$. Thus the largest r for which the triple (r, n, n) $(2 \parallel n)$ is admissible is 2. Indeed, writing $n = 2m$ we have

$$\begin{aligned} &(X_1^2 + X_2^2)(Y_1^2 + Y_2^2 + \ldots + Y_{2m-1}^2 + Y_{2m}^2) \\ &\quad = (X_1^2 + X_2^2)(Y_1^2 + Y_2^2) + (X_1^2 + X_2^2)(Y_3^2 + Y_4^2) + \ldots \\ &\quad = Z_1^2 + Z_2^2 + Z_3^2 + Z_4^2 + \ldots + Z_{2m-1}^2 + Z_{2m}^2, \end{aligned}$$

by the two-square identity. So $(2, n, n)$ is admisible when $2 \parallel n$, and 2 is the largest such integer as required.

In future, therefore, we shall suppose that $n/2$ is even (i.e. that $4|n$).

Step 3. Find a (complex) unitary matrix U (i.e. U satisfies $U\overline{U}' = I$) such that $D_{r-2} = UE_{r-2}U'$, where

$$E_{r-2} = \begin{pmatrix} 0 & I_{n/4} \\ -I_{n/4} & 0 \end{pmatrix}$$

and put $E_i = U^{-1}D_i\overline{U}$ $(i = 1, 2, \ldots, r-3)$. (This is possible by the corollary to Proposition 4 Appendix 1.) The $E_1, E_2, \ldots, E_{r-2}$ satisfy

$$\left.\begin{array}{ll} (1) & E_i + E_i' = 0 \\ (2) & E_i\overline{E}_i = -I_n/2 \\ (3) & E_i\overline{E}_j + E_j\overline{E}_i = 0\ (i \neq j) \end{array}\right\} \quad 1 \le i, j \le r-2 \qquad (10.6)$$

as may be easily checked from (10.5). Now write

$$E_i = \begin{pmatrix} P_i & Q_i \\ S_i & R_i \end{pmatrix} \quad \text{(all } n/4 \times n/4 \text{ blocks).}$$

Then by (3) of (10.6), with $j = r - 2$, we have, for $1 \le i \le r - 3$,

$$0 = \begin{pmatrix} P_i & Q_i \\ S_i & R_i \end{pmatrix} \begin{pmatrix} 0 & I \\ -I & 0 \end{pmatrix} + \begin{pmatrix} 0 & I \\ -I & 0 \end{pmatrix} \begin{pmatrix} \overline{P}_i & \overline{Q}_i \\ \overline{S}_i & \overline{R}_i \end{pmatrix}$$

$$= \begin{pmatrix} -Q_i & P_i \\ -R_i & S_i \end{pmatrix} + \begin{pmatrix} \overline{S}_i & \overline{R}_i \\ -\overline{P}_i & -\overline{Q}_i \end{pmatrix}.$$

It follows that $-Q_i + \overline{S}_i = 0,\ P_i + \overline{R}_i = 0$. Hence

$$E_i = \begin{pmatrix} P_i & Q_i \\ \overline{Q}_i & -\overline{P}_i \end{pmatrix}.$$

Furthermore, by (1) of (10.6), we find:

$$\left.\begin{aligned} P_i + P_i' &= 0 \\ Q_i + \overline{Q}_i' &= 0 \end{aligned}\right\} \tag{10.7}$$

Now decompose P_i, Q_i into real and imaginary parts:

$$P_i = -T_i^{(1)} - \imath T_i^{(2)},$$

$$Q_i = -T_i^{(3)} + \imath S_i.$$

By (10.7), we see that the S_i are symmetric, while the $T_i^{(1)}, T_i^{(2)}, T_i^{(3)}$ are all skew-symmetric.

We now enter into quaternion matrices. Put

$$F_i = S_i + \epsilon_1 T_i^{(1)} + \epsilon_2 T_i^{(2)} + \epsilon_3 T_i^{(3)} \quad (i = 1, 2, \ldots, r-3);$$

these are $n/4 \times n/4$ matrices. Then $\overline{F}_i$ is given by

$$\begin{aligned} \overline{F}_i &= S_i - \epsilon_1 T_i^{(1)} - \epsilon_2 T_i^{(2)} - \epsilon_3 T_i^{(3)} \\ &= S_i' + \epsilon_1 T_i^{(1)'} + \epsilon_2 T_i^{(2)'} + \epsilon_3 T_i^{(3)'} \\ &= F_i'. \end{aligned}$$

Now we have

$$\begin{aligned} F_i F_j &= (S_i + \epsilon_1 T_i^{(1)} + \epsilon_2 T_i^{(2)} + \epsilon_3 T_i^{(3)}) \times \\ &\qquad (S_j + \epsilon_1 T_j^{(1)} + \epsilon_2 T_j^{(2)} + \epsilon_3 T_j^{(3)}) \\ &= (S_i S_j - T_i^{(1)} T_j^{(1)} - T_i^{(2)} T_j^{(2)} - T_i^{(3)} T_j^{(3)}) \\ &\quad + \epsilon_1 (T_i^{(1)} S_j + S_i T_j^{(1)} + T_i^{(2)} T_j^{(3)} - T_i^{(3)} T_j^{(2)}) \\ &\quad + \epsilon_2 (T_i^{(2)} S_j + S_i T_j^{(2)} + T_i^{(3)} T_j^{(1)} - T_i^{(1)} T_j^{(3)}) \\ &\quad + \epsilon_3 (T_i^{(3)} S_j + S_i T_j^{(3)} + T_i^{(1)} T_j^{(2)} - T_i^{(2)} T_j^{(3)}), \end{aligned} \tag{10.8}$$

since

$$\epsilon_1^2 = \epsilon_2^2 = \epsilon_3^2 = -1,\ \epsilon_1 \epsilon_2 = -\epsilon_2 \epsilon_1 = \epsilon_3,\ \epsilon_2 \epsilon_3 = -\epsilon_3 \epsilon_2 = \epsilon_1$$

$$\epsilon_3 \epsilon_1 = -\epsilon_1 \epsilon_3 = \epsilon_2.$$

Then

$$E_i\overline{E}_j = \begin{pmatrix} P_i & Q_i \\ \overline{Q}_i & -\overline{P}_i \end{pmatrix} \begin{pmatrix} \overline{P}_j & \overline{Q}_j \\ Q_j & -P_j \end{pmatrix}$$

$$= \begin{pmatrix} P_i\overline{P}_j + Q_iQ_j & P_i\overline{Q}_j - Q_iP_j \\ \overline{Q}_i\overline{P}_j - \overline{P}_iQ_j & \overline{Q}_i\overline{Q}_j + \overline{P}_iP_j \end{pmatrix} = \begin{pmatrix} M & N \\ -\overline{N} & \overline{M} \end{pmatrix},$$

say. We have

$$\begin{aligned} M &= P_i\overline{P}_j + Q_iQ_j \\ &= (-S_iS_j + T_i^{(1)}T_j^{(1)} + T_i^{(2)}T_j^{(2)} + T_i^{(3)}T_j^{(3)}) \\ &\quad + \imath(-S_iT_j^{(3)} - T_i^{(3)}S_j - T_i^{(1)}T_j^{(2)} + T_i^{(2)}T_j^{(1)}) \end{aligned}$$

and

$$\begin{aligned} N &= P_i\overline{Q}_j - Q_iP_j \\ &= (-S_iT_j^{(2)} - T_i^{(2)}S_j - T_i^{(3)}T_j^{(1)} + T_i^{(1)}T_j^{(3)}) \\ &\quad + \imath(S_iT_j^{(1)} + T_i^{(1)}S_j + T_i^{(2)}T_j^{(3)} - T_i^{(3)}T_j^{(2)}) \end{aligned} \tag{10.9}$$

Comparing (10.8) and (10.9) we get

$$E_i\overline{E}_j = \begin{pmatrix} -(F_iF_j)_0 - \imath(F_iF_j)_3 & -(F_iF_j)_2 + \imath(F_iF_j)_1 \\ (F_iF_j)_2 + \imath(F_iF_j)_1 & -(F_iF_j)_0 + \imath(F_iF_j)_3 \end{pmatrix} \tag{10.10}$$

where $q = q_0 + \epsilon_1 q_1 + \epsilon_2 q_2 + \epsilon_3 q_3$ for any quaternion q.

It is now easy to check that the F's satisfy:

$$\left.\begin{aligned} &(1) \quad F_i = \overline{F}_i' \\ &(2) \quad F_i^2 = I_{n/4} \\ &(3) \quad F_iF_j + F_jF_i = 0 \ (i \neq j) \end{aligned}\right\} \quad 1 \le i,j \le 1,2,\ldots,r-3 \tag{10.11}$$

Indeed (1) has already been verified. For (3), we have (for $i \neq j$)

$$\begin{aligned} 0 &= E_i\overline{E}_j + E_j\overline{E}_i \quad \text{(by (3) of (10.6))} \\ &= \begin{pmatrix} -(F_iF_j + F_jF_i)_0 & -\imath(F_iF_j + F_jF_i)_3, \ \textit{etc.} \\ (F_iF_j + F_jF_i)_2 & +\imath(F_iF_j + F_jF_i)_1, \ \textit{etc.} \end{pmatrix} \end{aligned}$$

Hence, see (10.10),

$$\begin{aligned} &-(F_iF_j + F_jF_i)_0 = 0, \ -(F_iF_j + F_jF_i)_3 = 0 \\ &-(F_iF_j + F_jF_i)_2 = 0, \ -(F_iF_j + F_jF_i)_1 = 0, \end{aligned}$$

i.e. $F_iF_j + F_jF_i = 0$ as all four components are zero.

Finally we prove (2). By (2) of (10.6) we have

$$\begin{pmatrix} -I_{n/4} & 0 \\ 0 & -I_{n/4} \end{pmatrix} = -I_{n/2} = E_i\overline{E}_i = \begin{pmatrix} -(F_i^2)_0 - \imath(F_i^2)_3 & ,\ldots \\ (F_i^2)_2 + \imath(F_i^2)_1 & ,\ldots \end{pmatrix},$$

so $(F_i^2)_0 = I_{n/4}$ and $(F_i^2)_{1,2,3}$ are all $= 0$ i.e. $F_i^2 = I_{n/4}$ as required.

Step 4. By the corollary to Proposition 5 in Appendix 1, find a quaternion matrix O (the so-called *symplectic matrix*) such that $F_{r-3} = O\ G_{r-3}\overline{O}'$, where $O\overline{O}' = I_{n/4}$ and

$$G_{r-3} = \begin{pmatrix} I_a & 0 \\ 0 & I_{\frac{n}{4}-a}, \end{pmatrix},$$

where a is some non-negative integer depending on F_{r-3}.

Put $F_i = OG_i\overline{O}'$ $(i = 1,2,\ldots,r-4)$. Now check that the G_i $(i = 1,2,\ldots,r-3)$ satisfy

$$\left.\begin{array}{ll} (1) & G_i = \overline{G}_i' \\ (2) & G_i^2 = I_{n/4} \\ (3) & G_iG_j + G_jG_i = 0\ (i \neq j) \end{array}\right\} \quad 1 \leq i,j \leq 1,2,\ldots,r-3; \qquad (10.12)$$

this is seen to follow directly from (10.11) and the definition of the G's. Now write

$$G_i = \begin{pmatrix} Q_i & P_i \\ R_i & S_i \end{pmatrix} \quad \text{say } (i = 1,2,\ldots,r-4)$$

where the blocks have type compatible with (10.12); thus for the product G_iG_j to exist we must have, for example, the number of columns of Q_i equal to the number of rows of Q_j, so Q_i is a square matrix, and so on. Now (1) of (10.12) gives $Q_i = \overline{Q}_i'$, $S_i = \overline{S}_i'$, $P_i = \overline{P}_i'$. Putting $j = r-3$ in (3) of (10.12) we get, for $1 \leq i \leq r-4$,

$$\begin{pmatrix} Q_i & P_i \\ \overline{P}_i' & S_i \end{pmatrix} \begin{pmatrix} I_a & 0 \\ 0 & -I_{\frac{n}{4}-a} \end{pmatrix} + \begin{pmatrix} I_a & 0 \\ 0 & -I_{\frac{n}{4}-a} \end{pmatrix} \begin{pmatrix} Q_i & P_i \\ -\overline{P}_i' & S_i \end{pmatrix} = 0$$

since

$$G_i = \begin{pmatrix} Q_i & P_i \\ \overline{P}_i' & S_i \end{pmatrix},$$

using the relations got from (1) of (10.12) i.e.

$$\begin{pmatrix} Q_i & -P_i \\ \overline{P}_i' & -S_i \end{pmatrix} + \begin{pmatrix} Q_i & P_i \\ -\overline{P}_i' & -S_i \end{pmatrix} = 0.$$

Thus $Q_i = 0$, $S_i = 0$ and it follows that G_i is of the form

$$\begin{pmatrix} 0 & P_i \\ \overline{P}_i' & 0 \end{pmatrix} \quad (i = 1,2,\ldots,r-4)$$

where the P_i are $a \times (\frac{n}{4} - a)$ matrices.

Further by (2) of (10.12) we find

$$P_i\overline{P}_i' = I_a,\ \overline{P}_i'P_i = I_{\frac{n}{4}-a}$$

i.e. the matrix equation $P_iX = Y$ has a solution for an arbitrary choice of Y viz. $X = \overline{P}_i'Y$. Being a quaternion matrix, this gives $4a$ linear equations over $\mathbf{R}$ and the unknowns are $4(n/4 - a)$ in number. Hence $4a \leq 4(\frac{n}{4} - a)$. But now reversing the roles of X, Y i.e. writing the equation $P_iX = Y$ as $\overline{P}_i'Y = X$, we see that this latter equation has a solution in Y (viz.

$Y = P_i X$) for an arbitrary choice of X. Hence $4(\frac{n}{4} - a) \leq 4a$. It follows that $4a = 4(\frac{n}{4} - a)$ i.e. that $8a = n$, so $8|n$ or $\frac{n}{4}$ is even. Thus if $n/4$ is odd, equation (3) of (10.12) has only one element in it i.e. $r - 3 = 1$ or $r = 4$. Thus the largest r for which the triple (r, n, n) $(4 \parallel n)$ is admissible is 4. Indeed, writing $n = 4m$ (m odd) we have:

$$\begin{aligned}
&(X_1^2 + \ldots + X_4^2)(Y_1^2 + \ldots + Y_4^2 + \ldots + Y_{4m-3}^2 + \ldots + Y_{4m}^2) \\
&\quad = (X_1^2 + \ldots + X_4^2)(Y_1^2 + \ldots + Y_4^2) + \ldots \\
&\qquad + (X_1^2 + \ldots + X_4^2)(Y_{4m-3}^2 + \ldots + Y_{4m}^2) \\
&\quad = (Z_1^2 + \ldots + Z_4^2) + \ldots + (Z_{4m-3} + \ldots + Z_{4m}^2),
\end{aligned}$$

by the 4-square identity. Thus $(4, n, n)$ is admissible when $4 \parallel n$ and 4 is the largest r such that (r, n, n) is admissible.

In future therefore, we shall suppose that $n/4$ is even i.e. that $8|n$. Let $n = 8b$ and we examine the G_i's. We have

$$G_i = \begin{pmatrix} 0 & P_i \\ \overline{P}'_i & 0 \end{pmatrix} \quad (i = 1, 2, \ldots, r-4)$$

where all these are quaternion matrices and the P_i have type $\frac{n}{8} \times \frac{n}{8}$. These satisfy

$$\left.\begin{aligned}
&(1) \quad P_i \overline{P}'_i = I_{n/8} \\
&(2) \quad \overline{P}'_i P_i = I_{n/8} \\
&(3) \quad P_i \overline{P}'_j + P_j \overline{P}'_j = 0 \ (i \neq j) \\
&(4) \quad \overline{P}'_i P_j + \overline{P}'_j P_i = 0 \ (i \neq j)
\end{aligned}\right\} \ 1 \leq i, j \leq r-4 \qquad (10.13)$$

Here (1), (2) have already been checked earlier; for (3), (4) use (3) of (10.12).

Now compare (2), (4) of (10.13) with (10.2). They are similar, to say the least, so that the determination of these P_i should follow the same lines as that of the A_i from the equations (10.2). This is just what we now propose to do in our next step:

Step 5. (cf. Step 1). Put $Q = P_{r-4}^{-1} P_i$ $(i = 1, 2, \ldots, r-5)$. Then the Q_i satisfy

$$\left.\begin{aligned}
&(1) \quad Q_i + \overline{Q}'_i = 0 \\
&(2) \quad Q_i^2 = -I_{n/8} \\
&(3) \quad Q_i Q_j = Q_j Q_i \ (i \neq j)
\end{aligned}\right\} \ 1 \leq i, j \leq r-5 \qquad (10.14)$$

This easy to check (cf. Step 1).

Step 6. (cf. Step 2). By the corollary to Proposition 6 in Appendix 1, we find a quaternion matrix O such that

$$Q_{r-5} = O R_{r-5} \overline{O}'$$

where $O\overline{O}' = I_{n/8}$, $R_{r-5} = \epsilon_1 I_{n/8}$ and define R_i $(i = 1, 2, \ldots, r-6)$ by $Q_i = OR_i\overline{O}'$. These R_i $(i = 1, 2, \ldots, r-6, r-5)$ satisfy

$$\left.\begin{array}{ll} (1) & R_i + \overline{R}_i' = 0 \\ (2) & R_i^2 = -I_{n/8} \\ (3) & R_iR_j + R_jR_i = 0 \ (i \neq j) \end{array}\right\} \quad 1 \leq i, j \leq r-5 \qquad (10.15)$$

This is a routine and we leave it to the reader to check.

Now write

$$R_i = S_i^{(0)} + \epsilon_1 S_i^{(1)} + \epsilon_2 S_i^{(2)} + \epsilon_3 S_i^{(3)}$$
$$(i = 1, 2, \ldots, r-6)$$

and take $j = r-5$ in (3) of (10.15). We find

$$(S_i^{(0)} + \epsilon_1 S_i^{(1)} + \epsilon_2 S_i^{(2)} + \epsilon_3 S_i^{(3)})(\epsilon_1 I) + (\epsilon_1 I)(S_i^{(0)} + \epsilon_1 S_i^{(1)} + \epsilon_2 S_i^{(2)} + \epsilon_3 S_i^{(3)}) = 0$$

i.e.

$$\epsilon_i S_i^{(0)} - S_i^{(1)} - \epsilon_3 S_i^{(2)} + \epsilon_2 S_i^{(3)} + \epsilon_1 S_i^{(0)} - S_i^{(1)} + \epsilon_3 S_i^{(2)} - \epsilon_2 S_i^{(3)} = 0.$$

In other words $2(\epsilon_1 S_i^{(0)} - S_i^{(1)}) = 0$ or $S_i^{(0)} = S_i^{(1)} = 0$, so $R_i = \epsilon_2 S_i^{(2)} + \epsilon_3 S_i^{(3)}$. Furthermore by (1) of (10.15) we find

$$\epsilon_2 S_i^{(2)} + \epsilon_3 S_i^{(3)} - \epsilon_2 S_i^{(2)'} - \epsilon_3 S_i^{(3)'} = 0$$

i.e. $S_i^{(2)} = S_i^{(2)'}, S_i^{(3)} = S_i^{(3)'}$ or $S_i^{(2)}, S_i^{(3)}$ are real symmetric matrices.

Now put $T_i = S_i^{(2)} + \imath S_i^{(3)}$ $(i = 1, 2, \ldots, r-6)$. It is not difficult to check that the T_i satisfy

$$\left.\begin{array}{ll} (1) & T_i' = T_i \\ (2) & T_i\overline{T}_i = I_{n/8} \\ (3) & T_i\overline{T}_j + T_j\overline{T}_i = 0 \ (i \neq j) \end{array}\right\} \quad (1 \leq i, j \leq r-6) \qquad (10.16)$$

Indeed we have

$$\begin{aligned} R_iR_j &= (\epsilon_2 S_i^{(2)} + \epsilon_3 S_i^{(3)})(\epsilon_2 S_j^{(2)} + \epsilon_3 S_j^{(3)}) \\ &= -(S_i^{(2)} S_j^{(2)} + S_i^{(3)} S_j^{(3)}) - \epsilon_1 (S_i^{(3)} S_j^{(2)} - S_i^{(2)} S_j^{(3)}) \\ &= \mathcal{R}(T_i\overline{T}_j) - \epsilon_1 \mathcal{I}(T_i\overline{T}_j), \end{aligned}$$

for

$$\begin{aligned} T_i\overline{T}_j &= (S_i^{(2)} + \imath S_i^{(3)})(S_j^{(2)} - \imath S_j^{(3)}) \\ &= (S_i^{(2)} S_j^{(2)} + S_i^{(3)} S_j^{(3)}) + \imath (S_i^{(3)} S_j^{(2)} - S_i^{(2)} S_j^{(3)}) \end{aligned}$$

So by (3) of (10.15) we get

$$\begin{aligned} 0 &= R_iR_j + R_jR_i \\ &= -\mathcal{R}(T_i\overline{T}_j) - \epsilon_1\mathcal{I}(T_i\overline{T}_j) - \mathcal{R}(T_j\overline{T}_i) - \epsilon_1\mathcal{I}(T_j\overline{T}_i) \\ &= -\mathcal{R}(T_i\overline{T}_j + T_j\overline{T}_i) - \epsilon_1\mathcal{I}(T_i\overline{T}_j + T_j\overline{T}_i) \end{aligned}$$

i.e. $\mathcal{R}(T_i\overline{T}_j + T_j\overline{T}_i) = 0$ and $\mathcal{I}(T_i\overline{T}_j + T_j\overline{T}_i) = 0$ so (3) of (10.16) follows. (1) of (10.16) is trivial:

$$\begin{aligned} T_i' &= S_i^{(2)\prime} + \imath S_i^{(3)\prime} \\ &= S_i^{(2)} + \imath S_i^{(3)} \\ &= T_i, \end{aligned}$$

since $S_i^{(2)}, S_i^{(3)}$ are symmetric.

Finally $-I_{n/8} = R_i^2 = -\mathcal{R}(T_i\overline{T}_i) - \epsilon_1\mathcal{I}(T_i\overline{T}_i)$ so $\mathcal{I}(T_i\overline{T}_i) = 0$ and $\mathcal{R}(T_i\overline{T}_i) = I_{n/8}$. Hence, by definition, as required,

$$T_i\overline{T}_i = \mathcal{R}(T_i\overline{T}_i) + \mathcal{I}(T_i\overline{T}_i) = I_{n/8}.$$

Step 7. (cf. Step 3). By the corollary to Proposition 3 in Appendix 1, find O such that $T_{r-6} = O\,V_{r-6}O'$, where is O is a complex matrix - note that the T's are complex,

$$O\overline{O}' = I_{n/8},\ V_{r-6} = I_{n/8},$$

and where V_i is defined, for $i = 1, 2, \dots, r-7$, by

$$T_i = OV_iO'.$$

The V_i again satisfy equations similar to (10.16):

$$\left.\begin{array}{ll} (1) & V_i' = V_i \\ (2) & V_i\overline{V}_i = I_{n/8} \\ (3) & V_i\overline{V}_j + V_j\overline{V}_i = 0\ (i \neq j) \end{array}\right\} \quad 1 \leq i,j \leq r-6 \tag{10.17}$$

Now take $j = r-6$ in (3) of (10.17) and use (1) of (10.17) to find that $V_i = \imath W_i$ $(i = 1, 2, \dots, r-7)$, where the W_i are real symmetric matrices. These W_i then satisfy:

$$\left.\begin{array}{l} (1)\ W_i' = W_i \\ (2)\ W_i^2 = I_{n/8} \\ (3)\ W_iW_j + W_jW_i = 0\ (i \neq j) \end{array}\right\} \quad 1 \leq i,j \leq r-7 \tag{10.18}$$

Step 8. (cf. Step 4). By the corollary to Proposition 1 in Appendix 1, find a real orthogonal matrix O such that

$$W_{r-7} = OX_{r-7}O'$$

where $OO' = I_{n/8}$, and

$$X_{r-7} = \begin{pmatrix} I_b & O \\ O & -I_{\frac{n}{8}-b} \end{pmatrix}.$$

and define X_i $(i = 1, 2, \dots, r-8)$ by

$$W_i = OX_iO'.$$

Again check that the X_i satisfy

$$\left.\begin{array}{lll} (1) & X_i' = X_i \\ (2) & X_i^2 = I_{n/8} \\ (3) & X_iX_j + X_jX_i = 0 \ (i \neq j) \end{array}\right\} \quad 1 \leq i, j \leq r-7 \qquad (10.19)$$

Now finally put $j = r - 7$ in (3) of (10.19) and use (1) of (10.19) to get

$$X_i = \begin{pmatrix} 0 & Y_i \\ Y_i' & 0 \end{pmatrix}.$$

Indeed first put

$$X_i = \begin{pmatrix} P_i & Q_i \\ S_i & R_i \end{pmatrix}$$

as before and use (10.19). Since $X_i^2 = I_{n/8}$, it follows that $|X_i| \neq 0$ and so this is possible only if $b = \frac{n}{8} - b$. Thus if $n/8$ is odd, (10.19) has only one element i.e. $r - 7 = 1$ or $r = 8$ and indeed for $r = 8$, (r, n, n) $(8 \parallel n)$ is admissible; put $n = 8m$ (m odd). Then using the 8-square identity, we find that

$$\begin{aligned} (X_1^2 + \dots + X_8^2)(Y_1^2 + \dots + Y_{8m}^2) &= (X_1^2 + \dots + X_8^2)(Y_1^2 + \dots + Y_8^2) + \\ &\quad \dots + (X_1^2 + \dots + X_8^2)(Y_{8m-7}^2 + \dots + Y_{8m}^2) \\ &= Z_1^2 + \dots + Z_8^2 + \dots + Z_{8m}^2, \end{aligned}$$

as required. If $n/8$ is even, then $b = n/16$ and the condition on the Y_i becomes (the Y_i are $\frac{n}{16} \times \frac{n}{16}$ matrices)

$$\left.\begin{array}{ll} (1) & Y_iY_i' = I_{n/16} \\ (2) & Y_iY_j' + Y_jY_i' = 0 \ (i \neq j) \end{array}\right\} \quad 1 \leq i, j \leq r-8 \qquad (10.20)$$

These are the same as (10.2) substituting r with $r - 8$ and n with $n/16$

We have proved the following:

> (r, n, n) *is admissible iff* (10.2) *holds iff* (10.20) *holds iff* $(r - 8, \frac{n}{16}, \frac{n}{16})$ *is admissible.*

i.e. $\rho(n) = r$ iff $\rho(n/16) = r - 8 = \rho(n) - 8$.

We have thus proved the crucial formula

$$\rho(n) = \rho\left(\frac{n}{16}\right) + 8 \qquad (*)$$

To get the result in the form given in Theorem 10.1 we may now proceed as follows. Write

$$
\begin{aligned}
n &= 2^{4l+k} \cdot u \ (u \text{ odd, } k = 0,1,2,3) \\
&= 16^l \cdot 2^k \cdot u.
\end{aligned}
$$

Then

$$
\begin{aligned}
\rho(n) &= \rho(16^{l-1} \cdot 2^k \cdot u) + 8 \text{ (by } *) \\
&= \rho(16^{l-2} \cdot 2^k \cdot u) + 8 + 8 \text{ (by } * \text{ again)} \\
&\ldots\ldots\ldots\ldots\ldots\ldots\ldots\ldots\ldots\ldots \\
&= \rho(2^k \cdot u) + 8l.
\end{aligned}
$$

Now according as $k = 0,1,2,3$

$$
\begin{aligned}
\rho(2^k u) &= \rho(u),\ \rho(2u),\ \rho(4u),\ \rho(8u) \\
&= \ 1,\ 2,\ 4,\ 8
\end{aligned}
$$

respectively. So $\rho(n) = 1{+}8l,\ 2{+}8l,\ 4{+}8l,\ 8{+}8l$, according as $k = 0,1,2,3$. This is exactly what Theorem 10.1 says for in that theorem, $4l + k = m$ so $k = 0,1,2,3$ implies

$$m = 4l,\ 4l+1,\ 4l+2,\ 4l+3$$

hence $2m+1 = 8l+1,\ 2m = 8l+2,\ 2m = 8l+4,\ 2m+2 = 8l+8$.

This completes the proof at last. □

Remark 1. Other proofs of the Hurwitz-Radon theorem and their versions have been given over various fields by many authors, with methods involving the representation theory of groups or Clifford algebras; see e.g. Eckmann (1943) [E1], Wolf (1963) [W3], T.Y. Lam (1973) [L2], Shapiro (1977) [S3,4,5] and Yuzvinsky (1982) [Y2].

Remark 2. See also the excellent survey article by D.B. Shapiro (1984) [S6] for other references.

Notes on Chapter 10

Very little is known about the admissibility of the general triple (r, s, n). There is the following "bold" conjecture:

> *The admissibility of (r, s, n) over a field K is independent of K (char $K \neq 2$).*

There are some interesting sequences of admissible triples. The Hurwitz-Radon theorem provides examples of size $(\rho(n), n, n)$ giving the sequence

$$(2,2,2), (4,4,4), (8,8,8), (9,16,16), (10,32,32), \ldots$$

The first example, not a consequence of this, was one of size $(10, 10, 16)$ obtained by K.Y. Lam in 1966 [L1]. In 1975, Adem [A2] found another infinite family of examples, the first few terms being

$$(3,5,7),(10,10,16),(12,12,26),(13,13,28),(17,18,32),\ldots.$$

In 1982, Yuzvinski [Y2] constructed yet another infinite sequence beginning with

$$(10,10,16),(12,20,32),(14,40,64),\ldots$$

Since $(10, 10, 16)$ appears in both these infinite sequences we give it explicitly in Chapter 14. See [S6] for more information about this.

Some interesting results regarding the admisibility of the general triple (r, s, n) are known. Here are a few samples:

1. (Adem–1980,1981, Shapiro–1983)(see [A3, A4], [S5]). Let char $K \neq 2$ and suppose $(r, n-1, n)$ is admissible over K. We have the following:

(i) if n is even, then $r \leq \rho(n)$,
(ii) if n is odd, then $r \leq \rho(n-1)$.

2. (Yuzvinski–1983). If $n \equiv 3(4)$, then $(4, n-2, n)$ is not admissible over K.

For more information see Shapiro [S6].

3. We give an explicit solution of the Hurwitz-Radon equations (10.2) for $r = \rho(n)$. Once the $A_1, \ldots, A_r$ are available, the matrix A given by $A = A_1X_1 + \ldots + A_rX_r$ gives the (r, n, n) identity $(X_1^2 + \ldots + X_r^2)(Y_1^2 + \ldots + Y_n^2) = Z_1^2 + \ldots + Z_n^2$ where $\underline{Z} = A\underline{Y}$. This, then is a proof of the Hurwitz-Radon theorem one way round for all fields of characteristic 0.

In Chapter 14, we shall give the so-called "integral version" of the Hurwitz-Radon-Eckmann theorem due to Gabel. In the exercises to Chapter 14, we outline a construction of the bilinear functions $Z_1, Z_2, \ldots, Z_n$, with integer coefficients for the above identity via the Hurwitz-Radon equations (10.3), rather than (10.2).

(i) For $1 \leq b \leq 3$, identify $\mathbf{R}^{2^b}$ as the algebras of complex numbers, quaternions, and Cayley numbers respectively with basis $e_0 = 1, e_1, \ldots, e_{2^b-1}$ satisfying $e_i^2 = -1$ $(1 \leq i \leq 2^b - 1)$ and $e_ie_j = -e_je_i$ $(1 \leq i,\ j \leq 2^b - 1,$ $i \neq j)$. Let $E_{b,j}$ $(0 \leq j \leq 2^b - 1)$ be the matrix corresponding to the left multiplication by e_j. These then form a system of $\rho(2^b) = 2^b$ Hurwitz-Radon matrices of order 2^b satisfying (10.2) $(b = 1, 2, 3)$.

(ii) Let $E = \begin{pmatrix} 0 & 1 \\ -1 & 0 \end{pmatrix}$, $T = \begin{pmatrix} 1 & 0 \\ 0 & -1 \end{pmatrix}$. For $n = 16$ these are $\rho(16) = 9$

Hurwitz-Radon matrices given by

$$E_{4,0} = I_{16}$$
$$E_{4,i} = T \otimes E_{3,6} \qquad (1 \le i \le 7)$$
$$E_{4,8} = E \otimes I_8$$

(iii) Finally, let $K = \begin{pmatrix} 0 & 1 \\ 1 & 0 \end{pmatrix} \otimes I_8$ of order 16 and let $K^{\otimes h} = K \otimes K \otimes \ldots \otimes K$ (h factors). For $n = 2^{4a+b}(2m+1)$ $(0 \le b \le 3)$, let

$$A_0 = I_n$$
$$A_{8h+i} = I_d \otimes (K^{\otimes h}) \otimes E_{4,i} (d = n/2^{4h+4}, 0 \le h \le a-1,\ 1 \le i \le 8)$$
$$A_{8a+j} = I_{2m+1} \otimes (K^{\otimes a}) \otimes E_{b,j} (1 \le j \le 2^b - 1,\ 1 \le b \le 3).$$

Then A_k $(0 \le k \le 8a + 2^b - 1)$ form a system of $\rho(n) = 8a + 2^b$ Hurwitz-Radon matrices of order n.

Note. The tensor product $A \otimes B$, is sometimes called the Kronecker product, also written $A \times B$. (See Chapter 14 for the definition.)

Exercises

It will be of help if readers familiarize themselves with manipulations of the quaternions $\mathbf{H}$ and the octonions Ω. The quaternions are defined as

$$q = a_0 \cdot 1 + a_1 i + a_2 j + a_3 k = (a_0, a_1, a_2, a_3)$$

where $a_0, a_1, a_2, a_3 \in \mathbf{R}$, $i^2 = j^2 = k^2 = -1$, $ij = k = -ji$, etc. Then we can multiply $q_1 = (x_0, x_1, x_2, x_3)$ and $q_2 = (y_0, y_1, y_2, y_3)$. For $q \in \mathbf{H}$, the conjugate quaternion $\bar{q}$ is given by

$$\bar{q} = a_0 - a_1 i - a_2 j - a_3 k = (a_0, -a_1, -a_2, -a_3).$$

Then $q\bar{q} = a_0^2 + a_1^2 + a_2^2 + a_3^2$, the norm of q, written $||q||$. If $q \ne 0$, then $||q|| \ne 0$ and for such a q we have the inverse $q^{-1} = \frac{1}{||q||}\bar{q}$. Further, $\overline{q_1 q_2} = \bar{q}_2 \bar{q}_1$ (check), and the map $\alpha \to (\alpha, 0, 0, 0)$ $(\alpha \in \mathbf{R})$ gives an injective isomorphism of $\mathbf{R}$ into $\mathbf{H}$. We have $aq = qa$ for all $a \in \mathbf{R}$, $q \in \mathbf{H}$. $\mathbf{H}$ is a vector space of dimension 4 over $\mathbf{R}$ with basis 1, i, j, k (by abuse of language).

The octonions Ω are an 8-dimensional vector space over $\mathbf{R}$. They form a non-associative algebra over $\mathbf{R}$. We may write a general octonion as

$$x_0 i_0 + x_1 i_1 + \ldots + x_7 i_7 \qquad (x_j \in \mathbf{R}).$$

The multiplication table is given in Chapter 14 and it will be assumed to be known.

1. Prove

(i) Ω is not associative i.e. $x(yz) \ne (xy)z$ in general

(ii) for all x, $y \in \Omega$, $(xx)y = x(xy)$
and $(yx)x = y(xx)$.

We say Ω is *alternative.*

2. Write $[x, y, z] = (xy)z - x(yz)$, the ***associator***. Using $[x, y+z, y+z] = 0 = [y+z, y+z, x]$ show that $[x, y, z]+[x, z, y] = 0$ and $[z, y, x]+[y, z, x] = 0$. Deduce that $[x, y, z] = -[x, z, y] = [z, x, y] = -[z, y, x]$; i.e. that $[x, y, z]$ is skew-symmetric.

3. For $x = x_0 i_0 + \ldots + x_7 i_7$, define the conjugate $\bar{x} = x_0 i_0 - \ldots - x_7 i_7$. Prove

(i) the mapping $x \xrightarrow{I} \bar{x}$ is involutorial (i.e. $I^2 = \varepsilon$) and an anti-automorphism of Ω;
(ii) $\overline{xy} = \bar{y}\bar{x}$, $\bar{\bar{x}} = x$ for all x, $y \in \Omega$;
(iii) the trace of $x, t(x) = x + \bar{x} \in \mathbf{R}$ and the norm of x, $n(x) = x\bar{x} = \bar{x}x \in \mathbf{R}$.

4. Show that as a vector space $/\mathbf{R}$, $\Omega \simeq \mathbf{H} \oplus \mathbf{H}$. Show that
(i) if $x = (a_1, a_2)$ $(a_1, a_2 \in \mathbf{H})$, then $\bar{x} = (\bar{a}_1, -a_2)$.
(ii) if $x = (a_1, a_2)$, $y = (b_1, b_2)$ are in Ω where a_1, a_2, b_1, $b_2 \in \mathbf{H}$ then

$$xy = (a_1 b_1 - \bar{b}_2 a_2, b_2 a_1 + a_2 \bar{b}_1).$$

5. Prove that $(1, 0, 0, 0, 0, 0, 0, 0)$ is a 2-sided unit in Ω.

6. The mapping $\phi : \mathbf{R} \to \Omega$ given by

$$\phi(x) = (x, 0, 0, 0, 0, 0, 0, 0)$$

is an injective isomorphism of $\mathbf{R}$ into Ω.

7. For x_1, $x_2, \ldots \in \Omega$ write $\mathbf{R}(x_1, x_2, \ldots)$ for the subalgebra in Ω generated by $x_1, x_2, \ldots$ over $\mathbf{R}$ show that
(i) $\bar{x} \in \mathbf{R}(x)$ for any x;
(ii) $\mathbf{R}(x, y)$ is associative for any x, $y \in \Omega$;
(iii) if $xy = yx$, then $\mathbf{R}(x, y)$ is a field, isomorphic to $\mathbf{R}$ or to $\mathbf{C}$; so that then there exists a $z \in \Omega$ such that $\mathbf{R}(x, y) = \mathbf{R}(z)$; see [A5] for more elaborate properties and identities in Ω.

11

Introduction to quadratic form theory

So far we have hardly used even the most basic definitions and properties of quadratic forms and it is now time to develop the rudiments of this theory as we shall need it in the next two chapters.

So let K be a field with char$K \neq 2$. By a quadratic form $f(X_1,\ldots,X_n)$ over K we mean a homogeneous polynomial of degree 2 in the variables $X_1,\ldots,X_n$ with coefficients in K:

$$f(X_1,\ldots,X_n) = \sum_{i=1}^{n}\sum_{j=1}^{n} a_{ij}X_iX_j \qquad (a_{ij} \in K)$$

To render the coefficients symmetric it is customary to rewrite f as

$$\begin{aligned} f(X_1,\ldots,X_n) &= \sum\sum \frac{1}{2}(a_{ij} + a_{ji})X_iX_j \\ &= \sum\sum a'_{ij}X_iX_j \end{aligned}$$

where $a'_{ij} = \frac{1}{2}(a_{ij} + a_{ji})$, so that $a'_{ij} = a'_{ji}$. So we may suppose in the first place that $a_{ij} = a_{ji}$.

Definition 11.1. The symmetric matrix $A = (a_{ij})$ is called the *matrix* of the form f.

Definition 11.2. $d = \det A$ is called the *determinant* of f and written $\det f$.

Definition 11.3. If $d = 0$, we say f is *singular*; otherwise, *non-singular*.

Let $\underline{X} = \begin{pmatrix} X_1 \\ \vdots \\ X_n \end{pmatrix}$. Then $f = \underline{X}'A\underline{X}$. Suppose the variables $X_1, \ldots, X_n$ are replaced by $Y_1, \ldots, Y_n$ according to the substitution

$$X_i = \sum_{j=1}^{n} c_{ij} Y_j \qquad 1 \leq i \leq n, c_{ij} \in K$$

i.e. $\underline{X} = C\underline{Y}$ where $\underline{Y} = \begin{pmatrix} Y_1 \\ \vdots \\ Y_n \end{pmatrix}$ and $C = (c_{ij})$. On substitution, $f(X_1, \ldots X_n)$ becomes $g(Y_1, \ldots, Y_n)$: $\underline{X}'A\underline{X} \to \underline{Y}'(C'AC)\underline{Y}$ so that the matrix of g equals $C'AC$.

Definition 11.4. f and g are called *equivalent* if there exists a non-singular change of variables taking f to g as described above: we write $f \sim g$.

Corollary 1. *$\sim$ is an equivalence relation.* □

Corollary 2. *If $f \sim g$ then* $\det g / \det f \in K^{*^2}$.

Indeed $\det g / \det f = |C'||A||C|/|A| = |C|^2 \in K^{*^2}$. □

Definition 11.5. Let $\gamma \in K$. We say f *represents* γ over K if there exist elements $\alpha_1, \ldots, \alpha_n \in K$ such that

$$f(\alpha_1, \ldots, \alpha_n) = \gamma.$$

Definition 11.6. The set of all non-zero elements of K represented by f will be denoted by $V_f(K)$ or $G_f(K)$. The symbol $G_f(K)$ is generally reserved for the special form $f = X_1^2 + \ldots + X_n^2$ and then it is written $G_n(K)$.

Corollary 3. *If $f \sim g$ then $V_f(K) = V_g(K)$.*

Proof. Let A, B be the matrices of f, g so that $f(\underline{X}) = \underline{X}'A\underline{X}$, $g(\underline{Y}) = \underline{Y}'B\underline{Y}$, $B = C'AC$ where $\underline{X} = C\underline{Y}$. If $\gamma \in V_g(K)$ then

$$\gamma = g(\underline{\beta}) = \underline{\beta}'B\underline{\beta} = \underline{\beta}'C'AC\underline{\beta} = (C\underline{\beta})'A(C\underline{\beta}) = f(C\underline{\beta}) \in V_f(K)$$

So $V_g(K) \subseteq V_f(K)$. Similarly $V_f(K) \subseteq V_g(K)$. □

Definition 11.7. We say $f(\underline{X})$ represents 0 over K if there exist values $\alpha_1, \ldots, \alpha_n \in K$ *not all* 0 such that $f(\alpha_1, \ldots, \alpha_n) = 0$.

Definition 11.8. If f represents 0 over K we say f is *isotropic* $/K$; otherwise *anisotropic.*

Clearly $f \sim g \Rightarrow f$ isotropic iff g is isotropic.

Theorem 11.1. *Let f represent $\gamma \neq 0$ over K. Then f is equivalent to a form of the type $\gamma X_1^2 + g(X_2, \ldots, X_n)$, where g is a form in the $n-1$ variables $X_2, \ldots, X_n$.*

Proof. Let $\gamma = f(\alpha_1, \ldots, \alpha_n)$. Since $\gamma \neq 0$, so $(\alpha_1, \ldots, \alpha_n) \neq (0, \ldots, 0)$. So we can find a non-singular matrix C whose first column is $(\alpha_1, \ldots, \alpha_n)$. Now apply to f the linear map whose matrix is C. We get $f(\underline{X}) = \underline{X}'A\underline{X} \rightarrow \underline{Y}'C'AC\underline{Y} = \underline{Y}'B\underline{Y}$ say where $B = C'AC$. Then

$$\underline{Y}'B\underline{Y} = (Y_1, \ldots, Y_n) \begin{pmatrix} b_{11} & b_{12} & \ldots & b_{1n} \\ b_{21} & b_{22} & \ldots & b_{2n} \\ \ldots & \ldots & \ldots & \ldots \\ b_{n1} & b_{n2} & \ldots & b_{nn} \end{pmatrix} \begin{pmatrix} Y_1 \\ Y_2 \\ \vdots \\ Y_n \end{pmatrix}$$

$$= (Y_1, \ldots, Y_n) \begin{pmatrix} b_{11}\ Y_1 + b_{12}\ Y_2 + \ldots + b_{1n}\ Y_n \\ b_{21}\ Y_1 + b_{22}\ Y_2 + \ldots + b_{2n}\ Y_n \\ \ldots\ldots\ldots\ldots \\ b_{n1}\ Y_1 + b_{n2}\ Y_2 + \ldots + b_{nn}\ Y_n \end{pmatrix}.$$

Now the coefficient of Y_1^2 is

$$\begin{aligned} b_{11} &= \sum\sum c'_{1k} a_{kl} c_{l1} \quad (\text{since } B = C'AC) \\ &= \sum_k c_{k1}(a_{k1}c_{11} + a_{k2}c_{21} + \ldots + a_{kn}c_{n1}) \\ &= c_{11}(a_{11}c_{11} + a_{12}c_{21} + \ldots + a_{1n}c_{n1}) \\ &\quad + c_{21}(a_{21}c_{11} + a_{22}c_{21} + \ldots + a_{2n}c_{n1}) \\ &\quad + \ldots\ldots\ldots\ldots \\ &\quad + c_{n1}(a_{n1}c_{11} + a_{n2}c_{21} + \ldots + a_{nn}c_{n1}) \\ &= (c_{11}, c_{21}, \ldots, c_{n1}) \begin{pmatrix} a_{11} & a_{12} & \ldots & a_{1n} \\ a_{21} & a_{22} & \ldots & a_{2n} \\ \ldots & \ldots & \ldots & \ldots \\ a_{n1} & a_{n2} & \ldots & a_{nn} \end{pmatrix} \begin{pmatrix} c_{11} \\ c_{21} \\ \vdots \\ c_{n1} \end{pmatrix} \\ &= (\alpha_1, \alpha_2, \ldots, \alpha_n) A \begin{pmatrix} \alpha_1 \\ \alpha_2 \\ \vdots \\ \alpha_n \end{pmatrix}, \quad \text{since} \begin{pmatrix} c_{11} \\ c_{21} \\ \vdots \\ c_{n1} \end{pmatrix} = \begin{pmatrix} \alpha_1 \\ \alpha_2 \\ \vdots \\ \alpha_n \end{pmatrix}, \\ &= \underline{\alpha}' A \underline{\alpha} = f(\underline{\alpha}) = \gamma. \end{aligned}$$

The other terms containing Y_1 are:

$$Y_1(b_{12}Y_2 + \ldots + b_{1n}Y_n) + b_{21}Y_1Y_2 + b_{31}Y_1Y_3 + \ldots + b_{n1}Y_1Y_n.$$

So the complete set of terms involving Y_1 is

$$b_{11}Y_1^2 + Y_1Y_2(b_{12} + b_{21}) + Y_1Y_3(b_{13} + b_{31}) + \ldots + Y_1Y_n(b_{1n} + b_{n1}).$$

But $B = C'AC$ is symmetric so this equals

$$b_{11}\left(Y_1 + b_{12}\frac{Y_2}{b_{11}} + b_{13}\frac{Y_3}{b_{11}} + \ldots + \frac{b_{1n}Y_n}{b_{11}}\right)^2 + g(Y_2, \ldots, Y_n).$$

Now let

$$\begin{aligned} Y_1 &\to Y_1 + \frac{b_{12}Y_2}{b_{11}} + \ldots + \frac{b_{1n}Y_n}{b_{11}} \\ Y_2 &\to Y_2 \\ &\ldots\ldots\ldots \\ Y_n &\to Y_n. \end{aligned}$$

This has the non-singular matrix

$$\begin{pmatrix} 1 & b_{12}/b_{11} & \ldots & b_{1n}/b_{11} \\ 0 & 1 & \ldots & 0 \\ \ldots & \ldots & \ldots & \ldots \\ 0 & 0 & \ldots & 1 \end{pmatrix}.$$

It transforms the original form $\underline{X}'A\underline{X}$ to $Y_1^2 + g(Y_2, \ldots, Y_n)$ as required.

□

Definition 11.9. If the matrix of A is diagonal, we say A is *diagonal.*

Successive application of Theorem 11.1 gives the following theorem.

Theorem 11.2. *Any quadratic form $/K$ can be diagonalized by a non-singular linear substitution, i.e. $f \sim$ a diagonal form.*

In terms of matrices this simply means that if A is a symmetric $n \times n$ matrix, then there exists a non-singular matrix C such that $C'AC$ is diagonal.

Let $f(X_1, \ldots, X_n)$, $g(X_1, \ldots, X_m)$ be two forms.

Definition 11.10. The sum $f \oplus g$ is defined to be the form $f(\underline{X}) + g(\underline{Y})$ in $m + n$ variables $X_1, \ldots, X_n, Y_1, \ldots, Y_m$.

We must make sure the variables in f and g are taken different otherwise it could be confused with the usual sum $f + g$ of f and g: e.g. if

$$\begin{aligned} f(X_1, X_2) &= X_1^2 + 2X_2^2 \\ g(X_1, X_2) &= 2X_1^2 + X_1X_2 + 3X_2^2, \end{aligned}$$

then

$$f \oplus g = X_1^2 + 2X_2^2 + 2X_3^2 + X_3X_4 + 3X_4^2$$

whereas $f + g = 3X_1^2 + X_1X_2 + 5X_2^2$.

Corollary 4. $g \sim h \Rightarrow f \oplus g \sim f \oplus h$.

Proof. Let A be the martix of the linear map taking g to h and let n be the order of f; then $\begin{pmatrix} I_n & 0 \\ 0 & A \end{pmatrix}$ takes $f \oplus g$ to $f \oplus h$.

□

Theorem 11.3 (Witt's cancellation law). *Let f, g, h be non-singular quadratic forms over K. Then*

$$f \oplus g \sim f \oplus h \Rightarrow g \sim h.$$

Proof. Since f_0 is diagonal, $f \sim f_0$. Then by Corollary 4, $f \oplus g \sim f_0 \oplus g$ and $f \oplus h \sim f_0 \oplus h$, so $f_0 \oplus g \sim f_0 \oplus h$ so we may suppose f to be diagonal and it is enough to look at the case when $\dim f = 1$, i.e. say $f = aX^2$ $(a \neq 0)$.

Let A, B be the matrices of g, h. By hypothesis

$$aX^2 \oplus g \sim aX^2 \oplus h$$

So there exists a matrix $C = \begin{pmatrix} \gamma & S \\ T & D \end{pmatrix}$ say such that

$$C' \begin{pmatrix} a & 0 \\ 0 & A \end{pmatrix} C = \begin{pmatrix} a & 0 \\ 0 & B \end{pmatrix}.$$

Here S is a row matrix and T a column matrix. This gives the following equations:

$$\left.\begin{aligned} \gamma^2 a + T'AT &= a \\ \gamma a S + T'AD &= 0 \\ S'aS + D'AD &= B \end{aligned}\right\} \qquad (11.1)$$

We must show that there exists a non-singular matrix M such that $M'AM = B$. This matrix M will be found in the form $M = D + \xi TS$, where we shall choose ξ suitably later on.

By (11.1) we get

$$\begin{aligned} M'AM &= (D' + \xi S'T')A(D + \xi TS) \\ &= D'AD + \xi S'T'AD + \xi D'ATS + \xi^2 S'T'ATS \\ &= D'AD + a\{(1-\gamma^2)(\xi^2 - 2\gamma\xi)\}S'S \end{aligned}$$

and this equals B if $(1-\gamma^2)\xi^2 - 2\gamma\xi = 1$ (on using the last equation in (11.1)). That is, if $\xi^2 - (\gamma\xi + 1)^2 = 0$ which can be made so if ξ is chosen suitably viz. we want $\xi = \pm(\gamma\xi + 1)$; either sign will do. If $\gamma = 1$, take the negative sign and $\xi = 1/2$ will do. If $\gamma \neq +1$, take the positive sign and $\xi = 1/(1-\gamma)$ will do.

Finally since B is non-singular, so is M since $M'AM = B$. This completes the proof. □

Theorem 11.4. *If a non-singular form f represents 0 then it represents all elements of K.*

Proof. Without loss of generality, we take f to be diagonal (see Corollary 3), say, $a_1X_1^2 + \ldots + a_nX_n^2$. By hypothesis there exists $(\alpha_1, \ldots, \alpha_n) \neq \underline{0}$ such that $f(\alpha_1, \ldots, \alpha_n) = 0$, i.e. $a_1\alpha_1^2 + \ldots + a_n\alpha_n^2 = 0$, where without loss of generality we may suppose $a_1\alpha_1^2 \neq 0$. Then we have

$$-a_1 = a_2\left(\frac{\alpha_2}{\alpha_1}\right)^2 + \ldots + a_n\left(\frac{\alpha_n}{\alpha_1}\right)^2.$$

If now $\gamma \in K^*$ is any element of K^*, we wish to prove that f represents γ. We have

$$\begin{aligned}\gamma &= a_1\left(\frac{1+\gamma/a_1}{2}\right)^2 - a_1\left(\frac{1-\gamma/a_1}{2}\right)^2 \\ &= a_1\left(\frac{1+\gamma/a_1}{2}\right)^2 + a_2\left(\frac{\alpha_2}{\alpha_1}\right)^2\left(\frac{1-\gamma/a_1}{2}\right)^2 + \ldots \\ &\qquad \ldots + a_n\left(\frac{\alpha_n}{\alpha_1}\right)^2\left(\frac{1-\gamma/a_1}{2}\right)^2 \\ &= f\left(\frac{1+\gamma/a_1}{2}, \frac{\alpha_2}{\alpha_1}\cdot\frac{1-\gamma/a_1}{2}, \ldots, \frac{\alpha_n}{\alpha_1}\cdot\frac{1-\gamma/a_1}{2}\right)\end{aligned}$$

as required. □

Definition 11.11. A form representing all elements of K is called *universal.* For such a form we have

$$V_f(K) = K^*.$$

Corollary 5. *A non-singular f represents $\gamma(\neq 0)$ in K iff*

$$g(X_0, X_1, \ldots, X_n) = -\gamma \oplus f$$

represents 0.

Proof. If f represents γ, say $f(\alpha_1, \ldots, \alpha_n) = \gamma$, then

$$g(1, \alpha_1, \ldots, \alpha_n) = -\gamma + \gamma = 0$$

as required.

Conversely suppose $-\gamma\alpha_0^2 + f(\alpha_1, \ldots, \alpha_n) = 0$; then

$$\gamma = f(\alpha_1/\alpha_0, \ldots, \alpha_n/\alpha_0)$$

if $\alpha_0 \neq 0$ as required. If, however, $\alpha_0 = 0$, then since not all α_j are 0, we see that $f(X_1, \ldots, X_n)$ represents 0 over K; hence by Theorem 11.4, f is universal so in particular represents γ. □

Theorem 11.5. *Let f be non-singular and isotropic, then $f \sim Y_1Y_2 + g(Y_3, \ldots, Y_n)$.*

Proof. Since f is isotropic, it is universal, so in particular it represents 1: $f(\alpha_1, \ldots, \alpha_n) = 1$. Hence by Theorem 11.1, $f \sim X_1^2 + f_1(X_2, \ldots, X_n)$.

Since f is isotropic, so is $X_1^2 + f_1$, i.e. we can find $\beta_1, \ldots, \beta_n$ such that

$$\beta_1^2 + f_1(\beta_2, \ldots, \beta_n) = 0$$

i.e.

$$f_1(\beta_2/\beta_1, \ldots, \beta_n/\beta_1) = -1.$$

Again by Theorem 11.1, $f_1 \sim -X_2^2 + g(Y_3, \ldots, Y_n)$ so $f \sim X_1^2 - X_2^2 + g(Y_3, \ldots, Y_n)$. Now put $X_1 - X_2 = Y_1$, $X_1 + X_2 = Y_2$, so $f \sim Y_1Y_2 + g(Y_3, \ldots, Y_n)$. □

Remark. In Theorem 11.5, $g(Y_3, \ldots, Y_n)$ may or may not be isotropic. If it is, then as in Theorem 11.5, $g \sim Y_3Y_4 + h(Y_5, \ldots, Y_n)$. Repeating this we get to a stage when

$$f \sim Y_1Y_2 + \ldots + Y_{2s-1}Y_{2s} + \phi(Y_{2s+1}, \ldots, Y_n)$$

where ϕ is anisotropic. Thus in any representation of 0 by $Y_1Y_2 + \ldots + Y_{2s-1}Y_{2s} + \phi$, at least one of the variables Y_1, $Y_2, \ldots, Y_{2s}$ is non-zero.

Definition 11.12. If $f \sim Y_1Y_2 + \ldots + Y_{2s-1}Y_{2s}$ where $2s = n$, then we say f is *hyperbolic*. Equivalently, then,

$$f \sim X_1^2 - X_2^2 + \ldots + X_{2s-1}^2 - X_{2s}^2.$$

Theorem 11.6. *Let $|K| > 5$. If $f = a_1X_1^2 + \ldots + a_nX_n^2$ ($a_j \in K$) is isotropic $/K$, then there exists a representation of 0 by f in which all the variables take non-zero values.*

Proof. We first prove that if $a\xi^2 = \lambda \neq 0$, then for any $b \neq 0$, there exist non-zero elements α, β such that

$$a\alpha^2 + b\beta^2 = \lambda. \qquad (11.2)$$

Consider the identity

$$\frac{(t-1)^2}{(t+1)^2} + \frac{4t}{(t+1)^2} = 1.$$

Multiply this by $\lambda = a\xi^2$:

$$a\left(\xi \cdot \frac{t-1}{t+1}\right)^2 + at\left(\frac{2\xi}{t+1}\right)^2 = \lambda. \tag{11.3}$$

Now choose $a\gamma \neq 0$ in K such that $b\gamma^2/a \neq \pm 1$. This is possible since $b\gamma^2 = a$, $b\gamma^2 = -a$ has at most two solutions each in K and since $|K| > 5$, γ can be chosen to avoid 0 and these four solutions.

Now put $t = b\gamma^2/a$ in (11.3) to get (11.2) as required.

To prove 11.6 we now proceed as follows:

If the representation $a_1\xi_1^2 + \ldots + a_n\xi_n^2 = 0$ is such that $\xi_1 \neq 0, \ldots, \xi_r \neq 0$, $\xi_{r+1} = \ldots = \xi_n = 0$ $(r \geq 2)$, then by the above, we can find α, β both non-0 such that

$$a_r\xi_r^2 = a_r\alpha^2 + a_{r+1}\beta^2$$

This then yields a representation of 0 in which the number of non-zero variables is increased by one. Repeating this process we arrive at a representation in which all the variables have non-zero values. □

Definition 11.13. A quadratic form in two variables is called a *binary form.*

Corollary 6. *All non-singular isotropic binary forms are equivalent.*

Proof. By Theorem 11.5, any such form is $\sim Y_1Y_2$. □

Theorem 11.7.

(i) *Let f be a binary form of determinant $d \neq 0$. Then f is isotropic $/K$ iff $-d = \gamma^2$ for some $\gamma \in K^*$.*

(ii) *Let f, g be non-singular binary forms $/K$. Then $f \sim g$ iff*
   (a) $\det f / \det g = \delta^2$ *for some $\delta \in K$ and*
   (b) *there exists some non-zero element of K which is represented by both f and g.*

Proof. (i) First let $f \sim aX^2 + bY^2$ so that $-d = -ab = \gamma^2$. Then

$$f(\gamma, a) = a\gamma^2 + ba^2 = a(\gamma^2 + ab) = a \cdot 0 = 0.$$

Since a, γ are non-zero, this representation is non-trivial.

Conversely suppose $f \sim aX^2 + bY^2$ represents 0 in K, say $a\alpha^2 + b\beta^2 = 0$,

$\alpha,\ \beta \in K,\ \alpha\beta \neq 0$. Then, as required,

$$-d = -ab = a \cdot \frac{a\alpha^2}{\beta^2} = \left(\frac{a\alpha}{\beta}\right)^2.$$

(ii) If $f \sim g$ then (i) and (ii) follow by Corollaries 2 and 3 respectively.

Conversely let $\gamma \neq 0$ be an element of K represented by both f and g. Then by Theorem 11.1

$$f \sim \gamma X^2 + \beta Y^2,$$
$$g \sim \gamma X^2 + \beta' Y^2.$$

Now

$$\delta = \det f / \det g = \frac{-\beta\gamma}{-\beta'\gamma} = \frac{\beta}{\beta'},$$

so $\beta Y^2 \sim \beta' Y^2$ (trivial), i.e. $f \sim g$. □

Exercises.

(i) Show that f singular $\Rightarrow f$ isotropic.
(ii) Show that Theorem 11.4 is false for singular quadratic forms.
(iii) Show that Theorem 11.6 is false if $|K| \leq 5$.
(iv) Show that if $K = \mathsf{F}_q$ $(q \neq 2)$, and f has 3 or more variables, then f represents $0/K$.

We now enter deeper waters. What follows will only be needed in Chapters 12 and 13 and indeed only in the latter parts.

There are three so-called representation theorems of which one has already been proved as Theorem 11.1. It will be referred to as the *first representation theorem.*

The *second representation theorem* has essentially been covered in Corollary 3 of Chapter 2. However, we need it in a slightly greater generality and for that we need the lemma of Cassels (see Chapter 2) in greater generality too. We proceed to work towards this.

Theorem 11.8.

(i) $s(K) = s(K(X))$ *(X is independent indeterminate $/K$).*
(ii) *If $f(X_1, \ldots, X_n)$ is anisotropic $/K$, then it remains anisotropic $/K(X)$.*
(iii) $V_f(K(X)) \cap K^* = V_f(K)$.

Proof. (i) Let $s(K) = n$, $s(K(X)) = m$. Since $K \subset K(X)$ we have trivially $m \leq n$. Conversely suppose $f_1^2(X) + \ldots + f_{m+1}^2(X) = 0$, where on clearing denominators we may suppose $f_j(X) \in K[X]$. Equating coefficients of the highest power of X to 0 this gives

$$a_1^2 + \ldots + a_{m+1}^2 = 0 \qquad (a_j \in K)$$

so $n \leq m$.

(ii) Let

$$f(f_1(X), \ldots, f_n(X)) = 0 \tag{11.4}$$

in $K(X)$, where not all $f_j(X)$ are zero. Clearing denominators, we may suppose that X does not divide all the $f_j(X)$, for if it does, just divide out by it. Now put $X = 0$ in (11.4) to get $f(f_1(0), \ldots, f_n(0)) = 0$ where $f_j(0)$ are not all zero; so f represents $0/K$ - a contradiction.

(iii) First note that $V_f(K(X)) \supset V_f(K)$ and $K^* \supset V_f(K)$ so the left side of (iii) is a subset of $V_f(K)$. It remains to prove that $V_f(K)$ is a subset of the left side of (iii). So let d be an element of this left hand side, i.e. $d \in K^*$ and $d = f(r_1(X), \ldots, r_n(X))$ $r_j(X) \in K(X)$. Now consider the form

$$\phi(X_0, X_1, \ldots, X_n) = f(X_1, \ldots, X_n) - dX_0^2.$$

Suppose $d \notin V_f(K)$; then ϕ does not represent $0/K$. For if it does, with $X_0 \neq 0$, then d is represented by f/K, i.e. $d \in V_f(K)$ which is a contradiction; and if it does with $X_0 = 0$, then f represents $0/K$ and so is universal. Thus it represents d, i.e. $d \in V_f(K)$ a contradiction again.

Then by (ii), ϕ does not represent $0/K(X)$, i.e. f does not represent d in $K(X)$, another contradiction.

□

The generalized form of Cassels' Lemma (proved in Chapter 2) is usually known as the Cassels-Pfister theorem. The exact statement is the following:

Theorem 11.9. (Cassels-Pfister Lemma). *Let $f(X_1, \ldots, X_n)$ be a quadratic form $/K$ and let $p(X)$ be a polynomial in $K[X]$ which is represented by $f(X)/K(X)$, i.e. there exists $r_1(X), \ldots, r_n(X) \in K(X)$ such that $p(X) = f(r_1(X), \ldots, r_n(X))$. Then already $p(X)$ is represented by f over the polynomial ring $K[X]$.*

Proof. The proof exactly follows that of Cassels' Lemma of Chapter 2. It is a good exercise to write out the proof.

We note the following.

Corollary 7. *Let $f(Y_1, \ldots, Y_m)$ be a quadratic form $/K$, where the $Y_1, \ldots, Y_m$ are independent indeterminates $/K$. Let $p(X_1, \ldots, X_n) \in K(X_1, \ldots, X_n)$ be a rational function $/K$ such that $p(a_1, \ldots, a_n)$ is defined $(a_j \in K)$. If f represents $p(X_1, \ldots, X_n)$ over $K(X_1, \ldots, X_n)$, i.e. there* *exist rational functions* $r_1(X_1, \ldots, X_n), \ldots, r_m(X_1, \ldots, X_n)$ *such that*

$$f(r_1, \ldots, r_m) = p(X_1, \ldots, X_n),$$

then f *represents* $p(a_1, \ldots, a_n)$ *over* K.

Proof. Again, it follows exactly that of Corollary 1 of Chapter 2 following Cassels' Lemma.

We now state and prove the following.

Theorem 11.10. (The second representation theorem). *Let*

$$f(X_1, \ldots, X_n) = a_1 X_1^2 + \ldots + a_n X_n^2$$

be an anisotropic form $/K$ $(n \geq 2)$. *Let* $g(Y_2, \ldots, Y_n) = a_2 Y_2^2 + \ldots + a_n Y_n^2$ *and let* $d \in K^*$. *Then* $d + a_1 X^2$ *is represented by* $f/K(X)$ *iff* d *is represented by* g/K, *i.e.* $d + a_1 X^2 \in V_f(K(X))$ *iff* $d \in V_g(K)$.

Remark. If f were isotropic $/K$, then f would be isotropic $/K(X)$ and so universal $/K(X)$, i.e.

$$V_f(K(X)) = K(X) - \{0\},$$

but clearly $V_g(K)$ may not be equal to K^*; so Theorem 11.10 is definitely false without the hypothesis that f is anisotropic $/K$.

Proof of Theorem 11.10. First let $d \in V_g(K)$, say $d = a_2\alpha_2^2 + \ldots + a_n\alpha_n^2$. Then

$$\begin{aligned} a_1 X^2 + d &= a_1 X^2 + a_2\alpha_2^2 + \ldots + a_n\alpha_n^2 \\ &= f(X, \alpha_n, \ldots, \alpha_n) \in V_f(K(X)). \end{aligned}$$

Conversely suppose $d + a_1 X^2 \in V_f(K(X))$, i.e. $d + a_1 X^2$ (a polynomial $\in K[X]$) is represented by f over $K(X)$. So by the Cassels-Pfister Lemma, it is represented by f over $K[X]$, say:

$$d + a_1 X^2 = a_1 f_1^2(X) + \ldots + a_n f_n^2(X), \qquad f_j(X) \in K[X] \qquad (11.5)$$

Here each $f_j(X)$ is a linear polynomial in X for equating the highest coefficient of X on both sides we get $a_1 b_1^2 + \ldots + a_n b_n^2 = 0$ $(b_j \in K)$ - which is a contradiction since f is anisotropic $/K$. Write

$$f_1(X) = a + bX \qquad (a, b \in K).$$

Now one of $a + bX = \pm X$ is always solvable for if $b = 1$ then $a + X = -X$ is solvable, otherwise $a + bX = X$ is solvable. Let C be a solution. Putting $X = C$ in (11.5) we get

$$d + a_1 C^2 = a_1(\pm C)^2 + \sum_{j \geq 2} a_j f_j^2(C),$$

hence

$$d = \sum_{j\geq 2} a_j f_j^2(C),$$

i.e. d is represented by g over K. □

Taking $f = X_1^2 + \ldots + X_n^2$ we get the following.

Corollary 8. *Let K be a field with Stufe at least n (so that -1 is not a sum of $n-1$ squares in K). If $d \in K^*$ and $d + X^2$ is a sum of n squares in $K(X)$ then d is a sum of $n-1$ squares in K (see Corollary 3 of Chapter 2).*

By an obvious induction on n we obtain the following

Corollary 9. *Let K be a field with Stufe at least n. Then*
(i) $1 + X_1^2 + \ldots + X_n^2$ *is not a sum of n squares in* $K(X_1, \ldots, X_n)$
(ii) $X_1^2 + \ldots + X_n^2$ *is not a sum of $n-1$ squares in* $K(X_1, \ldots, X_n)$.

Note that (i) and (ii) are actually equivalent.

We now come to the *third representation theorem*, which is also known as the *subform theorem*. First we need the following

Definition 11.14. Let $f(Z_1, \ldots, Z_n)$ and $g(Y_1, \ldots, Y_m)$ be non-singular quadratic forms $/K$ of dimension n and m respectively. We say g *dominates* f (written $g \succ f$) or that f *submits to* g (written $f \prec g$) if, for indeterminates $X_1, \ldots, X_n/K$, we have

$$f(X_1, \ldots, X_n) \in V_g(K(X_1, \ldots, X_n)).$$

In other words $g(Y_1, \ldots, Y_m)$ represents $f(X_1, \ldots, X_n)$ over $K(X_1, \ldots, X_n)$, i.e. there exist $\gamma_1(X_1, \ldots, X_n), \ldots, \gamma_m(X_1, \ldots, X_n)$ such that

$$f(X_1, \ldots, X_n) = g(\gamma_1, \ldots, \gamma_m)$$

Remark. If g is isotropic $/K$, then

$$V_g(K(X_1, \ldots, X_n)) = K(X_1, \ldots, X_n) - \{0\}$$

and so g dominates any given form f. Thus the notion of dominance is significant only in the case when g is anisotropic $/K$.

Corollary 10. *Let $a \in K^*$. Then, by definition, $g \succ aX^2$ iff $aX^2 \in V_g(K(X))$ iff $a \in V_g(K(X))$, for $aX^2 \in V_g(K(X)$ which means there exist rational functions $\gamma_1(X), \ldots, \gamma_m(X)$ such that $aX^2 = g(\gamma_1(X), \ldots, \gamma_m(X))$. Thus, if for example, g is diagonal i.e. $g = a_1\gamma_1^2(X) + \ldots + a_m\gamma_m^2(X)$ so*

$$a = a_1\left(\frac{\gamma_1(X)}{X}\right)^2 + \ldots + a_m\left(\frac{\gamma_m(X)}{X}\right)^2,$$

and conversely if $a = a_1\gamma_1^2(X) + \ldots + a_m\gamma_m^2(X)$, *then*

$$aX^2 = a_1(X\gamma_1(X))^2 + \ldots + a_m(X\gamma_m(X))^2$$

iff $a \in V_g(K)$ *(by Theorem 11.8, (ii)).*

We now state and prove the main theorem.

Theorem 11.11 (The third representation theorem). *Let* $g(Y_1, \ldots, Y_m)$ *be an anisotropic form over* K *and let* $f(Z_1, \ldots, Z_n)$ *be any form of dimension* n *over* K. *Then* $g \succ f$ *iff* f *is equivalent to a subform of* g*; in other words, there exists a form* h *over* K *such that* $g \sim f \oplus h$. *In particular this implies* $m \geq n$.

Proof. First suppose $g \sim f \oplus h$, $f(X_1, \ldots, X_n) \in V_f(K(X_1, \ldots, X_n))$; for we must show that $f(Z_1, \ldots, Z_n)$ represents $f(X_1, \ldots, X_n)$ over $K(X_1, \ldots, X_n)$. To do this just take $Z_1 = X_1 \in K(X_1, \ldots, X_n), \ldots, Z_n = X_n \in K(X_1, \ldots, X_n)$. Then

$$V_f(K(X_1, \ldots, X_n)) \subset V_g(K(X_1, \ldots, X_n)),$$

i.e. $f \prec g$.

To prove the converse use induction on m, the dimension of g. For $m = 0$, there is nothing to prove. So suppose we know the result for $m - 1$.

Let $f \sim b_1Z_1^2 + \ldots + b_nZ_n^2$. By hypothesis

$$b_1X_1^2 + \ldots + b_nX_n^2 \in V_g(K(X_1, \ldots, X_n)) \tag{11.6}$$

So by Corollary 7, $b_1 \cdot 1^2 + b_2 \cdot 0^2 + \ldots + b_n \cdot 0^2 \in V_g(K)$ i.e. $b_1 \in V_g(K)$.

Hence by Theorem 11.1,

$$g \sim b_1Y_1^2 \oplus g'(Y_2, \ldots, Y_m) \tag{11.7}$$

for some form g' of dimension $m - 1$. Here g' is anisotropic $/K$ since g is. View g as a form over $K' = K(X_2, \ldots, X_n)$ and note that g is still anisotropic $/K(X_2, \ldots, X_n)$, by Theorem 11.8; and put $d = b_2X_2^2 + \ldots + b_nX_n^2 \in K'$. Then (11.6) becomes

$$b_1X_1^2 + d \in V_g(K(X_1, \ldots, X_n)) = V_g(K'(X_1)) = V_{b_1Y_1^2 \oplus g'(Y_2, \ldots, Y_m)}.$$

Hence by Theorem 11.10

$$d \in V_{g'(Y_2, \ldots, Y_m)}(K')$$

i.e.

$$b_2X_2^2 + \ldots + b_nX_n^2 \in V_{g'}(K(X_2, \ldots, X_n))$$

or

$$g' \succ b_2Z_2^2 + \ldots + b_nZ_n^2.$$

It follows by the induction hypothesis that, say,

$$g' \sim b_2Z_2^2 + \ldots + b_nZ_n^2 \oplus h,$$

So by (11.7)

$$\begin{aligned} g &\sim b_1Y_1^2 \oplus g'(Y_2, \ldots, Y_m) \\ &\sim b_1Z_1^2 + b_2Z_2^2 + \ldots + b_nZ_n^2 \oplus h \\ &\sim f \oplus h. \end{aligned}$$

□

Definition 11.15. Let $M_f(K) = \{c \in K^* | cf \sim f\}$. Such a c is called a *similarity factor* for f over K.

We have already met this set $M_f(K)$ in Chapter 2 where we proved some interesting properties about it. We now prove the following.

Theorem 11.12. *Let $f(Y_1, \ldots, Y_m)$ be an anisotropic form $/K$ and $\phi(Z_1, \ldots, Z_n)$ a form of dimension n over K.*

If $\phi(X_1, \ldots, X_n) \in M_f(K(X_1, \ldots, X_n))$ (i.e. $\phi(X_1, \ldots, X_n) \cdot f \sim f$ over $K(X_1, \ldots, X_n)$) then for any $a \in V_f(K)$, f contains a subform equivalent to $a\phi$ (i.e. $f \sim a\phi \oplus \psi$ for some ψ.)

Proof. Since a is represented over K by f so a^2 is represented over K by af, i.e. 1 is represented over K by af (just divide by a^2).

Now $\phi(X_1, \ldots, X_n) \cdot f \sim f$ by hypothesis, so $\phi(X_1, \ldots, X_n) \cdot af \sim af$ (over $K(X_1, \ldots, X_n)$). Since $1 \in V_{af}(K)$ we have $1 \in V_{af}(K(X_1, \ldots, X_n))$. Hence

$$\phi(X_1, \ldots, X_n) \in V_{af}(K(X_1, \ldots, X_n)),$$

i.e. $af \succ \phi$.

Thus by Theorem 11.11 $af \sim \phi \oplus \psi$ say, so $a^2 f \sim a\phi \oplus a\psi$, i.e. $f \sim a\phi \oplus a\psi$.

□

Remark. The first half of this chapter (i.e. up to the end of Theorem 11.7) has been beautifully developed in the Appendix of [B2].

Exercises

1. Let f be a quadratic form over K and let $a \in K^*$. Show that $M_f(K) = M_{af}(K)$.

2. Let f be as above and let $a, b \in K^*$. Show that $f \oplus aZ^2$ represents $-b$ iff $f \oplus bZ^2$ represents $-a$.

3. Let a, $b \in K$ be such that $a^2 + b^2 = c \neq 0$. Show that $X^2 + Y^2 - cZ^2 - cW^2 \sim XY + ZW$.

4. Show that the following conditions are equivalent:

(a) Every 4-dimensional form with determinant -1 is isotropic.
(b) Every even-dimensional form with determinant -1 is isotropic.
(c) Every 3-dimensional form represents its own determinant.
(d) Every odd-dimensional form represents its own determinant.

5.* (Legendre) Let a, b, c be square free relatively prime integers, not all of the same sign. Show that

$$aX^2 + bY^2 + cZ^2$$

is isotropic $/\mathbb{Q}$ iff $-bc$ is a square mod a, $-ca$ is a square mod b and $-ab$ is a square mod c.

6. p, q, r, s are distant odd primes such that $pqrs \not\equiv 1(8)$. Show that $pX^2 + qY^2 - rZ^2 - sW^2$ is isotropic $/\mathbb{Q}$.

7. Let f_1, f_2 be anisotropic $/K$. Then $f_1 \oplus X f_2$ is also anisotropic $/K(X)$.

8. Let $f \sim g$. Show that $f - g \sim$ a hyperbolic form.
Hint: Let a be represented by f and so by g too. Then $f \sim aX_1^2 + \phi(X_2, \ldots)$, $g \sim aY_1^2 + \psi(Y_2, \ldots)$. Since $f \sim g$ we have by Witt's cancellation theorem, $\phi \sim \psi$. Then

$$\begin{aligned} f \oplus -g &\sim a(X_1^2 - Y_1^2) \oplus \phi \oplus -\psi \\ &\sim X_1^2 - Y_1^2 \oplus \phi \oplus -\psi. \end{aligned}$$

where $\phi \sim \psi$. Now use induction.

12

Theory of multiplicative forms and of Pfister forms

The object of this chapter is to look at the third generalization of the identity (1.4) mentioned at the end of Chapter 1. Thus instead of looking at products of sums of squares, as in (1.4), we look for more general quadratic forms $q(X_1, \ldots, X_n)$ over the field K, which satisfy the identity

$$q(\underline{X}) \cdot q(\underline{Y}) = q(\underline{Z})$$

where Z_k is required to be bilinear in the X_i and the Y_j or indeed more generally we could let $Z_k \in K(X_1, \ldots, X_n, Y_1, \ldots, Y_n)$. This leads to the study of the so-called multiplicative forms and in particular the Pfister forms of which we have already had a flavour in Chapter 5.

Pfister forms have revealed various new facets in the theory of quadratic forms and promise to hold the key to other unknown areas in the study of Witt rings.

It is expedient to first introduce the general Pfister forms rather than look directly at the identity $q(\underline{X}) \cdot q(\underline{Y}) = q(\underline{Z})$. We have already met the 1-fold and 2-fold Pfister forms viz.

$$\Phi_1(X, Y) = \Phi_a(X, Y) = X^2 + aY^2$$

and

$$\Phi_2(X, Y, Z, T) = \Phi_{a,b}(Z, Y, Z, T) = X^2 + aY^2 + b(Z^2 + aT^2)\,(a, b \in K).$$

These satisfy the following two remarkable properties:

Theorem 12.1.

(i) *The totality of non-zero values of K (i.e. the sets $G_\Phi(K)$ or $V_\Phi(K)$*

in the earlier notation) represented by Φ_1 (or Φ_2) form a group under multiplication.

(ii) *If Φ_1 (or Φ_2) is isotropic, then it is hyperbolic, i.e.*

$$\Phi_1 \sim X^2 - Y^2 \sim XY$$
$$\Phi_2 \sim X^2 - Y^2 + Z^2 - T^2 \sim XY + ZT.$$

Proof. (i) The closure property follows from the following striking identities:

$$\text{(I)} \quad (X_1^2 + aX_2^2)(Y_1^2 + aY_2^2) = (X_1Y_1 + aX_2Y_2)^2 + a(X_1Y_2 - X_2Y_1)^2$$

$$\begin{aligned} &(X_1^2 + aX_2^2 + bX_3^2 + abX_4^2)(Y_1^2 + aY_2^2 + bY_3^2 + abY_4^2) \\ \text{(II)} \quad &= (X_1Y_1 + aX_2Y_2 + bX_3Y_3 + abX_4Y_4)^2 \\ &+ a(-X_1Y_2 + X_2Y_1 - bX_3Y_4 + bX_4Y_3)^2 \\ &+ b(-X_1Y_3 + X_3Y_1 + aX_2Y_4 - aX_4Y_2)^2 \\ &+ ab(-X_1Y_4 + X_4Y_1 - bX_2Y_3 + bX_3Y_2)^2. \end{aligned}$$

As for the inverse of $\alpha = X^2 + aY^2$, we have

$$\frac{1}{\alpha} = \frac{\alpha}{\alpha^2} = \frac{X^2 + aY^2}{\alpha^2} = \left(\frac{X}{\alpha}\right)^2 + a\left(\frac{Y}{\alpha}\right)^2 \in G_{\Phi_1}(K)$$

and similarly for $G_{\Phi_2}(K)$.

(ii) We first look at $\Phi_1 = X^2 + aY^2$. Since Φ_1 is isotropic, there exist α, β, not both zero, such that $\alpha^2 + a\beta^2 = 0$. Here both α, β are non-zero for $\alpha = 0$ if and only if $\beta = 0$. Hence $a = -(\alpha/\beta)^2$ and so, transforming $X \to X$ and $\left(\frac{\alpha}{\beta}Y\right) \to Y$, and noting that the determinant of this map is $\begin{vmatrix} 1 & 0 \\ 0 & \beta\backslash\alpha \end{vmatrix} = \beta/\alpha \neq 0$, we have $X^2 + aY^2 = X^2 - \left(\frac{\alpha Y}{\beta}\right)^2 \sim X^2 - Y^2$, as required.

For the form Φ_2, there exist α, β, γ, δ, not all zero, such that

$$\alpha^2 + a\beta^2 + b\gamma^2 + ab\delta^2 = 0.$$

Here if $\alpha \neq 0$, on dividing by it we see that -1 is represented by $aY^2 + bZ^2 + abT^2$ while if $\alpha = 0$, then $aY^2 + bZ^2 + abT^2$ represents 0 non-trivially in K and so is universal; so in particular represents -1. Thus $aY^2 + bZ^2 + abT^2$ always represents -1 over K and so is $\sim -Y^2 + \psi(Z,T) \sim -Y^2 + sZ^2 + tT^2$ say. Therefore $\Phi_2 \sim X^2 - Y^2 + sZ^2 + tT^2$. Comparing determinants we get $t = \frac{-1}{s} \times$ (a square) so $\Phi_2 \sim X^2 - Y^2 + s(Z^2 - T^2)$. But now on letting $Z + T \to (Z - T)/s$, $Z - T \to Z + T$ we see that

$$s(Z^2 - T^2) = s(Z + T)(Z - T) \to s\frac{Z - T}{s} \cdot (Z + T) = Z^2 - T^2.$$

Hence $\Phi_2 \sim X^2 - Y^2 + Z^2 - T^2$. □

We define the m-fold Pfister form by induction as

$$\Phi_m(X_1,\ldots,X_2m) = \Phi_{a_1,a_2,\ldots,a_m}(X_1,\ldots,X_2m) = \Phi_{m-1} \oplus a_m\Phi_{m-1}.$$

Our first aim is to prove that properties (i) and (ii) of Theorem 12.1 hold true for Φ_m. Before proving this we familiarize ourselves in playing around with easy identities. We prove the following.

Theorem 12.2.

(i) *$a(X^2 - Y^2) \sim X^2 - Y^2$ (any $a \in K^*$).*

(ii) *If some $a_i = -1$ then Φ_m is hyperbolic i.e. $\sim X_1^2 - X_2^2 + \ldots + X_{2^m-1}^2 - X_{2^m}^2$.*

(iii) *If $a_1 = 1$ then $\Phi_m(a_1,\ldots,a_m) \sim 2\Phi_{m-1}(a_2,\ldots,a_m)$.*

For 2-fold Pfister forms, we have

(iv) *If α, $\beta \in K$ are such that $\gamma = a_1\alpha^2 + \beta^2 \neq 0$, then*

$$X_1^2 + a_1X_2^2 + a_2(X_3^2 + a_1X_4^2) \sim X_1^2 + a_1X_2^2 + a_2\gamma(X_3^2 + a_1X_4^2).$$

(v) *If α, $\beta \in K$ are such that $\gamma = a_1\alpha^2 + a_2\beta^2 \neq 0$, then*

$$X_1^2 + a_1X_2^2 + a_2(X_3^2 + a_1X_4^2) \sim X_1^2 + \gamma X_2^2 + a_1a_2(X_3^2 + \gamma X_4^2).$$

Proof. (i) Let $X + Y \to \frac{1}{a}(X - Y)$, $X - Y \to X + Y$ and result follows.
(ii) If for example $a_1 = -1$, then

$$\Phi_{a_1,a_2,\ldots,a_m}(X_1,\ldots,X_{2^m}) = X_1^2 - X_2^2 + a_2(X_3^2 - X_4^2) + a_3(\ldots) + \ldots \sim X_1^2 - X_2^2 + X_3^2 - X_4^2 + \ldots$$

using (i) above repeatedly.
(iii) If $a_1 = 1$ then

$$\begin{aligned}
&\Phi_{a_1,a_2,\ldots,a_m}(X_1,\ldots,X_2m)\\
&= X_1^2 + X_2^2 + a_2(X_3^2 + X_4^2) + a_3(X_5^2 + X_6^2 + a_2(X_7^2 + X_8^2) + \ldots\\
&= X_1^2 + a_2X_3^2 + a_3(X_5^2 + a_2X_6^2) + \ldots + X_2^2 + a_2X_4^2\\
&\qquad + a_3(X_6^2 + a_2X_8^2) + \ldots\\
&= \Phi_{a_2,a_3,\ldots,a_m}(X_1,X_3,\ldots,X_{2^m-1}) + \Phi_{a_2,a_3,\ldots,a_m}(X_2,X_4,\ldots,X_{2^m})\\
&= 2\Phi_{a_2,\ldots,a_m}
\end{aligned}$$

as required.
(iv) By Witt's cancellation law, it is enough to prove that

$$a_2(X_3^2 + a_1X_4^2) \sim a_2\gamma(X_3^2 + a_1X_4^2).$$

Now the left side represents $a_2\beta^2 + a_1a_2\alpha^2 = a_2\gamma$ and so is $\sim a_2\gamma X_3^2 + cX_4^2$ say. Comparing determinants we get $a_2 \cdot a_1a_2 = a_2\gamma c\times$ (a square) or $c = a_1a_2'\times$ (a square), so the above is equivalent to

$$a_2\gamma X_3^2 + a_1a_2\gamma X_4^2 \sim a_2\gamma(X_3^2 + a_1X_4^2).$$

(v) $a_1X^2 + a_2Y^2$ represents $\gamma = a_1\alpha^2 + a_2\beta^2$ ($\neq 0$) and so is $\sim \gamma X^2 + cY^2$. Comparing determinants we get $c = a_1a_2\times$ (a square)/γ. Hence

$$a_1X^2 + a_2Y^2 \sim \gamma X^2 + a_1a_2\gamma Y^2.$$

It follows that

$$\begin{aligned} &X_1^2 + a_1X_2^2 + a_3(X_3^2 + a_1X_4^2) \\ &\quad \sim X_1^2 + a_1X_2^2 + a_2X_3^2 + a_1a_2X_4^2 \\ &\quad \sim X_1^2 + a_1a_2X_2^2 + a_1X_3^2 + a_2X_4^2 \quad \text{(remaining)} \\ &\quad \sim X_1^2 + a_1a_2X_2^2 + \gamma X_3^2 + a_1a_2\gamma X_4^2 \quad \text{(by the above)} \\ &\quad \sim X_1^2 + \gamma X_2^2 + a_1a_2(X_3^2 + \gamma X_4^2) \end{aligned}$$

□

Pfister showed that the two properties of Theorem 12.1 hold true for the m-fold Pfister form. From Theorems 12.1 and 12.2 we see that isotropic and anisotropic forms behave differently. However, the corresponding isotropic forms also satisfy identities similar to those satisfied by Φ_a and $\Phi_{a,b}$. Indeed,

$$(X_1^2 - X_2^2)(Y_1^2 - Y_2^2) = (X_1Y_1 + X_2Y_2)^2 - (X_1Y_2 + X_2Y_1)^2.$$

Since $X_1^2 - X_2^2 \sim X_1X_2$ we ought to have a simpler identity of the equivalent form X_1X_2. Write the above identity as

$$\begin{aligned} &(X_1 + X_2)(X_1 - X_2)(Y_1 + Y_2)(Y_1 - Y_2) \\ &= (X_1Y_1 + X_2Y_2 + X_1Y_2 + X_2Y_1)^2 \times \\ &\qquad (X_1Y_1 + X_2Y_2 - X_1Y_2 - X_2Y_1)^2, \end{aligned}$$

and apply the transformations

$$\left.\begin{aligned} X_1 &\to (X_1 + X_2)/2 \\ X_2 &\to (X_1 - X_2)/2 \end{aligned}\right\}, \quad \left.\begin{aligned} Y_1 &\to (Y_1 + Y_2)/2 \\ Y_2 &\to (Y_1 - Y_2)/2 \end{aligned}\right\}$$

which are supposed to take $X_1^2 - X_2^2$ to X_1X_2. We get $X_1X_2 \cdot Y_1Y_2 = Z_1Z_2$ where $Z_1 = X_1Y_1$, $Z_2 = X_2Y_2$ which is a trivial looking formula of course. Since this is simpler than the identity for $X_1^2 - X_2^2$, we look for the identity for the isotropic Pfister form in this shape and we indeed have:

$$(X_1X_2 + X_3X_4)(Y_1Y_2 + Y_3Y_4) = Z_1Z_2 + Z_3Z_4$$

where

$$\begin{aligned} Z_1 &= (X_1Y_2 + X_3Y_4), Z_2 = (X_2Y_1 + X_4Y_3) \\ Z_3 &= (X_1Y_3 - X_3Y_1), Z_4 = (X_2Y_4 - X_4Y_2). \end{aligned}$$

Actually in Chapter 5 we already proved the group property of the set $G_{\Phi_m}(K)$; but we shall now give a different proof of this to exhibit various methods employed in this set up. The crucial result in this game is the following.

Theorem 12.3. *If $b \neq 0$ is an element of K represented by Φ_m, then $\Phi_m \sim b\Phi_m$.*

Proof. We use induction on m. For $m = 1$, $\Phi_1 = X^2 + aY^2$, $b = u^2 + av^2 \neq 0$ is represented by Φ_1. Then

$$\begin{aligned} b\Phi_1 &= (u^2 + av^2)(X^2 + aY^2) = (uX + avY)^2 + a(uY - vX)^2 \\ &\sim X^2 + aY^2 \end{aligned}$$

using the non-singular transformation

$$\begin{aligned} uX + avY &\to X \\ -vX + uY &\to Y \end{aligned}$$

whose determinant equals $\begin{vmatrix} u & av \\ -v & u \end{vmatrix} = u^2 + av^2 \neq 0$.

So suppose the result is true for $m - 1 \geq 1$. We have $\Phi_m = \Phi_{m-1} \oplus a_m\Phi_{m-1}$. Since Φ represents b, we have $b = c + a_m d$, where c, $d \in K$ and are represented by Φ_{m-1}.

Case 1: $cd \neq 0$. Then by the induction hypothesis

$$c\Phi_{m-1} \sim \Phi_{m-1} \sim d\Phi_{m-1} \sim cd\Phi_{m-1} \tag{12.1}$$

(the last since $f \sim g \Rightarrow af \sim ag$ for any $a \in K^*$). Now

$$\begin{aligned} b\Phi_m &= (c + a_m d)\Phi_m = (c + a_m d)(\Phi_{m-1} \oplus a_m\Phi_{m-1}) \\ &\sim (c + a_m d)(\Phi_{m-1} \oplus cda_m\Phi_{m-1}) \\ &= (c + a_m d)\Phi_{m-1} \oplus cda_m(c + a_m d)\Phi_{m-1}. \end{aligned}$$

Now

$$(c + a_m d)(X^2 + cda_m Y^2) \sim cX^2 + a_m dY^2 \tag{12.2}$$

for the right side represents $c + a_m d$ (put $X = Y = 1$) so it is equivalent to $(c + a_m d)X^2 + \alpha Y^2$. Comparing determinants we get $ca_m d = (c + a_m d)\alpha \times$ (a square) so the right side is

$$\begin{aligned} &\sim (c + a_m d)X^2 + \frac{ca_m d}{(c + a_m d) \times (\text{a square})} \cdot Y^2 \\ &= (c + a_m d)(X^2 + cda_m Y^2) = \quad \text{left side.} \end{aligned}$$

So the above (starting from $b\Phi_m$) is, using (12.2) repeatedly,

$$\begin{aligned}
&\sim (c + a_m d)[X_1^2 + a_1 X_2^2 + a_2(X_3^2 + a_1 X_4^2) + \ldots \\
&\qquad + cda_m(X_1'^2 + a_1 X_2'^2 + a_2(X_3'^2 + a_1 X_4'^2) + \ldots)] \\
&= (c + a_m d)[(X_1^2 + cda_m X_1'^2) + a_1(c + a_m d)(X_2^2 + cda_m X_2'^2) + \ldots \\
&\sim (cX_1^2 + a_m dX_1'^2 + a_1(cX_2^2 + a_m dX_2'^2) + \ldots \\
&= c\Phi_{m-1} \oplus a_m d\Phi_{m-1} \\
&\sim \Phi_{m-1} \oplus a_m \Phi_{m-1} \quad \text{(by (12.1))} \\
&= \Phi_m
\end{aligned}$$

as required.

Case 2: $cd = 0$. If $c = 0$ then $b = a_m d$ where Φ_{m-1} represents d. Hence

$$\begin{aligned}
b\Phi_m &= a_m d(\Phi_{m-1} \oplus a_m \Phi_{m-1}) \\
&\sim a_m d\Phi_{m-1} \oplus d\Phi_{m-1} \\
&\sim a_m \Phi_{m-1} \oplus \Phi_{m-1} \quad \text{(by (12.1))} \\
&= \Phi_m
\end{aligned}$$

as required. Finally if $d = 0$, then $b = c$ and

$$\begin{aligned}
b\Phi_m &= c(\Phi_{m-1} \oplus a_m \Phi_{m-1}) \\
&\sim c\Phi_{m-1} \oplus ca_m \Phi_{m-1} \sim \Phi_{m-1} \oplus a_m \Phi_{m-1} \quad \text{(by (12.1))} \\
&= \Phi_m
\end{aligned}$$

as required. □

Remark. The property of Theorem 12.3 is expressed by saying that the form Φ_m is "*round*".

Going back to the notation of Chapter 2, we have

$$M_{\Phi_m}(K) = \{c \in K^* \mid c\Phi_m \sim \Phi_m\},$$

the group of similarity factors of Φ_m over K. We prove now that $M_{\Phi_m}(K) = V_{\Phi_m}(K)$: the totality of non-zero elements of K represented by Φ_m (also denoted by $G_{\Phi_m}(K)$).

Now Φ_m represents 1 (since it begins with X_1^2 and putting $X_1 = 1$, $X_2 = \ldots = 0$ gives 1). Noting this let $\alpha \in V$ so Φ_m represents α; hence $\alpha\Phi_m \sim \Phi_m$ by Theorem 12.3, i.e. $\alpha \in M$ or $V \subseteq M$.

Conversely let $c \in M$ so that $c\Phi_m \sim \Phi_m$. Since Φ_m represents 1 so $c\Phi_m$ represents c. But $c\Phi_m \sim \Phi_m$ so Φ_m represents c, i.e. $c \in V$ so $M \subseteq V$.

It follows that $M = V$ as claimed.

We have thus proved the following.

Theorem 12.4. *Let Φ_m be the m-fold Pfister form; then $V_{\Phi_m}(K) = M_{\Phi_m}(K)$ is a group under multiplication.*

This proves property (i) of Theorem 12.1 for the general Pfister form Φ_m. We now deal with property (ii) and prove

Theorem 12.5. *Let Φ_m be isotropic; then Φ_m is hyperbolic.*

Proof. We use induction on m. For $m = 1, 2$, Theorem 12.1 (ii) gives the result, so suppose the result is true for $m-1$ and write

$$\Phi_m = \Phi_{m-1}(X_1, \ldots, X_{2^{m-1}}) \oplus a_m \Phi_{m-1}(X_{2^{m-1}+1}, \ldots, X_{2^m}).$$

Case 1. Φ_{m-1} is isotropic. Then by the induction hypothesis Φ_{m-1} is hyperbolic, say $\Phi_{m-1} \sim X_1^2 - X_2^2 + \ldots + X_{2^{2m-1}-1} - X_{2^{m-1}}^2$. Then by Theorem 12.2 (i), $a_m \Phi_{m-1} \sim \Phi_{m-1}$.

So $\Phi_m \sim$ what is required.

Case 2. Φ_{m-1} is not isotropic. Since Φ_m is isotropic, we get an equation $\alpha + a_m \beta = 0$, where, since Φ_{m-1} is not isotropic so $\alpha\beta \neq 0$. Here α, β are represented by Φ_{m-1}. So by Theorem 12.4, $\alpha\beta$ is represented by Φ_{m-1}, but $\alpha\beta = -a_m\beta^2$; hence $-a_m\beta^2$ and $-a_m$ is represented by Φ_{m-1}. Theorem 12.3 now gives $-a_m\Phi_{m-1} \sim \Phi_{m-1}$.

Hence

$$\begin{aligned}
\Phi_m &= \Phi_{m-1} \oplus a_m \Phi_{m-1} \sim \Phi_{m-1} \oplus -\Phi_{m-1} \\
&= X_1^2 + a_1 X_2^2 + a_2 X_3^2 + \ldots - X_{2^{m-1}+1}^2 - a_1 X_{2^{m-1}+2}^2 \\
&\qquad - a_2 X_{2^{m-1}+3}^2 - \ldots \\
&= X_1^2 - X_{2^{m-1}+1}^2 + a_2(X_2^2 - X_{2^{m-1}+2}^2) + a_2(X_3^2 - X_{2^{m-1}+3}^2) + \ldots \\
&\sim X_1^2 - X_{2^{m-1}+1}^2 + (X_2^2 - X_{2^{m-1}+2}^2) + (X_3^2 - X_{2^{m-1}+3}^2) + \ldots
\end{aligned}$$

by Theorem 12.2 (i). This completes the proof. □

The group property in Theorem 12.4 has been derived without the use of the famous identity satisfied by the Pfister form Φ_m, proved in Chapter 5. We now again prove this identity using Theorem 12.4. We have

Theorem 12.6. *Let $X_1, \ldots, X_{2^m}; Y_1, \ldots, Y_{2^m}$ be two sets of independent indeterminants over K and let $L = K(X_1, \ldots, X_{2^m}, Y_1, \ldots, Y_{2^m})$. Then there exist rational functions $Z_1, \ldots, Z_{2^m} \in L$ such that*

$$\Phi_m(X_1, \ldots, X_{2^m}) \cdot \Phi_m(Y_1, \ldots, Y_{2^m}) = \Phi_m(Z_1, \ldots, Z_{2^m}) \qquad (12.3)$$

Proof. By Theorem 12.4, $G_{\Phi_m}(L)$ is a group under multiplication. Denote the variables in Φ_m by $u_1, \ldots, u_{2^m}$. Then $\Phi_m(u_1, \ldots, u_{2^m})$ represents the elements $\Phi_m(X_1, \ldots, X_{2^m})$ and $\Phi_m(Y_1, \ldots, Y_{2^m}) \in L$ (just put $u_j = X_j$ and Y_j respectively), so it represents their product, i.e. there exist $Z_1, \ldots, Z_{2^m} \in L$ satisfying (12.3). □

Are there other quadratic forms which satisfy similar identities? To formalize the property expressed by the identity (12.3), we make the following

Definition 12.1. Let K be a field and $q(X_1, \ldots, X_n)$ a quadratic form defined over K having a non-singular matrix. We say q is *multiplicative* if there exists a formula

$$q(X_1, \ldots, X_n)q(Y_1, \ldots, Y_n) = q(Z_1, \ldots, Z_n) \tag{12.3$'$}$$

where $Z_j \in K(X_1, \ldots, X_n, Y_1, \ldots, Y_n)$.

We have the following.

Theorem 12.7. *Let q be a quadratic form defined over K. Then q is multiplicative iff for any field extension $L \supset K$ the set $Vq(L)$ is a group under multiplication. In particular q represents 1, the identity of V (over K).*

Proof. Let $L = K(X_1, \ldots, X_n, Y_1, \ldots, Y_n)$. By hypothesis $Vq(L)$ is a group, so q is multiplicative (see the proof of this fact for Pfister forms given above). This proves sufficiency.

Conversely, let q be multiplicative, so we have the identity (12.3)$'$. We have to prove that $Vq(L)$ is a group for any extension field L/K.

Let $q(c_1, \ldots, c_n)$, $q(d_1, \ldots, d_n)$ be two non-zero values taken by q over L (i.e. $c_j, d_j \in L$). By (12.3)$'$, q represents the polynomial $q(X_1, \ldots, X_n) \cdot q(Y_1, \ldots, Y_n)$ over $K(X_1, \ldots, X_n, Y_1, \ldots, Y_n)$ and hence also over $L(X_1, \ldots, X_n, Y_1, \ldots, Y_n)$ (X_j, Y_j remain indeterminates over L). Hence by the Cassels-Pfister Lemma, Theorem 11.9, q represents

$$q(c_1, \ldots, c_n)q(d_1, \ldots, d_n) \text{ over } L.$$

So $V_q(L)$ is a group under multiplication. □

Example 1. If q is a Pfister form, then by Theorem 12.6, q is multiplicative.

Example 2. If q is any isotropic form defined over K, then q is isotropic over any extension field L/K and so $V_q(L) = L^*$, a group, so again q is multiplicative. Thus any form q isotropic $/K$ is multiplicative. There are no further examples of multiplicative forms because we have the following remarkable theorem.

Theorem 12.8 (Pfister). *Let q be an anisotropic multiplicative form defined over K. Then q must be a Pfister form.*

Proof. Let q be of dimension n. Since q is multiplicative, q represents 1, by Theorem 12.7, and so

$$q \sim X^2 \oplus f,$$

say. If $n = 1$, then $q \sim X^2$ which is the zero-fold Pfister form as required. So suppose $n \geq 2$. Let m be the largest positive integer such that q contains (as a direct summand) a subform equivalent to an m-fold Pfister form $\Phi_m = \Phi_{(a_1,a_2,\ldots,a_m)}(X_1,\ldots,X_{2^m})$: say,

$$q = \Phi_m \oplus f.$$

We claim that $n = 2^m$. If not let $n > 2^m$ and let

$$L = K(X_1,\ldots,X_{2^m},Y_1,\ldots,Y_{2^m})$$

where the X_j, Y_j are independent indeterminates. Since Φ_m is multiplicative, we have a formula

$$\Phi_m(X_1,\ldots,X_{2^m})\Phi_m(Y_1,\ldots,Y_{2^m}) = \Phi_m(Z_1,\ldots,Z_{2^m})$$

with $Z_1,\ldots,Z_{2^m} \in L$.

Since $n > 2^m$, the equation $q = \Phi_m \oplus f$ shows that $\dim f > 0$. Let $c \in V_f(K)$ be any element of K^* represented by f. In the field L we have

$$A = \Phi_m(X_1,\ldots,X_{2^m}) + c\Phi_m(Y_1,\ldots,Y_{2^m}) \neq 0,$$

for otherwise we would have an algebraic equation connecting the X_j and the Y_j with coefficients in K.

Write $\underline{X} = (X_1,\ldots,X_{2^m})$, $\underline{Y} = (Y_1,\ldots,Y_{2^m})$, $\underline{Z} = (Z_1,\ldots,Z_{2^m})$, so that $\Phi_m(\underline{X}) \cdot \Phi_m(\underline{Y}) = \Phi_m(\underline{Z})$. Then

$$\begin{aligned} A &= \frac{\Phi_m(\underline{Z})}{\Phi_m(\underline{Y})} + c\Phi_m(\underline{Y}) \\ &= \Phi_m(\underline{Y})\left(\frac{\Phi_m(\underline{Z})}{(\Phi_m(\underline{Y}))^2} + c\right) \\ &= \Phi_m(\underline{Y})[\Phi_m(\underline{Z}/\Phi_m(\underline{Y})) + c]. \end{aligned}$$

Here the first factor $\Phi_m(\underline{Y}) \in V_{\Phi_m}(L) \subset V_q(L)$, while the second factor $\in V_{\Phi_m} \oplus f(L) \subset V_q(L)$ too. But q is multiplicative, so $V_q(L)$ is a group. It follows that $A \in V_q(L)$.

Since q is isotropic $/K$, we can apply the third representation theorem (Theorem 11.11) and conclude that over the field K, q contains a subform equivalent to A, i.e. to $\Phi_m \oplus c\Phi_m$. But $\Phi_m \oplus c\Phi_m$ is the $(m+1)$-fold Pfister form $\Phi_{m+1} = \Phi_{(a_1,\ldots,a_m,c)}$ and this is a contradiction to the maximal choice of m. Thus $n = 2^m$ and in the equation $q = \Phi_m \oplus f$, comparing dimensions we get $\dim f = 0$ so $q = \Phi_m$ as required. □

In the identity (12.3) satisfied by all Pfister forms, we saw in Chapter 5, that we could take Z_i to be linear polynomials in the Y_j with coefficients in $K(X_1, \dots, X_{2^m})$. We now give a direct proof of this fact in the spirit of this chapter. We ask if this result is true for all multiplicative forms. Since only isotropic forms remain, we state the full result in the form of the following.

Theorem 12.9. *Let q be a multiplicative quadratic form of dimension n over a field K, so that q satisfies $q(\underline{X}) \cdot q(\underline{Y}) = q(\underline{Z})$, $\underline{Z} \in (K(\underline{X}, \underline{Y}))^n$. Then*

(i) *If q is isotropic $/K$, we can choose the Z_k as polynomials in both the X_i and the Y_j $(1 \le i, j, k \le n)$,*

(ii) *If q is anisotropic $/K$, so that q is an m-fold Pfister form Φ_m and $n = 2^m$, then the Z_k can be chosen homogeneous linear polynomials in $Y_1, \dots, Y_n$ with coefficients in $K(X_1, \dots, X_n)$.*

Proof. (i) Since q is isotropic $/K$, we have

$$q(u_1, \dots, u_n) \sim u_1^2 - u_2^2 + f(u_3, \dots, u_n) \qquad (12.4)$$

Then

$$q(\underline{X}) \cdot q(\underline{Y}) = \left(\frac{q(\underline{X}) \cdot q(\underline{Y}) + 1}{2}\right)^2 - \left(\frac{q(\underline{X}) \cdot q(\underline{Y}) - 1}{2}\right)^2$$

so by taking $u_1 = (q(\underline{X}) \cdot q(\underline{Y}) + 1)/2$, $u_2 = (q(\underline{X}) \cdot q(\underline{Y}) - 1)/2$ and $u_3 = \dots = u_n = 0$ in (12.4), we see that $q(u_1, \dots, u_n)$ represents $q(\underline{X}) \cdot q(\underline{Y})$ and the u's can be chosen as polynomials in the X_i and the Y_j with coefficients in K since the equivalence (12.4) is linear $/K$; i.e. in (12.3)$'$ the Z_k got from the u_k using the equivalence (12.4) are polynomials in the X_i and the Y_j. This proves (i).

(ii) Let A be the $n \times n$ matrix $/K$ corresponding to the Pfister form $\Phi_m(u_1, \dots, u_{2^m})$ so that

$$\Phi_m(\underline{u}) = \underline{u} A \underline{u}' \qquad (\underline{u} \text{ a row vector}) \qquad (12.5)$$

Now let $L = K(\underline{X}, \underline{Y})$. $\Phi_m(\underline{u})$ represents the element $\Phi_m(\underline{X}) \in K(\underline{X})$ (just put $u_j = X_j$), i.e. $\Phi_m(\underline{X}) \in V_{\Phi_m(\underline{u})}(K(X))$ which is $= M_{\Phi_m(\underline{u})}(K(\underline{X}))$. Hence by Theorem 12.3, $\Phi_m(\underline{u}) \sim \Phi_m(\underline{X}) \cdot \Phi_m(\underline{u})$ over $K(\underline{X})$ ($\Phi_m(\underline{X})$ is the b of Theorem 12.3), i.e. the matrices A and $\Phi_m(\underline{X})A$ are congruent $/K(\underline{X})$, i.e. there exists a matrix equation

$$\Phi_m(\underline{X}) \cdot A = BAB' \qquad (12.6)$$

where B is an $n \times n$ matrix $/K(\underline{X})$.

Now in the field L we have

$$\Phi_m(\underline{X}) \cdot \Phi_m(\underline{Y}) = \Phi_m(\underline{X}) \cdot \underline{Y} \cdot A\underline{Y}' \quad \text{(see (12.5))}$$

$$
\begin{aligned}
&= \underline{Y}(\Phi_m(\underline{X})A)\underline{Y}' \quad (\Phi_m(\underline{X}) \text{ is a scalar of } K(\underline{X}))\\
&= \underline{Y}BAB'\underline{Y}' \quad \text{by (12.6)}\\
&= (\underline{Y}B)A(\underline{Y}B)'\\
&= \Phi_m(\underline{Y}B) = \Phi_m(\underline{Z}),
\end{aligned}
$$

say, where $\underline{Z} = \underline{Y}B$ and so each Z_j is linear in the Y_i with coefficients from the matrix B, i.e. in $K(\underline{X})$. This proves (ii). □

Remark. The converse of Theorem 12.9 holds. Namely, if q is any multiplicative form (isotropic or anisotropic) then $q(\underline{X}) \in M_{q(\underline{u})}(K(\underline{X}))$ i.e. $q(\underline{X})$ is a similarity factor of $q(\underline{u})$ over the field $K(\underline{X})$.

Proof. The matrix argument in the proof of (ii) above is reversible. □

Finally we have the following.

Definition 12.2. q is called ***strongly multiplicative*** if there is a formula $q(\underline{X}) \cdot q(\underline{Y}) = q(\underline{Z})$ with Z_k linear in Y_j coefficients in $K(\underline{X})$

We have the following.

Theorem 12.10.

(i) *Let q be anisotropic. Then q is strongly multiplicative iff q is multiplicative, iff q is a Pfister form.*

(ii) *Let q be isotropic. Then q is strongly multiplicative iff q is hyperbolic.*

Proof. (i) q strongly multiplicative $\Rightarrow$ q multiplicative (trivial) $\Rightarrow$ q is a Pfister form (Theorem 12.8) $\Rightarrow$ q is strongly multiplicative (Theorem 12.9 (ii)).

(ii) Let q be hyperbolic say $q \sim u_1^2 - u_2^2 + \dots$. Let $r(\underline{X}) \in K(\underline{X}) - \{0\}$. Regarding q as a form $/K(\underline{X})$ we have $r(\underline{X}) \cdot q(\underline{u}), \sim q(\underline{u})$, over $K(\underline{X})$ (Theorem 12.2 (i)). Since this is true for all $r(\underline{X}) \in K(\underline{X})$, it is true, in particular, for $q(\underline{X})$. Thus $q(\underline{X}) \cdot q(\underline{u}) \sim q(\underline{u})$ over $K(\underline{X})$. Now proceeding as in the proof of Theorem 12.9 (ii), we see that q is strongly multiplicative.

Conversely let q be strongly multiplicative (and isotropic). Decompose q according to the remark following Theorem 11.5:

$$
\begin{aligned}
q(u_1, \dots, u_n) &\sim u_1^2 - u_2^2 + \dots + u_{2s-1}^2 - u_{2s}^2 + \gamma(u_{2s+1}, \dots, u_n)\\
&= \mathsf{H} \oplus \gamma,
\end{aligned}
$$

say, where $\dim \mathsf{H} > 0$ and γ is anisotropic. We must show that $\dim \gamma = 0$. Suppose to the contrary that $\dim \gamma > 0$.

Now q is multiplicative so by the remark following Theorem 12.9, we have an equivalence

$$q(\underline{X})(\mathsf{H} \oplus \gamma) \sim \mathsf{H} \oplus \gamma \quad \text{over} \quad K(\underline{X}).$$

But $q(\underline{X}) \cdot \mathsf{H} \sim \mathsf{H}$ (Theorem 12.2 (i)). By Witt's cancellation theorem we get $q(\underline{X}) \cdot \gamma \sim \gamma$ over $K(\underline{X})$. Hence for any $a \in V_\gamma(K)$ we see, by Theorem 11.12 that γ contains a subform equivalent to aq:

$$\gamma \sim aq \oplus \gamma'.$$

It follows that $q \sim \mathsf{H} \oplus \gamma \sim \mathsf{H} \oplus aq \oplus \gamma'$. Comparing dimensions, we get a contradiction.

Thus $\dim \gamma = 0$, i.e. $q \sim \mathsf{H}$. □

The Venn diagram (given by Lam in [L2], p. 288) summarizes all the results about multiplicative, strongly multiplicative, isotropic, anisotropic and Pfister forms. We use the following abbreviations; the Venn diagram is given on the right:

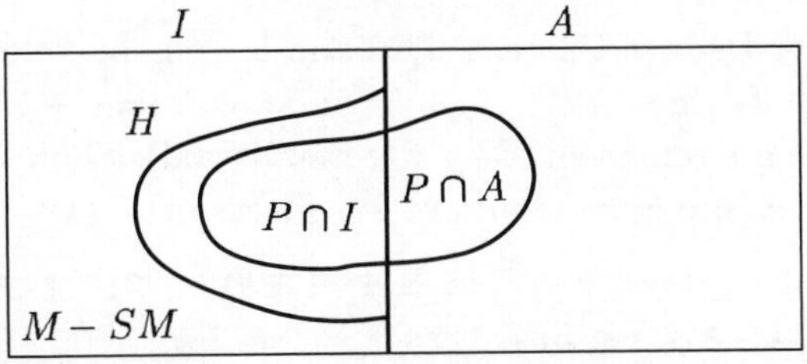

I = Isotropic forms

A = Anistropic forms

H = Hyperbolic forms

P = Pfister forms

M = Multiplicative forms

SM = Strongly Multiplicative forms

I A H $P \cap I$ $P \cap A$ $M - SM$

$$H = SM \cap I \qquad M = P \cup I \qquad SM = P \cup H$$

Exercises

1. Prove that if $q = aX^2 + bY^2 + cZ^2$ represents $-abc$ then q is isotropic. ($q \sim -abcX^2 + \alpha Y^2 + \beta Z^2$. Comparing determinants $abc = -abc\alpha\beta$ so $\beta = -1/\alpha$, i.e.

$$q \sim -abcX^2 + \alpha Y^2 - \frac{1}{\alpha} Z^2$$

which represents 0 non-trivially with $X = 0$, $Y = 1$, $Z = \alpha$).

Moral: a 3-dimensional form is universal iff it is isotropic.

2. (a) Let ϕ be a 5-dimensional form. Show that we can find scalars a, b, c, d in K such that

$$d\phi \sim X^2 + aY^2 + bZ^2 + abW^2 + CT^2.$$

(b) Deduce that ϕ is universal iff ϕ is isotropic.

3. Let ϕ be a 4-dimensional form $/K$ of unit determinant. Show that the number of cosets of K^{*^2} contained in $G_\phi(K)$ is either infinite or a power of 2.

4. (a) Let Φ be a Pfister form $/K$ and write $\Phi \sim X^2 \oplus \Phi'$. Show that the equivalence class of Φ' is uniquely determined by Φ, i.e. if also $\Phi \sim X^2 \oplus \Phi''$ then $\Phi' \sim \Phi''$. Φ' is called the pure subform of Φ.
(b) Let K be pythagorean in (a). Show that if $1 \in G_{\Phi'}(K)$ then $G_\Phi(K) = G_{\Phi'}(K)$.

5. Show that the four linear forms on the right side of the Pfister identity (II) given in the proof of Theorem 12.1 are linearly independent over K.

6. Deduce Pfister's Theorem 5.1 on the following lines:

Let $q = X_1^2 + \ldots + X_m^2$ and let $Y = a + bZ$, where $Z \in K$ is fixed and a, b are represented by q (zero allowed). Show that $1 + Y$ is of the same form: let a_i denote a sum of 2^i squares in K $(1 \leq i \leq m)$ so that

$$Y = e_1^2 + e_2^2 + a_1 + a_2 + \ldots + a_{m-1} + a_m Z. \tag{1}$$

Let, $1 \leq i \leq m-1$,

$$\left.\begin{aligned} b_i &= \begin{cases} a_i & \text{if } a_i \neq 0 \\ 1 & \text{if } a_i = 0 \end{cases} \\ b_m &= \begin{cases} Za_m & \text{if } a_m \neq 0 \\ Z & \text{if } a_m = 0 \end{cases} \end{aligned}\right\} \tag{2}$$

and let Φ_m be the Pfister form $\Phi_{b_1,b_2,\ldots,b_m}(u_1, u_2, \ldots, u_2m)$ so that Φ_m represents $e_1^2 + e_2^2$. Inserting this representation in (1) and using (2), show that

$$\begin{aligned} 1 + Y = 1 + X_1^2 + a_1(1 + X_2^2) + a_2(1 + X_3^2 + a_1 X_4^2) + \ldots \\ + a_m Z(1 + X_{2m-1+1}^2 + \ldots + a_1 a_2 \ldots a_{m-1} X_{2m}^2) \ldots \end{aligned} \tag{3}$$

Show that each bracket has a sum containing respectively $2^1, 2^2, \ldots, 2^m$ squares (note that each a_i, and so $a_1 a_2 \ldots a_i$, is a sum of 2^i squares). Deduce that (3) is of the form $u + vZ$ where u, v are represented by q as required.

13

The rational admissibility of the triple (r, s, n) and the Hopf condition

We know by Hurwitz's theorem (Theorem 1.1) that the triple (n, n, n) is admissible over a field K, i.e. the identity

$$(X_1^2 + \ldots + X_n^2)(Y_1^2 + \ldots + Y_n^2) = Z_1^2 + \ldots + Z_n^2 \tag{13.1}$$

with each Z_k bilinear in the X_i and the Y_j with coefficients in K, holds, iff $n = 1, 2, 4, 8$.

By allowing the Z_k to be rational functions of the X_i, Y_j with coefficients in K, (13.1) is seen to hold iff n is a power 2^m of 2 (see Theorem 2.1).

Very little is known regarding the admissibility of the general triple (r, s, n), i.e. about the identity

$$(X_1^2 + \ldots + X_r^2)(Y_1^2 + \ldots + Y_s^2) = Z_1^2 + \ldots + Z_n^2 \tag{13.2}$$

with the Z_k bilinear in the X_i, Y_j with coefficients in K.

For $r = s$, the Radon-Hurwitz theorem (Theorem 10.1) tells us that (r, n, n) is admissible over any K iff $r \leq \rho(n)$.

Definition 13.1. We define *rational admissibility* of the triple (r, s, n) to mean the existence of (13.2) with the Z_k rational functions of the X_i, Y_j with coefficients in K.

The object of this chapter is to give some very striking necessary and sufficient conditions for the rational admissibility of (r, s, n) (and related results) – e.g. the Hopf conditions. Here is the definition and some basic properties of the Hopf condition.

Definition 13.2. If all the binomial coefficients $\binom{n}{k}$ $(n - r < k < s)$ are even, we say the triple (r, s, n) *satisfies the Hopf condition* or that $\mathcal{H}(r, s, n)$ holds.

Some easy properties of the binomial coefficients give us the following.

Theorem 13.1.

(i) *If* $n-r+1 > s-1$ *(i.e.* $r+s < n+2$*) then* $\mathcal{H}(r,s,n)$ *holds vacuously,*
(ii) $\mathcal{H}(r,s,n) \Rightarrow \mathcal{H}(r,s,n+1)$,
(iii) $\mathcal{H}(n,n,n)$ *holds iff* $n = 2^m$,
(iv) *If* $n = 2^m \cdot u$ *(u odd) then* $\mathcal{H}(r,s,n)$ *holds iff* $r \leq 2^m$,
(v) $\mathcal{H}(r,s,n)$ *holds iff* $\mathcal{H}(s,r,n)$ *holds.*

Proof. (i) If $r + s < n + 2$ then there are no terms in the sequence $\binom{n}{n-r+1}, \ldots, \binom{n}{s-1}$.

(ii) We use the identity

$$\binom{n+1}{k} = \binom{n}{k} + \binom{n}{k-1} \tag{13.3}$$

We have $\mathcal{H}(r,s,n)$ holds if and only if $\binom{n}{n-r+1}, \ldots, \binom{n}{s-1}$ are all even while $\mathcal{H}(r,s,n+1)$ holds ifand only if $\binom{n+1}{n+1-r+1}, \ldots, \binom{n+1}{s-1}$ are all even.

But by (13.3),

$$\begin{array}{rcl} \binom{n+1}{n-r+2} & = & \binom{n}{n-r+2} + \binom{n}{n-r+1}, \\ \cdots\cdots & & \\ \binom{n+1}{s-1} & = & \binom{n}{s-1} + \binom{n}{s-2}. \end{array}$$

Here all the coefficients on the right side are even by $\mathcal{H}(r,s,n)$, hence so are all the ones on the left side, i.e. iff $\mathcal{H}(r,s,n+1)$ holds.

The moral of this result is that for a given r, s, the minimal n for which $\mathcal{H}(r,s,n)$ holds is the relevant value of n; then $\mathcal{H}(r,s,m)$ holds for all $m \geq n$.

(iii) $\mathcal{H}(n,n,n)$ holds iff $\binom{n}{1}, \binom{n}{2}, \ldots, \binom{n}{n-1}$ are all even, i.e. iff $n = 2^m$, which is easy to check if $n = 2^m$; conversely if $n = 2^r \cdot u$ (u odd) then for $k = 2^r$, $\binom{n}{k} = \binom{2^r \cdot u}{2^r}$ is odd unless $n = 1$.

(iv) We have $(1+x)^2 \equiv 1 + x^2 \pmod 2$ and by induction $(1+x)^n \equiv 1 + x^n$, whenever n is a power of 2. Now simply compare coefficients.

Remark. More generally let $e(1) < e(2) < \ldots < e(r)$ be the positions of the 1's in the expansion of n to the base 2; thus $n = \sum 2^{e(j)}$ and

$$(1+x)^n \equiv \prod (1 + x^{e(j)}) \qquad (\text{mod } 2).$$

It follows that the binomial coefficient $\binom{n}{m}$ is odd precisely when the places of the 1's in the dyadic expansion of m are a subset of the places of the 1's in the expansion of n.

(v) We have $\binom{n}{k} = \binom{n}{n-k}$ and so $\mathcal{H}(r, s, n)$ holds iff

$$\binom{n}{n-r+1}, \binom{n}{n-r+2}, \ldots, \binom{n}{s-1}$$

are all even, i.e. iff

$$\binom{n}{r-1}, \binom{n}{r-2}, \ldots, \binom{n}{n-s+1}$$

are all even, i.e. iff $\mathcal{H}(s, r, n)$ holds. □

Another integer that provides interesting links with rational admissibility of (r, s, n) is denoted by $r_{\circ}s$. To define it, we go back to the sets

$$G_t(K) = \{a \in K^* \mid a = a_1^2 + \ldots + a_t^2, \quad a_j \in K\}.$$

We wish to know whether or not there is a value $n = n(r, s)$ such that $G_r(K) \cdot G_s(K) = G_n(K)$ for all fields K. We have the following.

Definition 13.3. Denote by $r_{\circ}s$ the smallest natural number t such that for all fields K, we have

$$G_r \cdot G_s \subseteq G_t.$$

Thus by definition (of $\circ$), $G_\alpha \cdot G_\beta \subseteq G_\gamma \Rightarrow \alpha_{\circ}\beta \leq \gamma$. We have the following.

Theorem 13.2.

(i) $r_{\circ}(s_1 + s_2) \leq r_{\circ}s_1 + r_{\circ}s_2$

(ii) *Let* $r \leq 2^m$ *and let* t *be an arbitrary natural number; then* $r_{\circ}(t \cdot 2^m) \leq t \cdot 2^m$.

(iii) $G_r \cdot G_s = G_{r_{\circ}s}$

(iv) $r_{\circ}s \leq r + s - 1$.

(v) $r_{\circ}s = q \cdot 2^\alpha + \lambda_{\circ}r$, *where* $r \leq s$ *and* α, λ *are determined by* $2^{\alpha-1} < r \leq 2^\alpha$, $s = q \cdot 2^\alpha + \lambda$, $(q \geq 0,\ 1 \leq \lambda \leq 2^\alpha)$.

Proof. (i)

$$
\begin{aligned}
G_r \cdot G_{s_1+s_2} &= \left\{ \sum_1^r x_i^2 \Big(\sum_1^{s_1} y_i^2 + \sum_1^{s_2} z_i^2\Big) \right\} \quad \text{(neither factor equal to zero)} \\
&= \left\{ \sum x_i^2 \cdot \sum y_i^2 + \sum x_i^2 \sum z_i^2 \right\} \quad (\neq 0) \\
&\subseteq G_r G_{s_1} + G_r G_{s_2} \qquad \text{(0 can occur now)} \\
&\subseteq G_{r_{\circ}s_1} + G_{r_{\circ}s_2} \\
&= \{(x_1^2 + \ldots + x_{r_{\circ}s_1}^2) + (y_1^2 + \ldots + y_{r_{\circ}s_2}^2)\} \\
&\qquad\qquad \text{(neither bracket equal to 0)} \\
&\subseteq G_{r_{\circ}s_1 + r_{\circ}s_2} \cup \{0\}.
\end{aligned}
$$

But 0 is not an element of the left side, i.e. of $G_r \cdot G_{s_1+s_2}$, so in fact $G_r \cdot G_{s_1+s_2} \subseteq G_{r_0 s_1 + r_0 s_2}$, i.e.

$$r_0(s_1+s_2) \leq r_0 s_1 + r_0 s_2.$$

(ii)

$$\begin{aligned} G_r \cdot G_{t \cdot 2^m} &\subseteq G_{2^m}(G_{2^m} + \ldots + G_{2^m}) \\ &\subseteq G_{2^m} \cdot G_{2^m} + \ldots + G_{2^m} \cdot G_{2^m} \\ &= G_{2^m} + \ldots + G_{2^m} \\ &= G_{t \cdot 2^m} \cup \{0\}. \end{aligned}$$

But the left side does not contain zero, so

$$G_r \cdot G_{t \cdot 2^m} \subseteq G_{t \cdot 2^m},$$

hence $r_0 t \cdot 2^m \leq t \cdot 2^m$.

(iii) By definition, $G_r \cdot G_s \subseteq G_{r_0 s}$, so it remains to show that $G_{r_0 s} \subseteq G_r \cdot G_s$. Here we may suppose, without loss of generality, that $r \leq s$. We use induction on r. For $r = 1$ we have, for all s, $1_0 s = s$ for $G_1 \cdot G_s \subseteq G_s$ so $1_0 s \leq s$; also $G_1 \cdot G_s \not\subseteq G_{s-1}$ for at least one field K, e.g. $K = \mathbf{R}(X_1, \ldots, X_s)$, for otherwise $X_1^2 + \ldots + X_s^2$ would be a sum of $s-1$ squares in K which is false by Corollary 4 of Chapter 2, so $1_0 s > s-1$, i.e. $1_0 s \geq s$.

So, for all s, suppose the result is true for all pairs (p, s), $p \leq s$, with $p < r$ and we prove it for the pair (r, s), i.e. we prove that $G_{r_0 s} \subseteq G_r \cdot G_s$.

Let $2^{m-1} < r \leq 2^m$ and divide s by 2^m giving $s = t \cdot 2^m + l$, $t \geq 0$, $1 \leq l \leq 2^m$. Then we have the following:

$$\begin{aligned} G_{r_0 s} &= G_{r_0(t \cdot 2^m + l)} \subseteq G_{r_0 t \cdot 2^m + r_0 l} \quad \text{(by (i))} \\ &\subseteq G_{t \cdot 2^m + r_0 l} \quad \text{(by (ii))} \\ &= \{a_1^2 + \ldots + a_{2^m}^2 + \ldots + a_{t \cdot 2^m}^2 + b_1^2 + \ldots + b_{r_0 l}^2\} \\ &= \{a_1^2 + \ldots + a_{t \cdot 2^m} + (c_1^2 + \ldots + c_r^2)(d_1^2 + \ldots + d_l^2)\}, \end{aligned}$$

since $G_{r_0 l} = G_{l_0 r} = G_l G_r$, by the induction hypothesis (since $l < r$). Hence, with $c_1^2 + \ldots + c_r^2 \neq 0$ of course,

$$\begin{aligned} G_{r_0 s} &= \Bigg\{(c_1^2 + \ldots + c_r^2)\left[\frac{a_1^2 + \ldots + a_{2^m}^2}{c_1^2 + \ldots + c_r^2} + \ldots \right. \\ &\qquad \left. + \frac{a_{(t-1)2^m+1}^2 + \ldots + a_{t \cdot 2^m}^2}{c_1^2 + \ldots + c_r^2} + d_1^2 + \ldots + d_l^2\right]\Bigg\} \\ &= \Big\{(c_1^2 + \ldots + c_r^2)\Big[a_1'^2 + \ldots + a_{t \cdot 2^m}'^2 + \ldots + d_1^2 + \ldots + d_l^2\Big]\Big\} \\ &\subseteq G_r \cdot G_{t \cdot 2^m + l} = G_r \cdot G_s \end{aligned}$$

as required, where we have used the facts that $r \leq 2^m$ and G_{2^r} is a group.

Finally if $c_1^2 + \ldots + c_r^2 = 0$, then as above

$$G_{r_0 s} \subseteq \{a_1^2 + \ldots + a_{t\cdot 2^m}^2 + (c_1^2 + \ldots + c_r^2)(d_1^2 + \ldots + d_\ell^2)\}$$
$$\subset G_{t\cdot 2^m} \subset G_{t\cdot 2^m + l} = G_r \cdot G_s.$$

(iv) It is enough to prove that $G_r \cdot G_s \subseteq G_{r+s-1}$. We use induction on $r+s$.

If $r+s=2$, then $r=s=1$, so $G_r G_s = G_1 G_1 = G_1 \subseteq G_1$, i.e. $1_0 1 \leq 1$ and so $1_0 1 = 1$. Now let r, s be positive integers, $r+s>2$ and without loss of generality let $r \leq s$. Let m be the least power of 2 such that $s \leq 2^m$ so that $r \leq s \leq 2^m$. Then

$$G_r G_s \subseteq G_{2^m} \cdot G_{2^m} = G_{2^m}.$$

If $2^m \leq r+s-1$ then $G_{2^m} \subseteq G_{r+s-1}$ and the result follows. So suppose $r+s-1 < 2^m$, i.e. $r+s \leq 2^m$. Since $r \leq s$ this gives $r \leq 2^{m-1} \leq s$ by the choice of m.

Hence now

$$\begin{aligned} G_r G_s &= \{(a_1^2 + \ldots + a_r^2)(b_1^2 + \ldots + b_s^2)\} \\ &= \{(a_1^2 + \ldots + a_r^2)(b_1^2 + \ldots + b_{2^{m-1}}^2 + b_{2^{m-1}+1}^2 + \ldots + b_s^2)\} \\ &= \{(a_1^2 + \ldots + a_r^2)(b_1^2 + \ldots + b_{2^{m-1}}^2) \\ &\qquad + (a_1^2 + \ldots + a_r^2)(b_{2^{m-1}+1}^2 + \ldots + b_s^2)\} \\ &\subseteq G_{2^{m-1}} + G_r \cdot G_{s 2^{m-1}} \\ &\subseteq G_{2^{m-1}} + G_{r+s-2^{m-1}-1} \quad \text{(by the induction hypothesis)} \\ &\subseteq G_{2^{m-1}+r+s-2^{m-1}-1} \cup \{0\}. \end{aligned}$$

But 0 is not an element of the left side so $G_r G_s \subseteq G_{r+s-1}$, giving $r_0 s \leq r+s-1$.

(v) Using (i)

$$\begin{aligned} r_0 s = r_0(q \cdot 2^\alpha + \lambda) &\leq r_0 q \cdot 2^\alpha + r_0 \lambda \\ &\leq q \cdot 2^\alpha + r_0 \lambda \text{ by (ii)} \\ &= q \cdot 2^\alpha + \lambda_0 r. \end{aligned}$$

Conversely we shall show that $G_{q\cdot 2^\alpha + r_0\lambda} \subset G_{r_0 s}$ whence we get the reverse inequality as required.

We have

$$\begin{aligned} &(\alpha_1^2 + \cdots + \alpha_{r_0\lambda}^2) + \beta_1^2 + \cdots + \beta_{k\cdot 2^\alpha} \\ &\quad = (u_1^2 + \cdots + u_r^2)(v_1^2 + \cdots + v_\lambda^2) \\ &\qquad + \frac{(u_1^2 + \cdots + u_r^2)\{(u_1^2 + \cdots + u_r^2)(\beta_1^2 + \cdots + \beta_{q\cdot 2^\alpha}^2)\}}{(u_1^2 + \cdots + u_r^2)^2} \quad \text{by (iii)} \\ &\quad = (u_1^2 + \cdots + u_r^2)\{v_1^2 + \cdots + v_\lambda^2 + w_1^2 + \cdots + w_{q\cdot 2^\alpha}^2\} \end{aligned}$$

(by q successive applications of the 2^α-identity). It follows that

$$G_{r_0\lambda+q\cdot 2^\alpha} \subset G_r \cdot G_{\lambda+q\cdot 2^\alpha} = G_r \cdot G_s \subset G_{r_0 s}$$

which proves (v). □

Remark. If $2^{\alpha-1} < \lambda$, it turns out that $\lambda_0 r$ (in (v) of Theorem 13.2) can be replaced by 2^α so that we get a more satisfactory expression for $r_0 s$. To see this we note that $G_{2^\alpha} = G_{2^{\alpha-1}} \cdot G_{2^{\alpha-1}+1}$ (see Exercise 13.6), which gives:

$$\begin{aligned} G_{2^\alpha} &\subset G_\lambda \cdot G_r \text{ (since } 2^{\alpha-1} < \lambda, 2^{\alpha-1}+1 \le r) \\ &\subset G_{2^\alpha} \cdot G_{2^\alpha} \text{ (since } \lambda < 2^\alpha, r < 2^\alpha) \\ &= G_{2^\alpha} \end{aligned}$$

It follows that $2^\alpha \le \lambda_0 r \le 2^\alpha$ (see after Definition 13.3). □

Theorem 13.3.

(i) $r_0 s = s_0 r$,
(ii) $1_0 s = s$,
(iii) $2^m_0 2^m = 2^m$,
(iv) $r \le r' \Rightarrow r_0 s \le r'_0 s$,
(v) $r < 2^m \Rightarrow r_0(s+2^m) = r_0 s + 2^m$.

Remark. These properties are in fact sufficient to define $r_0 s$ for all r, $s \in \mathbb{N}$. We shall prove this after we have proved the theorem.

Proof. (i) $G_r \cdot G_s = G_s \cdot G_r$; so (1) follows.
(ii) $G_1 \cdot G_s \subseteq G_s$ so $1_0 s \le s$. However, $G_1 G_s \not\subset G_{s-1}$ for every field K, e.g. if $K = \mathbb{R}(X_1, \ldots, X_s)$ then by Corollary 4, Chapter 2, we know that $X_1^2 + \ldots + X_s^2$ is not a sum of $s-1$ squares in K; so $1_0 s \not< s$.
(iii) $G_{2^m} \cdot G_{2^m} \subseteq G_{2^m}$ (G_{2^m} being a group), so

$$2^m_0 2^m \le 2^m.$$

Here again $G_{2^m} \cdot G_{2^m} \not\subset G_{2^m-1}$ for at least one K, e.g. take $K = \mathbb{R}(X_1, \ldots, X_{2^m})$, then

$$(1^2 + 0^2 + \ldots + 0^2)(X_1^2 + \ldots + X_{2^m}^2) \notin G_{2^m-1}.$$

(iv) $r \le r' \Rightarrow G_r \subseteq G_{r'}$ so $G_r G_s \subseteq G'_r G_s \subseteq G_{r'_0 s}$, i.e. $r_0 s \le r'_0 s$ as required.
(v)

$$\begin{aligned} G_r G_{s+2^m} &= \{(x_1^2 + \ldots + x_r^2)(y_1^2 + \ldots y_s^2 + z_1^2 + \ldots z_{2^m}^2)\} \\ &= \{(x_1^2 + \ldots + x_r^2)(y_1^2 + \ldots + y_s^2) \\ &\qquad + (x_1^2 + \ldots + x_r^2)(z_1^2 + \ldots + z_{2^m}^2)\} \\ &\subseteq (G_r G_s \cup \{0\}) + (G_r G_{2^m} \cup \{0\}) \end{aligned}$$

$$\begin{aligned} &\subseteq (G_{r_0 s} \cup \{0\})(G_{2^m} \cup \{0\}) \quad (\text{since } r < 2^m) \\ &\subset G_{r_0 s + 2^m} \cup \{0\}. \end{aligned}$$

Now since the left side is free of 0, it follows that $G_r G_{s+2^m} \subseteq G_{r_0 s+2^m}$, giving $r_0(s+2^m) \leq r_0 s + 2^m$. Conversely, since $G_r G_s = G_{r_0 s}$,

$$\begin{aligned} G_{r_0 s+2^m} &= \{x_1^2 + \ldots + x_{r_0 s}^2 + y_1^2 + \ldots + y_{2^m}^2\} \\ &= \{(x_1'^2 + \ldots + x_r'^2)(z_1^2 + \ldots + z_s^2) + y_1^2 + \ldots + y_{2^m}^2\} \\ &= \left\{(x_1'^2 + \ldots + x_r'^2)\left[z_1^2 + \ldots + z_s^2 + \frac{y_1^2 + \ldots + y_{2^m}^2}{x_1'^2 + \ldots + x_r'^2}\right]\right\} \\ &\subseteq G_r \cdot G_{s+2^m} \end{aligned}$$

since $r < 2^m$ so this quotient is an element of $G_{2^m} \subseteq G_{r_0}(s+2^m)$. Hence $(r_0 s) + 2^m \leq r_0(s+2^m)$. □

Proof of the Remark after Theorem 13.3. Take for example, $2^{m-1} < r$, $s \leq 2^m$; then $r_0 s \leq 2^m_0 2^m = 2^m$, while

$$r_0 s \geq 2^m_0(1+2^{m-1}) = (2^{m-1}_0 1) + 2^{m-1} = 2^{m-1} + 2^{m-1} = 2^m.$$

so $r_0 s = 2^m$. □

Similarly we can calculate exactly the value of $r_0 s$ for all r, $s \in \mathbb{N}$. Pfister provided the following more direct way to calculate $r_0 s$.

Theorem 13.4. *Write $r - 1 = \sum_{i\geq 0} r_i 2^i$, $s - 1 = \sum_{i \geq 0} s_i 2^i$ in their binary scale, so that $r_i, s_i = 0$ or 1. Then*

$$r_0 s = \begin{cases} \sum_{i \geq l}(r_i + s_i)2^i & \text{if } r_l = s_l = 1,\ r_i s_i = 0 \text{ for all } i > l \\ \\ r + s - 1 & \text{if } r_i s_i = 0 \text{ for all } i\ , \end{cases}$$

in particular $r_0 s \leq r + s - 1$ with equality iff $r_i s_i = 0$ for all i.

Proof. Define α, λ as in (v) of Theorem 13.2, so that

$$r_i = \begin{cases} 0 & \text{if } i \geq \alpha \\ 1 & \text{if } i = \alpha - 1 \end{cases} \tag{i}$$

Write $\lambda - 1 = \Sigma \lambda_i 2^i$. Then it follows by the relation between λ and s, that

$$\lambda_i = \begin{cases} s_i & \text{if } i < \alpha \\ 0 & \text{if } i \geq \alpha \end{cases} \tag{ii}$$

Case 1: $\lambda > 2^{\alpha-1}$. Then $s_{\alpha-1} = 1$ and the right hand side of the theorem equals (note that $\ell = \alpha - 1$)

$$\sum_{i \geq \alpha - 1} (r_i + s_i)2^i = \sum_{i \geq \alpha}(r_i + s_i)2^i + 2^{\alpha-1}(r_{\alpha-1} + s_{\alpha-1})$$

$$
\begin{aligned}
&= \sum_{i \geq \alpha} r_i 2^i + \sum_{i \geq \alpha} s_i 2^i + 2^{\alpha-1}(1+1) \\
&= 0 + \sum_{\text{all } i} s_i 2^i - \sum_{i<\alpha} s_i 2^i + 2^\alpha \quad \text{(by (i))} \\
&= (s-1) - (\lambda - 1) + 2^\alpha \text{ (by (ii))} \\
&= q \cdot 2^\alpha + 2^\alpha
\end{aligned}
$$

as required. (Note that the case "$r_i s_i = 0$ for all i" does not occur here since $r_{\alpha-1} = s_{\alpha-1} = 1$). So, by the remark at the end of Theorem 13.2,

$$\sum_{i \geq \alpha-1} (r_i + s_i)2^i = r \circ s.$$

Case 2: $\lambda \leq 2^{\alpha-1}$. We have in this case

$$
\begin{aligned}
s - 1 = \sum_{\text{all } i} s_i \cdot 2^i &= \sum_{1 \leq i \leq \alpha-1} \lambda_i \cdot 2^i + \sum_{i > \alpha-1} s_i \cdot 2^i \\
&= \lambda - 1 + \sum_{i > \alpha-1} s_i \cdot 2^i;
\end{aligned}
$$

hence

$$\sum_{i > \alpha-1} s_i \cdot 2^i = s - \lambda = q \cdot 2^\alpha \tag{iii}$$

So now the right hand side of Theorem 13.4 is

$$
\begin{aligned}
\sum_{i \geq \ell} (r_i + s_i)2^i &= \sum_{\ell \leq i < \alpha} (r_i + s_i)2^i + \sum_{\ell < \alpha \leq i} (r_i + s_i)2^i \\
&= \sum_{\ell \leq i < \alpha} (r_i + \lambda_i)2^i + \sum_{\ell < \alpha \leq i} s_i 2^i \\
&= \sum_{\alpha-1 \geq i \geq \ell} (r_i + s_i)2^i + q \cdot 2^\alpha \text{ (by (iii) and (ii))}
\end{aligned} \tag{iv}
$$

But now the left hand side of the theorem is $r \circ s = q \cdot 2^\alpha + \lambda \circ r$ (by (v) of Theorem 13.2), so by the induction hypothesis, since in Case 2, $\lambda \leq 2^{\alpha-1} < r \leq s$ i.e. $\lambda < s$

$$
r \circ s = \begin{cases} q \cdot 2^\alpha + \sum_{i \geq \ell} (r_i + \lambda_i) \cdot 2^r & \text{if } r_\ell = \lambda_\ell = 1,\ r_i \lambda_i = 0 \text{ if } i > \ell, \\ q \cdot 2^\alpha + r + \lambda - 1 & \text{if } r_i \lambda_i = 0 \text{ for all } i \end{cases}
$$

$$= q \cdot 2^\alpha + \sum_{\ell \leq i \leq \alpha-1} (r_i + s_i)2^i$$

by (iii) if $r_\ell = \lambda_\ell = 1, r_i \lambda_i = 0$ if $i > \ell$. This equals $r + s - 1$ (since $s - 1 = q \cdot 2^\alpha + \lambda - 1$)) if $r_i s_i = 0$ for all i. Comparing with (iv), Theorem 13.4 follows. □

We now prove a few easy consequences of what we have been doing.

Proposition 1 (Cassels) (cf. Corollary 4, Chapter 2). $1+X_1^2+\ldots+X_n^2$ *is not a sum of* n *squares in* $\mathbf{R}(X_1,\ldots,X_n)$.

Proof. Use induction on n. For $n=1$, $1+X_1^2$ is not a single square in $\mathbf{R}(X_1)$. This is trivial. Suppose $1+X_1^2+\ldots+X_{n-1}^2$ is not a sum of $n-1$ squares in $\mathbf{R}(X_1,\ldots,X_{n-1})=K$ say. Then if $1+X_1^2+\ldots+X_{n-1}^2+X_n^2$ is a sum of n squares in $\mathbf{R}(X_1,\ldots,X_{n-1},X_n)=K(X_n)$, we would get $X_n+(1+X_1^2+\ldots+X_{n-1}^2)$ is a sum of n squares in $K(X_n)$ and so by Corollary 3 of Chapter 2, since -1 is not at all a sum of squares in K, we find $d=1+X_1^2+\ldots+X_{n-1}^2$ is a sum of $n-1$ squares in K, which contradicts the induction hypothesis. □

Proposition 2. *Let* K *be a field with* $s(K)\geq n$*; then*

$$G_n(K(X_1,\ldots,X_n))\subsetneqq G_{n+1}(K(X_1,\ldots,X_n)).$$

Proof. It is easy to check that $s(K(X))\geq n$ too, for if it were not, then $0=f_1^2(X)+\ldots+f_n^2(X)$, where by clearing the denominators, we suppose $f_j(X)\in K[X]$. Equating coefficients of the highest power of X to zero we get $0=a_1^2+\ldots+a_n^2$ ($a_j\in K$, not all $a_j=0$), hence $s(K)<n$, a contradiction; so $s(K(X))\geq n$.

It follows that $s(K(X_1,\ldots,X_j))\geq n$ (for all j). Now as in the proof of Proposition 1, $1+X_1^2+\ldots+X_n^2$ is not a sum of n squares in $K(X_1,\ldots,X_n)$. So the element $1+X_1^2+\ldots+X_n^2\in G_{n+1}$ but is not in G_n. □

Proposition 3. *If* (r,s,n) *is admissible* $/K$, *then for any field* $L\supseteq K$, $G_r(L)\cdot G_s(L)\subseteq G_n(L)$.

Proof. Let

$$(X_1^2+\ldots+X_r^2)(Y_1^2+\ldots+Y_s^2)=Z_1^2+\ldots+Z_n^2, \tag{13.4}$$

Z_k bilinear in X_i, Y_j with coefficients in K, be the (r,s,n)-identity assumed to hold. Let

$$a=a_1^2+\ldots+a_r^2\in G_r(L),$$
$$b=b_1^2+\ldots+b_s^2\in G_s(L).$$

Put $X_i=a_i$, $Y_j=b_j$ in (13.4) ($1\leq i\leq r$, $1\leq j\leq s$) to get $a\cdot b\in G_n(L)$, i.e. $G_r(L)\cdot G_s(L)\subseteq G_n(L)$. □

Proposition 4. *Let* $s(K)\geq(r\circ s)-1$. *If* (r,s,n) *is admissible over* K, *then* $r\circ s\geq n$.

Proof. By Theorem 13.2 (iii) and Proposition 3, $G_{r_0 s}(K) = G_r(K) \cdot G_s(K) \subseteq G_n(K)$.

Indeed this is true for any field $L \supseteq K$:

$$G_{r_0 s}(L) \subseteq G_n(L) \quad (\text{for any} \quad L \supseteq K) \tag{13.5}$$

Suppose, to the contrary, that $n < r_0 s$. By Proposition 1 find a field K' with Stufe at least n such that $G_n(K') \subsetneqq G_{n+1}(K')$. Since $n < r_0 s$ it follows that $n+1 \leq r_0 s$ so $G_{n+1}(K') \subseteq G_{r_0 s}(K')$ and $G_n(K') \subsetneqq G_{r_0 s}(K')$, contradicting (13.5). □

Proposition 5 (Köhnen (1978)). *$r_0 s \leq n$ iff $\mathcal{H}(r,s,n)$ holds.*

Proof. Let $g(r,s) = \min\left(n \mid \mathcal{H}(r,s,n) \text{ holds}\right)$. We show that $g(r,s) = r_0 s$ and to do this we need to check that the function $g(r,s)$ satisfies the conditions of Theorem 13.3. This we verify step by step.

(i) $g(r,s) = g(s,r)$ for $\min\left(n|\mathcal{H}(r,s,n) \text{ holds}\right) = \min\left(n|\mathcal{H}(s,r,n) \text{ holds}\right)$; by (v) of Theorem 13.1.

(ii) $g(1,s) = \min\left(n|\binom{n}{k} \text{ is even for } n-1 < k < s\right)$. Now for $n = s-1$, $k = s-1$ is a value of k for which $\binom{n}{k} = \binom{s-1}{s-1} = 1$ is *not* even so $n \geq s$.

For $n = s$, the condition is vacuously true, so $n = s$ is the required minimum n, i.e. $g(1,s) = s$.

(iii) $r \leq r' \Rightarrow g(r,s) \leq g(r',s)$, since $\min\left(n|\mathcal{H}(r,s,n) \text{ holds}\right) = \min\left(n|\binom{n}{k} \text{ is even for } n-r < k < n\right)$, and $\min\left(n|\mathcal{H}(r',s,n) \text{ holds}\right) = \min\left(n|\binom{n}{k} \text{ is even for } n-r' < k < n\right)$. Clearly the second follows from the first since the range of k in the second is a subset of the range of k in the first.

(iv) $g(2^m, 2^m) = 2^m$, because the left side equals

$$\min\left(n|\mathcal{H}(2^m,2^m,n) \text{ holds}\right) = \min\left(n|\binom{n}{k} \text{ is even if } n-2^m < k < 2^m\right).$$

Now if $n < 2^m$, the range of k takes negative values which is not allowed and if $n = 2^m$, then $\mathcal{H}(2^m, 2^m, n)$ holds (by Theorem 13.1, (iv)), so $n = 2^m$ is the required minimum.

(v) Let $r \leq 2^m$. We must prove $g(r, s+2^m) = g(r,s) + 2^m$. This reduces to the statement: if $k < 2^m$ then $\binom{n}{k} \equiv \binom{n+2^m}{n} \pmod 2$, which follows immediately from the remark given after the proof of Theorem 13.1, (iv).

Now $r_0 s = \min n$ such that $G_r \cdot G_s = G_n \subseteq G_N$ for all N with $n \leq N$ and $g(r,s) = \min n$ such that $\mathcal{H}(r,s,n)$ holds, i.e. $\mathcal{H}(r,s,n)$ holds for all N with $n \leq N$.

To prove Proposition 5, first let $r_0 s \leq n$, i.e. $g(r,s) \leq n$. Since $\mathcal{H}((r,s,g(r,s))$ holds so $\mathcal{H}(r,s,n)$ holds.

Conversely let $\mathcal{H}(r,s,n)$ hold and let $g(r,s)$ be the minimum n such that $\mathcal{H}(r,s,m)$ holds, so $g(r,s) \leq n$ i.e. $r_0 s \leq n$. □

Proposition 6. *If $\mathcal{H}(r,s,n)$ holds, then $(X_1^2+\ldots+X_r^2)(Y_1^2+\ldots+Y_s^2)$ is a sum of n squares in $L = K(X_1,\ldots,X_r,Y_1,\ldots,Y_s)$.*

Proof. Let $\mathcal{H}(r,s,n)$ hold. Then by Proposition 5, $r_0 s \le n$, so $G_{r_0}s(L) \subseteq G_n(L)$ for all fields L (definition of r_0), so $G_r(L)G_s(L) \subseteq G_n(L)$ for all L. Taking $L = K(X_1,\ldots,X_r,Y_1,\ldots,Y_s)$, this says that

$$(X_1^2+\ldots+X_r^2)(Y_1^2+\ldots+Y_s^2) \in G_n(L)$$

i.e. that the left side is a sum of n squares in L. □

We now come to our main result.

Theorem 13.5. *The following statements are equivalent.*

(i) $\mathcal{H}(r,s,n)$ *holds.*
(ii) $r_0 s \le n$.
(iii) $G_r(K)\cdot G_s(K) \subseteq G_n(K)$ *for every field* K.

Furthermore, if K is a field with Stufe $K \ge n$, then the following are also equivalent to (i), (ii), (iii)*:*

(iv) $G_r(L)\cdot G_s(L) \subseteq G_n(L)$ *for* $L = K(X_1,\ldots,X_r,Y_1,\ldots,Y_s)$.
(v) *There is a formula* $(X_1^2+\ldots+X_r^2)(Y_1^2+\ldots+Y_s^2) = Z_1^2+\ldots+Z_n^2$, *where* $Z_k \in K(X_1,\ldots,X_r,Y_1,\ldots,Y_s)$.
(vi) *There is such a multiplication formula as in* (v) *with each Z_k linear functions of the $Y_1,\ldots,Y_s$ with coefficients in $K(X_1,\ldots,X_r)$.*

Proof. (i)⇔(ii) is Proposition 5.

(ii)⇔(iii): by the definition of $r_0 s$, $G_r(K)\cdot G_s(K) = G_{r_0}s(K)$. Now $r_0 s \le n \Leftrightarrow G_{r_0}s(K) \subseteq G_n(K)$, so we see that $r_0 s \le n \Leftrightarrow G_r(K)\cdot G_s(K) \subseteq G_n(K)$.

(iii)⇒(iv) is trivial since L is special in (iv).

(iv)⇒(iii): let $s(K) \ge n$ so that by (iv)

$$G_r(L)\cdot G_s(L) \subseteq G_n(L) \text{ for } L = K(X_1,\ldots,X_r,Y_1,\ldots,Y_s) \tag{13.6}$$

and we have to prove that $G_r(F)\cdot G_s(F) \subseteq G_n(F)$ for *all* fields F.

Suppose for some F, $G_r(F)\cdot G_s(F) \supsetneqq G_n(F)$. Then $r_0 s > n$, i.e. $r_0 s \ge n+1$ and so

$$G_{r_0}s \text{ (any field } A) \supseteq G_{n+1}(A) \tag{13.7}$$

Now

$$\begin{aligned} G_r(L)\cdot G_s(L) = G_{r_0}s(L) &\quad L \text{ as in (13.6))} \\ \subseteq G_n(L) &\quad \text{(by (13.6))} \\ \subsetneqq G_{n+1}(L) &\quad \text{(by Proposition 2)} \end{aligned}$$

which is a contradiction to (13.7). Note that $n+1 \le r_0 s \le r+s-1$, by

Theorem 13.2, (iv), so $n+2 \le r+s$ say $r+s = n+2+a$. Now $s(K) \ge n$ so $s(K(X)) \ge n$ and indeed $s(K(X_1, \dots, X_{2+a})) \ge n$. Let

$$F_1 = K(X_1, \dots, X_{2+a}).$$

Then $L = F_1(X_{2+a+1}, \dots, X_r, Y_1, \dots, Y_s)$ $(s+r-(2+a)$ an equation with n indeterminates). Then by Proposition 2, $G_n(L) \subsetneqq G_{n+1}(L)$ as required.

(iv)⇒(v): this is trivial.

(vi)⇒(v) is also trivial but we don't need to mention it here.

It now remains to prove that (v)⇒(vi)⇒(ii).

(v)⇒(vi): K is a given field with Stufe $K \ge n$ and we are given a formula

$$(X_1^2 + \dots + X_r^2)(Y_1^2 + \dots + Y_s^2) = Z_1^2 + \dots + Z_n^2,$$

$Z_k \in K(X_1, \dots, X_r, Y_1, \dots, Y_s)$. Put $X = X_1^2 + \dots + X_r^2$. Thus $X(Y_1^2 + \dots Y_s^2)$ is a sum of n squares in the field

$$K(X_1, \dots, X_r, Y_1, \dots, Y_s) = K'(Y_1, \dots, Y_s),$$

where $K' = K(X_1, \dots, X_r)$, i.e. $XY_1^2 + \dots + XY_s^2 \in G_n(K'(Y_1, \dots, Y_s))$.

Now, see Theorem 11.8, $s(K') = s(k) \ge n$, so $Z_1^2 + \dots + Z_n^2$ is anisotropic $/K'$ and the subform theorem implies that

$$\begin{aligned} Z_1^2 + \dots + Z_n^2 &\sim XY_1^2 + \dots + XY_s^2 \oplus g \\ &\sim XY_1^2 + \dots + XY_s^2 + \lambda_1 W_1^2 + \dots + \lambda_t W_t^2 \end{aligned}$$

where we assumed, without loss of generality, that g is diagonal. Then $n = s+t$ and this equivalence means that there exist linear substitutions (over K'):

$$\begin{aligned} Z_1 &= a_{11}Y_1 + \dots + a_{1s}Y_s + b_{1,s+1}W_1 + \dots + b_{1,s+t}W_t \\ Z_2 &= a_{21}Y_1 + \dots + a_{2s}Y_s + b_{2,s+1}W_1 + \dots + b_{2,s+t}W_t \\ &\dots\dots\dots\dots\dots\dots\dots\dots\dots\dots \\ Z_n &= a_{n1}Y_1 + \dots + a_{ns}Y_s + b_{n,s+1}W_1 + \dots + b_{n,s+t}W_t \end{aligned}$$

i.e.

$$\begin{pmatrix} Z_1 \\ \vdots \\ Z_n \end{pmatrix} = [A \mid B] \begin{pmatrix} Y_1 \\ \vdots \\ Y_s \\ W_1 \\ \vdots \\ W_t \end{pmatrix},$$

(where A, B have n rows, A has s columns and B has t columns), taking the form $Z_1^2 + \dots + Z_n^2$ to $XY_1^2 + \dots + XY_s^2 + \lambda_1 W_1^2 + \dots + \lambda_t W_t^2$. So

$$Z_1^2 + \dots + Z_n^2 = (Z_1, \dots, Z_n) \begin{pmatrix} Z_1 \\ \vdots \\ Z_n \end{pmatrix} =$$

$$= (Y_1,\ldots,Y_s,W_1,\ldots,W_t)\left(\frac{A'}{B'}\right)(A|B)\begin{pmatrix}Y_1\\ \vdots\\ Y_s\\ W_1\\ \vdots\\ W_t\end{pmatrix}$$

$$= [A \mid B]\begin{pmatrix}Y_1\\ \vdots\\ Y_s\\ W_1\\ \vdots\\ W_t\end{pmatrix},$$

$$= (Y_1,\ldots,Y_s,W_1,\ldots,W_t)\begin{pmatrix}XI_s & & & 0\\ & \lambda_1 & & \\ & & \ddots & \\ 0 & & & \lambda_t\end{pmatrix}\begin{pmatrix}Y_1\\ \vdots\\ Y_s\\ W_1\\ \vdots\\ W_t\end{pmatrix}.$$

It follows that

$$\begin{pmatrix}A'A & A'B\\ B'A & B'B\end{pmatrix} = \begin{pmatrix}XI_s & & & 0\\ & \lambda_1 & & \\ & & \ddots & \\ 0 & & & \lambda_t\end{pmatrix}$$

and so in particular $A'A = XI_s$.

Now we have

$$(X_1^2+\ldots+X_r^2)(Y_1^2+\ldots+Y_s^2) = X(Y_1,\ldots,Y_s)\begin{pmatrix}Y_1\\ \vdots\\ Y_s\end{pmatrix}$$

$$= (Y_1,\ldots,Y_s)XI_s\begin{pmatrix}Y_1\\ \vdots\\ Y_s\end{pmatrix}$$

$$= (Y_1,\ldots,Y_s)A'A\begin{pmatrix}Y_1\\ \vdots\\ Y_s\end{pmatrix}$$

$$= (Z_1,\ldots,Z_n)\begin{pmatrix}Z_1\\ \vdots\\ Z_n\end{pmatrix},$$

where

$$\begin{pmatrix} Z_1 \\ \vdots \\ Z_n \end{pmatrix} = A \begin{pmatrix} Y_1 \\ \vdots \\ Y_s \end{pmatrix} = Z_1^2 + \ldots + Z_n^2.$$

Thus Z_k are linear forms in $Y_1, \ldots, Y_s$ with coefficients in $K' = K(X_1, \ldots, X_r)$ as required.

Finally we prove (vi)$\Rightarrow$(ii): we are given a field K with Stufe $K \geq n$ and a multiplication formula:

$$(X_1^2 + \ldots + X_r^2)(Y_1^2 + \ldots + Y_s^2) = Z_1^2 + \ldots + Z_n^2,$$

where Z_k are linear functions of the $Y_1, \ldots, Y_s$ with coefficients in $K(X_1, \ldots, X_r)$. We have to prove that $r_0 s \leq n$. For this it is enough to prove that $G_{r_0 s}(F) \subseteq G_n(F)$ for the field $F = K(t_1, \ldots, t_n)$, for once this is done, then by Proposition 2, $G_n(F) \subsetneqq G_{n+1}(F)$, i.e. $G_{r_0 s}(F) \subsetneqq G_{n+1}(F)$ and so clearly $r_0 s < n+1$. In other words, $r_0 s \leq n$ as required.

Now pick any $\beta \in G_s(F)$, i.e. $\beta = b_1^2 + \ldots + b_s^2$ $(b_j \in F)$. In the multiplication formula given above, since each Z_k is linear in the Y_j, we can substitute b_j for Y_j to obtain

$$(X_1^2 + \ldots + X_r^2)\beta = \hat{Z}_1^2 + \ldots + \hat{Z}_n^2 \quad (\hat{Z}_j \in F(X_1, \ldots, X_r))$$

i.e. $\beta X_1^2 + \ldots + \beta X_r^2 \in G_n(F(X_1, \ldots, X_r))$.

Now $s(F) = s(K) \geq n$ so $W_1^2 + \ldots + W_n^2$ is anisotropic $/F$ and so by the subform theorem

$$W_1^2 + \ldots + W_n^2 \sim \beta X_1^2 + \ldots + \beta X_r^2 \oplus g \quad (\text{over } F).$$

Hence $\beta \cdot G_r(F) \subseteq G_n(F)$. Since this is true for every $\beta \in G_s(F)$, it follows that $G_s(F) \cdot G_r(F) \subseteq G_n(F)$, i.e. $G_{r_0 s}(F) \subseteq G_n(F)$ as required.

This completes the proof of Theorem 13.5. □

Finally we have a very beautiful theorem of Shapiro. Let $n = 2^m$. Then in the n-square identity

$$(X_1^2 + \ldots + X_n^2)(Y_1^2 + \ldots + Y_n^2) = Z_1^2 + \ldots + Z_n^2$$

we know that Z_k may be taken as linear functions in the Y_j's with coefficients in $K(X_1, \ldots, X_n)$. Here Pfister has noted that Z_1 can also be taken linear in the X_i's as well; in fact he takes $Z_1 = X_1Y_1 + \ldots + X_nY_n$ (see Chapter 2, just before Theorem 2.3). This raises the question of how many more of these Z_k can also be taken linear in the X_i's too. The answer is the following.

Theorem 13.6 (Shapiro - 1978). *Suppose $n = 2^m$ and let K be any field. Then in the n-square identity*

$$(X_1^2 + \ldots + X_n^2)(Y_1^2 + \ldots + Y_n^2) = Z_1^2 + \ldots + Z_n^2 \qquad (13.8)$$

with the Z_k's linear in the Y_j's with coefficients in $K(X_1, \ldots, X_n)$, the $Z_1, \ldots, Z_r$ can also be taken linear in the X_i's iff $r \leq \rho(n)$, the Radon function.

Proof. Before giving the proof we note that in (13.8) above we can easily arrange a formula where 8 of the Z_k are bilinear (when $n \geq 8$). To do this start with the known $(8, 8, 8)$ bilinear identity and apply the "doubling" process given in Pfister's theorem of Chapter 2. Indeed write the 8-square identity twice over, once for the variables $X_1, \ldots, X_8$; $Y_1, \ldots, Y_8$; $Z_1, \ldots, Z_8$ and once for $X_9, \ldots, X_{16}$; $Y_9, \ldots, Y_{16}$; $Z_9, \ldots, Z_{16}$. Thus

$$\begin{pmatrix} Z_1 \\ Z_2 \\ \vdots \\ Z_8 \end{pmatrix} =$$

$$\begin{pmatrix} X_1 & -X_2 & -X_3 & -X_4 & -X_5 & -X_6 & -X_7 & -X_8 \\ X_2 & X_1 & -X_4 & X_3 & -X_6 & X_5 & X_8 & -X_7 \\ \cdots & \cdots & \cdots & \cdots & \cdots & \cdots & \cdots & \cdots \\ X_8 & X_7 & -X_6 & -X_5 & X_4 & X_3 & -X_2 & X_1 \end{pmatrix} \begin{pmatrix} Y_1 \\ Y_2 \\ \vdots \\ Y_8 \end{pmatrix}$$

and

$$\begin{pmatrix} Z_9 \\ Z_{10} \\ \vdots \\ Z_{16} \end{pmatrix} =$$

$$\begin{pmatrix} X_9 & -X_{10} & -X_{11} & -X_{12} & -X_{13} & -X_{14} & -X_{15} & -X_{16} \\ X_{10} & X_9 & -X_{12} & X_{11} & -X_{14} & X_{13} & X_{16} & -X_{15} \\ \cdots & \cdots & \cdots & \cdots & \cdots & \cdots & \cdots & \cdots \\ X_{16} & X_{15} & -X_{14} & -X_{13} & X_{12} & X_{11} & -X_{10} & X_9 \end{pmatrix} \begin{pmatrix} Y_9 \\ Y_{10} \\ \vdots \\ Y_{16} \end{pmatrix},$$

(simply read off the identity (1.3)); say $\underline{Z}_1 = S_1\underline{Y}_1$ and $\underline{Z}_2 = S_2\underline{Y}_2$ for short. Then

$$\begin{pmatrix} \underline{Z}_1 \\ \underline{Z}_2 \end{pmatrix} = \begin{pmatrix} S_1 & S_2 \\ S_2 & -S_2^{t^{-1}} S_1^t S_2 \end{pmatrix} \begin{pmatrix} \underline{Y}_1 \\ \underline{Y}_2 \end{pmatrix}$$

by Pfister's Theorem 2.1. We see that $Z_1, Z_2, \ldots, Z_8$ are bilinear in the X_i and the Y_j as claimed. The process can be repeated for $32, 64, \ldots$ variables; the 8 bilinear terms will persist.

But of course even for $n = 16$, Theorem 13.6 is stronger than the above method as it gives us nine fully bilinear terms.

This problem was posed by Baeza and solved by Shapiro in a letter to Baeza in 1976.

To prove the theorem recall that the fully bilinear identity

$$(X_1^2+\ldots+X_r^2)(Y_1^2+\ldots+Y_s^2)=(Z_1^2+\ldots+Z_n^2) \tag{13.9}$$

is equivalent to the system of Hurwitz matrix equations.

Write $\underline{Z} = A\underline{Y}$ (where $\underline{Y}, \underline{Z}$ are column matrices and A is an $n \times s$ matrix whose entries are linear forms in $X_1, \ldots, X_r$). The composition formula (13.9) becomes

$$\left(\sum_1^r X_i^2\right)\underline{Y}^t\underline{Y} = \underline{Z}^t\underline{Z} = \underline{Y}^t A^t A\underline{Y}$$

from which we get (see above (10.2))

$$A^tA = \left(\sum_1^r X_i^2\right) I_s.$$

Now write $A = X_1A_1+\ldots+X_rA_r$ and we get the Hurwitz matrix equations.

The same argument applies to Pfister type composition of size $n \times n$, but now A is allowed to have entries in $K(X_1, \ldots, X_n)$, the function field of rational functions over K.

Now suppose there is a Pfister composition for n squares for which $Z_1, \ldots, Z_r$ are linear in the $X_1, \ldots, X_n$ also. Then we have an $n \times n$ matrix A satisfying

$$A^tA = \left(\sum_1^n X_i^2\right) I_n \tag{13.10}$$

where A has entries in $K(X_1, \ldots, X_n)$ *and* where the first r rows of A have entries which are linear forms in $X_1, \ldots, X_n$.

Write $A = \left(\frac{L}{N}\right)$ in blocks, where L is the $r \times n$ linear part and N the $(n-r) \times n$ remaining part. Expanding the formula (13.10) for A, we get

$$L^tL + N^tN = \left(\sum_1^n X_i^2\right) I_n$$

which is not useful. However we also know that

$$AA^t = \left(\sum_1^n X_i^2\right) I_n. \tag{13.11}$$

To see this, note that since A is non-singular, we may multiply (13.10) by A^{-1} on the right and $(A^t)^{-1}$ on the left to get

$$I_n = (A^t)^{-1}A^{-1}\left(\sum_1^n X_i^2\right) = (AA^t)^{-1}\left(\sum_1^n X_i^2\right)$$

giving (13.11).

Then

$$AA^t = \begin{pmatrix} L \\ N \end{pmatrix}(L^tN^t) = \begin{pmatrix} LL^t & LN^t \\ NL^t & NN^t \end{pmatrix}$$

and this equals $(\sum_1^n X_i^2)\begin{pmatrix} I_r & 0 \\ 0 & I_{n-r} \end{pmatrix}$. It follows that

$$LL^t = \left(\sum_1^n X_i^2\right) I_r$$

and the $n \times r$ matrix L^t fulfills the Hurwitz condition for a composition formula of size (n, r, n). But the "r" and "s" parts of a composition are obviously interchangeable; so we get an (r, n, n) formula and applying the Hurwitz-Radon Theorem (Theorem 10.1), valid for any field K with char $K \neq 2$, we conclude that

$$r \leq \rho(n).$$

Conversely we need to *construct* Pfister identities for $n = 2^m$ squares having $r = \rho(n)$ bilinear terms. To do this we need the following matrix-theoretic

Lemma. *Suppose B_1 is an $n \times r$ matrix over a field F, with $B_1^t B_1 = \mu I_r$. Then we can enlarge B_1 to an $n \times n$ matrix $B = (B_1|B_2)$ (B_2 an $n \times (n-r)$ matrix) such that $B^t B = \mu I_n$.*

Proof. The condition $B_1^t B_1 = \mu I_r$ simply means that the r columns of B_1 form a determinant of r mutually orthogonal vectors in F^n, each of "length" μ. Write

$$B_1 = (\underline{b}_1, \underline{b}_2, \ldots, \underline{b}_r)$$

so that the $\underline{b}_j$ are orthogonal $n \times 1$ column vectors. By the Gram–Schmidt process, we can find columns $\underline{c}_{r+1}, \ldots, \underline{c}_n$ such that the complete set

$$\underline{b}_1, \ldots, \underline{b}_r, \underline{c}_{r+1}, \ldots, \underline{c}_n$$

is a linearly independent set of orthogonal vectors.

Let $M = (\underline{b}_1, \ldots, \underline{b}_r, \underline{c}_{r+1}, \ldots, \underline{c}_n = (B_1|C)$ say. Then

$$M^t M = \begin{pmatrix} B_1^t \\ C^t \end{pmatrix}(B_1|C) = \begin{pmatrix} B_1^t B_1 & B_1^t C \\ C^t B_1 & C^t C \end{pmatrix}.$$

But $M^t M$ is a diagonal matrix, the columns of M being orthogonal vectors; say $M^t M = \begin{pmatrix} \mu I_r & 0 \\ 0 & D \end{pmatrix}$ (note that $B_1^t B_1 = \mu I_r$) where

$$D = \operatorname{diag}(d_{r+1}, \ldots, d_n) \; (d_j \neq 0).$$

Hence $B_1^t C = 0$ and $C^t C = D$. Further

$$X_1^2 + \ldots + X_n^2 \sim \mu X_1^2 + \ldots + \mu X_r^2 + d_{r+1} X_{r+1}^2 + \ldots + d_n x_n^2$$

via the transformation $\underline{X} = M\underline{X}$. But n is a power, 2^m, of 2 and so the form $X_1^2 + \ldots + X_n^2$ is the Pfister form $\langle\langle 1, 1, \ldots, 1\rangle\rangle$; we know from Theorem 12.3 that it is "round" i.e. that $X_1^2 + \ldots + X_n^2 \sim \mu(X_1^2 + \ldots + X_n^2)$. Hence

$$\mu(X_1^2 + \ldots + X_r^2) + d_{r+1} X_{r+1}^2 + \ldots + d_n X_n^2 \sim \mu(X_1^2 + \ldots + X_n^2).$$

Cancelling, by Witt's cancellation lemma (Theorem 11.3) we get

$$d_{r+1}X_{r+1}^2 + \ldots + d_nX_n^2 \sim \mu(X_{r+1}^2 + \ldots + X_n^2).$$

Hence there exists a non-singular $(n-r)\times(n-r)$ matrix U such that

$$U^tDU = \mu I_{n-r}$$

and then we have

$$\begin{aligned}(B_1|CU)^t(B_1|CU) &= \begin{pmatrix} B_1^t \\ U^tC^t \end{pmatrix}(B_1|CU) \\ &= \begin{pmatrix} B_1^tB_1 & B_1^tCU \\ U^tC^tB_1 & U^tC^tCU \end{pmatrix} \\ &= \begin{pmatrix} \mu I_r & 0 \\ 0 & U^tDU \end{pmatrix} \\ &= \begin{pmatrix} \mu I_r & 0 \\ 0 & \mu I_{n-r} \end{pmatrix} = \mu I_n\end{aligned}$$

and taking $B_2 = CU$ completes the proof of the lemma. □

Now return to our bilinear terms construction. Let $n = 2^m$, $r = \rho(n)$. Since there is a bilinear composition of size (r, n, n), there is one of size (n, r, n) and we find an $n \times r$ matrix B_1 whose entries are linear forms in $X_1, \ldots, X_n$ satisfying

$$B_1^tB_1 = \left(\sum_1^n X_i^2\right) I_r.$$

By the lemma above, there exists $B = (B_1|B_2)$ over $F = K(X_1, \ldots, X_n)$ with $B^tB = (\sum_1^n X_i^2)I_n$. Again since $\sum_1^n X_i^2$ is a non-zero 'scalar', we have

$$BB^t = \left(\sum_1^n X_i^2\right) I_n.$$

Now use the matrix $A = \left(\frac{L}{N}\right)$ where $A = B^t$, $L = B_1^t$, $N = B_2^t$ to give a Pfister composition for n squares for which $Z_1, \ldots, Z_r$ are linear in $X_1, \ldots, X_n$ too. □

Exercises

1. Prove a result similar to Theorem 13.6 for the m-fold Pfister forms that is set

$$\Phi_{a_1,\ldots,a_m} = \Phi_m = \langle\langle a_1, a_2, \ldots, a_m\rangle\rangle,$$

be the m-fold Pfister form over a field K, and let $n = 2^m$. Then in the identity

$$\begin{aligned}\Phi_{a_1,\ldots,a_m}&(X_1, \ldots, X_n)\cdot\Phi_{a_1,\ldots,a_m}(Y_1, \ldots, Y_n) \\ &= \Phi_{a_1,\ldots,a_m}(Z_1, \ldots, Z_n),\end{aligned}$$

where $Z_1, \ldots, Z_n (\in K(X_1, \ldots, X_n)[Y_1, \ldots, Y_n])$ are all linear forms in the Y_j, we can make the Z_k linear in the X_i's too if and only if $k \le \rho(n)$.

We give below certain definitions followed by exercises related to them. These will come in handy when some of Lam's [L1] or Adem's [A2] constructions of normed maps are to be understood. These normed maps lead to exciting (r, s, n)-identities.

Definitions. A mapping $f : K^r \times K^s \to K^n$ is called

(i) *bilinear* if for $\mathbf{a}, \mathbf{a}' \in K^r, \mathbf{b}, \mathbf{b}' \in K^s$, we have

$$f(\mathbf{a}, \mathbf{b} + \mathbf{b}') = f(\mathbf{a}, \mathbf{b}) + f(\mathbf{a}, \mathbf{b}') \text{ and}$$
$$f(\mathbf{a} + \mathbf{a}', b) = f(\mathbf{a}, \mathbf{b}) + f(\mathbf{a}', b).$$

(ii) *non singular* if $f(\mathbf{a}, \mathbf{b}) = 0 \Rightarrow \mathbf{a} = \mathbf{0}$ or $\mathbf{b} = \mathbf{0}$

(iii) *normed* if $\|f(\mathbf{a}, \mathbf{b})\|^2 = \|\mathbf{a}\|^2 \cdot \|\mathbf{b}\|^2$ for all $\mathbf{a} \in K^r, \mathbf{b} \in K^s$, where if $X = (X_1, \ldots, X_m) \in K^m$, then the norm $\|\mathbf{X}\|^2$ of $\mathbf{X}$ is defined to be $X_1^2 + X_2^2 + \cdots + X_m^2$

(iv) *bi-skew* if

$$f(-\mathbf{X}, \mathbf{Y}) = f(\mathbf{X}, -\mathbf{Y}) = -f(\mathbf{X}, \mathbf{Y}) \text{ for all } \mathbf{X} \in K^r, \mathbf{Y} \in K^s.$$

We have already defined $r_* s$ to be the least n for which the triple (r, s, n) is admissible over $\mathbf{R}$ (see Chapter 10).

(v) We define $r \# s$ to be the least n for which there is a non-singular bilinear map $f : \mathbf{R}^r \times \mathbf{R}^s \to \mathbf{R}^n$.

2. Prove

(i) If f is normed then f is non-singular.

(ii) The triple (r, s, n) is admissible over K if and only if there is a bilinear normed mapping

$$f : K^r \times K^s \to K^n.$$

(iii) f bilinear $\Rightarrow f$ bi-skew (Hint: $\mathbf{0} = f(\mathbf{X}, \mathbf{0}) = f(\mathbf{X}, \mathbf{Y} - \mathbf{Y}) = f(\mathbf{X}, \mathbf{Y}) + f(\mathbf{X}, -\mathbf{Y})$.)

3. Prove that

(i) $\max(r, s) \le r \# s \le r \# s$

(ii) $(r + r')_* s \le (r_* s) + (r'_* s)$

(iii) $(r + r') \# s \le r \# s + r' \# s$

(iv) $r \# s \le r + s - 1$

Hint: If $f : \mathbf{R}^r \times \mathbf{R}^s \to \mathbf{R}^n, g : \mathbf{R}^{r'} \times \mathbf{R}^s \to \mathbf{R}^{n'}$ are bilinear maps, define the direct sum

$$h : \mathbf{R}^{r+r'} \times \mathbf{R}^s \to \mathbf{R}^{n+n'}$$

by $h((\mathbf{X},\mathbf{X}'),\mathbf{Y}) = (f(\mathbf{X},\mathbf{Y}), g(\mathbf{X}',\mathbf{Y}))$.

If f, g are non-singular, so is h and if f, g are normed, so is h. This proves (ii), and (iii) is similar.

For (iv), define $f(X,Y) : \mathbf{R}^r \times \mathbf{R}^s \to r+s-1$ as the coefficients of $1, t, t^2, \ldots, t^{r+s-2}$ in the product

$$\left(\sum_{i=1}^{r} X_i t^{i-1}\right)\left(\sum_{j=1}^{s} Y_j t^{j-1}\right)$$

in the ring $\mathbf{R}[t]$.

4. Prove that the map $\mathbf{R}^2 \times \mathbf{R}^2 \xrightarrow{f} \mathbf{R}^2$ (got from multiplication of complex numbers) given by

$$f(X_1, X_2), (Y_1, Y_2)) = (X_1Y_1 - X_2Y_2, X_1Y_2 + X_2Y_1)$$

is bilinear, bi-skew, non-singular and normed.

Show similarly that the maps got by multiplication of quaternions and octonions are also non-singular, bilinear, bi-skew and normed.

5. By definition $r\#s \leq n$ if there is a non-singular bilinear map $f : \mathbf{R}^r \times \mathbf{R}^s \to \mathbf{R}^n$. In particular if (r, s, n) is admissible $\mathbf{R}$ then $r\#s \leq n$. Prove that if (r, s, n) is admissible $\mathbf{C}$ (a weaker condition than admissibility $\mathbf{R}$) then $r\#s \leq n$.

Hint: (r, s, n) admissible $\mathbf{C} \Rightarrow$ the existence of a formula $(X_1^2 + \cdots + X_r^2)(Y_1^2 + \cdots + Y_s^2) = Z_1^2 + \cdots + Z_n^2$, with Z_k bilinear in X_i, Y_j with coefficients in $\mathbf{C}$:

$$Z_k = \sum_{i=1}^{r}\sum_{j=1}^{s} \alpha_{ij} X_i Y_j \quad (\alpha_{ij} \in \mathbf{C}).$$

Writing $\alpha_{ij} = a_{ij} + \sqrt{-1}\, b_{ij}$ we see that $Z_k = u_k + \sqrt{-1}\, v_k$ whereu_k, v_k are bilinear in X_i, Y_j with coefficients in $\mathbf{R}$. Now comparing real parts in the (r, s, n)-identity we get

$$(X_1^2 + \cdots + X_r^2)(Y_1^2 + \cdots + Y_s^2) = u_1^2 - v_1^2 + \cdots + u_n^2 - v_n^2 \qquad (*)$$

Now define the map $f : \mathbf{R}^r \times \mathbf{R}^s \to \mathbf{R}^n$ by

$$\begin{aligned} f(a_1, \ldots, a_r; b_1, \ldots, b_s) &= f(\mathbf{a}, \mathbf{b}) \\ &= [u_1^2(\mathbf{a},\mathbf{b}) - v_1^2(\mathbf{a},\mathbf{b}), \ldots, u_n^2(\mathbf{a},\mathbf{b}) - v_n^2(\mathbf{a},\mathbf{b})] \end{aligned}$$

and check that f is bilinear and non-singular (use $(*)$).

6. Prove that if G_n is a group. That is, if $n = 2^m$, then

$$G_n G_{n+1} = G_{2n};$$

we have

$$u_1^2 + \cdots + u_{2n}^2 = (u_1^2 + \cdots + u_n^2)\left[1 + \frac{(u_1^2 + \cdots + u_n^2)(u_{n+1}^2 + \cdots + u_{2n}^2)}{(u_1^2 + \cdots + u_n^2)^2}\right]$$
$$\in G_n G_{n+1},$$

and conversely we have

$$(u_1^2 + \cdots + u_n^2)(v_1^2 + \cdots + v_{n+1}^2) = (u_1^2 + \cdots + u_n^2)(v_1^2 + \cdots + v_n^2) + (u_1^2 + \cdots + u_n^2)v_{n+1}^2$$
$$\in G_{2n}.$$

7. Define

$$\Gamma_n = \{a \in K |\ a \text{ is a sum of at most } m \text{ squares in } K\}$$
$$= G_n \cup \{0\}.$$

For integers $m, n \geq 0$, show that there is a function

$$m_0 n = n_0 m$$

with the property that $\Gamma_m(K).\Gamma_n(K) = \Gamma_{m_0 n}(K)$ for all fields K. (Follow the steps given below.)

(I) $2^t_0 2^t = 2^t$ (this is just Pfister's theorem).

(II) Suppose that $2^t \leq m < 2^{t+1}, 2^t < n \leq 2^{t+1}$ for some t. Then $m_0 n = 2^{t+1}$.

(III) Suppose that $m \leq 2^t < n$ for some t and that $m_0(n - 2^t)$ is already known. Then $m_0 n = m_0(n - 2^t) + 2^t$.

(IV) Now use induction on $m + n$ to define $m_0 n$ as required.

Remarks:

1. The equality in Exercise 7 makes two assertions:

(a) Given $u \in \Gamma_m(K), v \in \Gamma_n(K)$, then $uv \in \Gamma_{m_0 n}(K)$

(b) Given $w \in \Gamma_{m_0 n}(K)$, there are $u \in \Gamma_m(K), v \in \Gamma_n(K)$ such that $w = uv$.

2. There are fields K (e.g. $\mathbf{R}(t_1, t_2, \ldots, t_j, \ldots)$) where the t_j are independent variables, such that the $\Gamma_m(K)$ are distinct. Hence $m_0 n$, if it exists, is certainly unique.

Hints: For (II) first let $u \in \Gamma_m, v \in \Gamma_n$. Then $u, v \in \Gamma_{2^{t+1}}$ and so $uv \in \Gamma_{2^{t+1}}$, by (I).

Next let $w \in \Gamma_{2^{t+1}}$. If $w = 0$, we have $w = 0 \cdot 0$, $0 \in \Gamma_m, \Gamma_n$. Otherwise $w = u + x$, where $u, x \in \Gamma_{2^t}, u \neq 0$. Then, by (I),

$$y = x/u = xu/u^2 \in \Gamma_{2^t}.$$

Hence v (say) $= 1 + y \in \Gamma_{2^t+1}$, and

$$w = uv; u \in \Gamma_{2^t} \subset \Gamma_m, \ v \in \Gamma_{2^t+1} \subset \Gamma_n$$

as required.
(III). First let $u \in \Gamma_m, v \in \Gamma_n$. Then $v = v_1 + v_2$ where $v_1 \in \Gamma(n - 2^t), v_2 \in \Gamma_{2^t}$, and $uv_1 \in \Gamma_{m_0(n-2^t)}$. Also $u \in \Gamma_m \subset \Gamma_{2^t}$; so $uv_2 \in \Gamma_{2^t}$, by (I). Hence

$$uv = uv_1 + uv_2 \in \Gamma_{m_0(n-2^t)} + \Gamma_{2^t}$$
$$= \Gamma_{m_0(n-2^t)+2^t}$$

Conversely let

$$w \in \Gamma_{m_0(n-2^t)+2^t}$$

If $w = 0$, there is nothing to prove. Otherwise $w = w_1 + w_2$, where $w_1 \neq 0$, $w_1 \in \Gamma_{m_0(n-2^t)}$, $w_2 \in \Gamma_{2^t}$. Then $w_1 = uv_1$ where $u \in \Gamma_m, v_1 \in \Gamma_{n-2^t}$. Put $v_2 = w_2/u = w_2u/u^2$. Since $u \in \Gamma_m \subset \Gamma_{2^t}$, $w_2 \in \Gamma_{2^t}$, we have $v_2 \in \Gamma_{2^t}$. Then $w = u(v_1 + v_2)$, where $v_1 + v_2 \in \Gamma_{n-2^t} + \Gamma_{2^t} = \Gamma_n$ as required.

14

Some interesting examples of bilinear identities and a theorem of Gabel

If we wish to find the quadratic forms $q(\underline{X})$ which satisfy the identity

$$q(\underline{X})q(\underline{Y}) = q(\underline{Z})$$

where the $Z_k \in K(\underline{X}, \underline{Y})$, then we have seen in Chapter 12 that two essentially different cases arise:

I. *q is anisotropic.* In this case q is of necessity a Pfister form and we know all about the identities satisfied by them.

II. *q is isotropic.* In this case $q \simeq$ a hyperbolic form and apart from the two cases

(a) $q \simeq \mathsf{H} = X_1X_2$
(b) $q \simeq 2\mathsf{H} = X_1X_2 + X_3X_4$,

we have not looked at any further forms and the identities they satisfy. The first new case here is the 6-variable one (the first time not a power of 2). So let $q \simeq 3\mathsf{H} = X_1X_2 + X_3X_4 + X_5X_6$, then

$$q(\underline{X}) \cdot q(\underline{Y}) = q(\underline{Z}),$$

where

$$Z_1 = X_2Y_1 + X_3Y_4 + X_5Y_6,$$
$$Z_2 = X_1Y_2 + X_4Y_3 + X_6Y_5,$$
$$Z_3 = X_2Y_3 - X_3Y_2 - X_6\left(\frac{X_3Y_5 - X_5Y_3}{X_1}\right),$$

$$Z_4 = X_1Y_4 - X_4Y_1,$$
$$Z_5 = X_2Y_5 - X_5Y_2 + X_4\left(\frac{X_3Y_5 - X_5Y_3}{X_1}\right),$$
$$Z_6 = X_1Y_6 - X_6Y_1;$$

This has 4 bilinear terms.

We have the three classical 2-, 4-, 8-, identities and we know that the 16-identity fails as a bilinear identity. We have already remarked that the minimum value of n for which $(16, 16, n)$ is admissible (over $\mathbb{R}$) is not known to date. Trivially, of course, $n \leq 32$ (use the 8-identity four times). However the maximal r for which $(r, 16, 16)$ is admissible is, by the Hurwitz-Radon theorem, equal to 9. We thus have the following identity giving the admissibility of the triple $(9, 16, 16)$:

$$(X_1^2 + X_2^2 + \ldots + X_9^2)(Y_1^2 + Y_2^2 + \ldots + Y_{16}^2) = Z_1^2 + Z_2^2 + \ldots + Z_{16}^2,$$

where

$$Z_1 = X_1Y_1 + X_2Y_2 + X_3Y_3 + X_4Y_4 + X_5Y_5 + X_6Y_6 + X_7Y_7 + X_8Y_8 + X_9Y_9$$
$$Z_2 = -X_2Y_1 + X_1Y_2 + X_4Y_3 - X_3Y_4 + X_6Y_5 - X_5Y_6 - X_8Y_7 + X_7Y_8 + X_9Y_{10}$$
$$Z_3 = -X_3Y_1 - X_4Y_2 + X_1Y_3 + X_2Y_4 + X_7Y_5 + X_8Y_6 - X_5Y_7 - X_6Y_8 + X_9Y_{11}$$
$$Z_4 = -X_4Y_1 + X_3Y_2 - X_2Y_3 + X_1Y_4 + X_8Y_5 - X_7Y_6 + X_6Y_7 - X_5Y_8 + X_9Y_{12}$$
$$Z_5 = -X_5Y_1 - X_6Y_2 - X_7Y_3 - X_8Y_4 + X_1Y_5 + X_2Y_6 + X_3Y_7 + X_4Y_8 + X_9Y_{13}$$
$$Z_6 = -X_6Y_1 + X_5Y_2 - X_8Y_3 + X_7Y_4 - X_2Y_5 + X_1Y_6 - X_4Y_7 + X_3Y_8 + X_9Y_{14}$$
$$Z_7 = -X_7Y_1 + X_8Y_2 + X_5Y_3 - X_6Y_4 - X_3Y_5 + X_4Y_6 + X_1Y_7 - X_2Y_8 + X_9Y_{15}$$
$$Z_8 = -X_8Y_1 - X_7Y_2 + X_6Y_3 + X_5Y_4 - X_4Y_5 - X_3Y_6 + X_2Y_7 + X_1Y_8 + X_9Y_{16}$$
$$Z_9 = -X_9Y_1 + X_1Y_9 - X_2Y_{10} - X_3Y_{11} - X_4Y_{12} - X_5Y_{13} - X_6Y_{14} - X_7Y_{15} - X_8Y_{16}$$
$$Z_{10} = -X_9Y_2 + X_2Y_9 + X_1Y_{10} - X_4Y_{11} + X_3Y_{12} - X_6Y_{13} + X_5Y_{14} + X_8Y_{15} - X_7Y_{16}$$
$$Z_{11} = -X_9Y_3 + X_3Y_9 + X_4Y_{10} + X_1Y_{11} - X_2Y_{12} - X_7Y_{13} - X_8Y_{14} + X_5Y_{15} + X_6Y_{16}$$
$$Z_{12} = -X_9Y_4 + X_4Y_9 - X_3Y_{10} + X_2Y_{11} + X_1Y_{12} - X_8Y_{13} + X_7Y_{14} - X_6Y_{15} + X_5Y_{16}$$
$$Z_{13} = -X_9Y_5 + X_5Y_9 + X_6Y_{10} + X_7Y_{11} + X_8Y_{12} + X_1Y_{13} - X_2Y_{14} - X_3Y_{15} - X_4Y_{16}$$
$$Z_{14} = -X_9Y_6 + X_6Y_9 - X_5Y_{10} + X_8Y_{11} - X_7Y_{12} + X_2Y_{13} + X_1Y_{14} + X_4Y_{15} - X_3Y_{16}$$
$$Z_{15} = -X_9Y_7 + X_7Y_9 - X_8Y_{10} - X_5Y_{11} + X_6Y_{12} + X_3Y_{13} - X_4Y_{14} + X_1Y_{15} + X_2Y_{16}$$
$$Z_{16} = -X_9Y_8 + X_8Y_9 + X_7Y_{10} - X_6Y_{11} - X_5Y_{12} + X_4Y_{13} + X_3Y_{14} - X_2Y_{15} + X_1Y_{16}$$

or in the matrix form the following: $\underline{Z} = A\underline{Y}$ ($\underline{Z}$, $\underline{Y}$ columns), and the

16×16 matrix $A =$

$$\begin{pmatrix}
X_1 & X_2 & X_3 & X_4 & X_5 & X_6 & X_7 & X_8 & X_9 & 0 & 0 & 0 & 0 & 0 & 0 & 0 \\
-X_2 & X_1 & X_4 & -X_3 & X_6 & -X_5 & -X_8 & X_7 & 0 & X_9 & 0 & 0 & 0 & 0 & 0 & 0 \\
-X_3 & -X_4 & X_1 & X_2 & X_7 & X_8 & -X_5 & -X_6 & 0 & 0 & X_9 & 0 & 0 & 0 & 0 & 0 \\
-X_4 & X_3 & -X_2 & X_1 & X_8 & -X_7 & X_6 & -X_5 & 0 & 0 & 0 & X_9 & 0 & 0 & 0 & 0 \\
-X_5 & -X_6 & -X_7 & -X_8 & X_1 & X_2 & X_3 & X_4 & 0 & 0 & 0 & 0 & X_9 & 0 & 0 & 0 \\
-X_6 & X_5 & -X_8 & X_7 & -X_2 & X_1 & -X_4 & X_3 & 0 & 0 & 0 & 0 & 0 & X_9 & 0 & 0 \\
-X_7 & X_8 & X_5 & -X_6 & -X_3 & X_4 & X_1 & -X_2 & 0 & 0 & 0 & 0 & 0 & 0 & X_9 & 0 \\
-X_8 & -X_7 & X_6 & X_5 & -X_4 & -X_3 & X_2 & X_1 & 0 & 0 & 0 & 0 & 0 & 0 & 0 & X_9 \\
-X_9 & 0 & 0 & 0 & 0 & 0 & 0 & 0 & X_1 & -X_2 & -X_3 & -X_4 & -X_5 & -X_6 & -X_7 & -X_8 \\
0 & -X_9 & 0 & 0 & 0 & 0 & 0 & 0 & X_2 & X_1 & -X_4 & X_3 & -X_6 & X_5 & X_8 & -X_7 \\
0 & 0 & -X_9 & 0 & 0 & 0 & 0 & 0 & X_3 & X_4 & X_1 & -X_2 & -X_7 & -X_8 & X_5 & X_6 \\
0 & 0 & 0 & -X_9 & 0 & 0 & 0 & 0 & X_4 & -X_3 & X_2 & X_1 & -X_8 & X_7 & -X_6 & X_5 \\
0 & 0 & 0 & 0 & -X_9 & 0 & 0 & 0 & X_5 & X_6 & X_7 & X_8 & X_1 & -X_2 & -X_3 & -X_4 \\
0 & 0 & 0 & 0 & 0 & -X_9 & 0 & 0 & X_6 & -X_5 & X_8 & -X_7 & X_2 & X_1 & X_4 & -X_3 \\
0 & 0 & 0 & 0 & 0 & 0 & -X_9 & 0 & X_7 & -X_8 & -X_5 & X_6 & X_3 & -X_4 & X_1 & X_2 \\
0 & 0 & 0 & 0 & 0 & 0 & 0 & -X_9 & X_8 & X_7 & -X_6 & -X_5 & X_4 & X_3 & -X_2 & X_1
\end{pmatrix}$$

The beauty of this formula is, I think undisputed.

At the end of Theorem 14.1, we have explained how this $(9, 16, 16)$ identity can be derived from Theorem 14.1.

Another interesting identity promised in Chapter 10 is the $(10, 10, 16)$ bilinear identity given by

$$(X_1^2 + \ldots + X_{10}^2)(Y_1^2 + \ldots + Y_{10}^2) = Z_1^2 + \ldots + Z_{16}^2,$$

where the $Z_1, \ldots, Z_{16}$ are given by the following (note that it is known that $10 * 10 = 16$ (see Remark 2.10, page 243 of [S6])):

$$Z_1 = X_1Y_1 - X_2Y_2 - X_3Y_3 - X_4Y_4 - X_5Y_5 - X_6Y_6 - X_7Y_7 - X_8Y_8 - X_9Y_9 - X_{10}Y_{10}$$

$$Z_2 = X_1Y_2 + X_2Y_1 + X_3Y_4 - X_4Y_3 + X_5Y_6 - X_6Y_5 - X_7Y_8 + X_8Y_7 + X_9Y_{10} - X_{10}Y_9$$

$$Z_3 = X_1Y_3 + X_3Y_1 - X_2Y_4 + X_4Y_2 + X_5Y_7 - X_7Y_5 + X_6Y_8 - X_8Y_6$$

$$Z_4 = X_1Y_4 + X_4Y_1 + X_2Y_3 - X_3Y_2 + X_5Y_8 - X_8Y_5 - X_6Y_7 + X_7Y_6$$

$$Z_5 = X_1Y_5 + X_5Y_1 - X_2Y_6 + X_6Y_2 - X_3Y_7 + X_7Y_3 - X_4Y_8 + X_8Y_4$$

$$Z_6 = X_1Y_6 + X_6Y_1 + X_2Y_5 - X_5Y_2 - X_3Y_8 + X_8Y_3 + X_4Y_7 - X_7Y_4$$

$$Z_7 = X_1Y_7 + X_7Y_1 + X_2Y_8 - X_8Y_2 + X_3Y_5 - X_5Y_3 - X_4Y_6 + X_6Y_4$$

$$Z_8 = X_1Y_8 + X_8Y_1 - X_2Y_7 + X_7Y_2 + X_3Y_6 - X_6Y_3 + X_4Y_5 - X_5Y_4$$

$$Z_9 = X_1Y_9 + X_9Y_1 - X_2Y_{10} + X_{10}Y_2$$

$$Z_{10} = X_1Y_{10} + X_{10}Y_1 + X_2Y_9 - X_9Y_2$$

$$Z_{11} = X_3Y_9 - X_9Y_3 - X_4Y_{10} + X_{10}Y_4$$

$$Z_{12} = X_3Y_{10} - X_{10}Y_3 + X_4Y_9 - X_9Y_4$$

$$Z_{13} = X_5Y_9 - X_9Y_5 - X_6Y_{10} + X_{10}Y_6$$

$$Z_{14}=X_5Y_{10}-X_{10}Y_5+X_6Y_9-X_9Y_6$$

$$Z_{15}=X_7Y_9-X_9Y_7+X_8Y_{10}-X_{10}Y_8$$

$$Z_{16}=-X_7Y_{10}+X_{10}Y_7+X_8Y_9-X_9Y_8$$

In the matrix notation, this becomes $\underline{Z} = A\underline{Y}$, where $\underline{Z}$, $\underline{Y}$ are column vectors with 16 and 10 components respectively. We have written out A below in full for the readers to appreciate and, as for the $(9,16,16)$ identity, we would like to say where this $(10,10,16)$ identity comes from.

$$\begin{pmatrix}
X_1 & -X_2 & -X_3 & -X_4 & -X_5 & -X_6 & -X_7 & -X_8 & -X_9 & -X_{10} \\
X_2 & X_1 & -X_4 & X_3 & -X_6 & X_5 & X_8 & -X_7 & -X_{10} & X_9 \\
X_3 & X_4 & X_1 & -X_2 & -X_7 & -X_8 & X_5 & X_6 & 0 & 0 \\
X_4 & -X_3 & X_2 & X_1 & -X_8 & X_7 & -X_6 & X_5 & 0 & 0 \\
X_5 & X_6 & X_7 & X_8 & X_1 & -X_2 & -X_3 & -X_4 & 0 & 0 \\
X_6 & -X_5 & X_8 & -X_7 & X_2 & X_1 & X_4 & -X_3 & 0 & 0 \\
X_7 & -X_8 & -X_5 & X_6 & X_3 & -X_4 & X_1 & X_2 & 0 & 0 \\
X_8 & X_7 & -X_6 & -X_5 & X_4 & X_3 & -X_2 & X_1 & 0 & 0 \\
X_9 & X_{10} & 0 & 0 & 0 & 0 & 0 & 0 & X_1 & -X_2 \\
X_{10} & -X_9 & 0 & 0 & 0 & 0 & 0 & 0 & X_2 & X_1 \\
0 & 0 & -X_9 & X_{10} & 0 & 0 & 0 & 0 & X_3 & -X_4 \\
0 & 0 & -X_{10} & -X_9 & 0 & 0 & 0 & 0 & X_4 & X_3 \\
0 & 0 & 0 & 0 & -X_9 & X_{10} & 0 & 0 & X_5 & -X_6 \\
0 & 0 & 0 & 0 & -X_{10} & -X_9 & 0 & 0 & X_6 & X_5 \\
0 & 0 & 0 & 0 & 0 & 0 & -X_9 & -X_{10} & X_7 & X_8 \\
0 & 0 & 0 & 0 & 0 & 0 & X_{10} & -X_9 & X_8 & -X_7
\end{pmatrix}$$

The identity can be squeezed out of Theorem 1(iv) of K.Y. Lam's paper [L1]. In the notation therein, take

$$u = (X_1, X_2) = (\alpha_1 u_0 + \alpha_2 u_1 + \ldots + \alpha_8 u_7, \alpha_9 u_0 + \alpha_{10} u_1) \in K^2,$$

where we restrict the element X_2 of K to the subspace $\mathbb{C}$ of complex numbers of K. Similarly take

$$\begin{aligned} v &= (Y_1, Y_2) \\ &= (\beta_1 u_0 + \ldots + \beta_8 u_7, \beta_9 u_0 + \beta_{10} u_1). \end{aligned}$$

Then

$$\begin{aligned} f(u,v) &= (x_1 y_1 - \overline{y}_2 x_2, y_2 x_1 + x_2 \overline{y}_1, x_2 y_2 - y_2 x_2) \\ &= (x_1 y_1 - \overline{y}_2 x_2, y_2 x_1 + x_2 \overline{y}_1, 0) \qquad (*) \end{aligned}$$

since x_2, $y_2 \in \mathbb{C}$ and so commute.

Hence the image $f(u,v) \in K^2$ rather than K^3. Now multiply out the Cayley numbers in $(*)$ to give

$$f(u,v) = (r_1 u_0 + \ldots + r_8 u_7, r'_1 u_0 + \ldots + r'_8 u_7).$$

If this bilinear f happens to be norm-preserving then we would indeed have

$$(\alpha_1^2 + \cdots + \alpha_8^2 + \alpha_9^2 + \alpha_{10}^2)(\beta_1^2 + \cdots \beta_8^2 + \beta_9^2 + \beta_{10}^2) = \gamma_1^2 + \cdots + \gamma_8^2 + \gamma_1'^2 + \cdots + \gamma_8'^2.$$

As it turns out, f *is* norm-preserving, although Lam does not say so anywhere (in fact not all the maps (i)–(viii) seem to be norm-preserving; for example (i) is not for otherwise it would give a (16, 16, 23) bilinear identity contradicting the fact that $16 *_{\mathbf{Z}} 16 \geq 25$ mentioned in [Y1]; indeed $16 *_{\mathbf{Z}} 16 \geq 29$). Changing the notation to $\mathbf{x}, \mathbf{y}, \mathbf{z}$ we get the required (10, 10, 16) identity.

Note that [Y1] also gives a method of getting the identity (10, 10, 16). For the convenience of the reader we give here the multiplication table of Cayley numbers.

	u_1	u_2	u_3	u_4	u_5	u_6	u_7
u_1	$-u_0$	u_3	$-u_2$	u_5	$-u_4$	$-u_7$	u_6
u_2	$-u_3$	$-u_0$	u_1	u_6	u_7	$-u_4$	$-u_5$
u_3	u_2	$-u_1$	$-u_0$	u_7	$-u_6$	u_5	$-u_4$
u_4	$-u_5$	$-u_6$	$-u_7$	$-u_0$	u_1	u_2	u_3
u_5	u_4	$-u_7$	u_6	$-u_1$	$-u_0$	$-u_3$	u_2
u_6	u_7	u_4	$-u_5$	$-u_2$	u_3	$-u_0$	$-u_1$
u_7	$-u_6$	u_5	u_4	$-u_3$	$-u_2$	u_1	$-u_0$

The u_0 above acts like 1 so that $u_0 u_j = u_j u_0 = u_j$ $(j = 1, 2 \ldots, 7)$. See [A5], p.137. Finally note that this (10, 10, 16) identity apparently goes back to Kirkman over 140 years ago! See [K5].

Although there is an (8, 8, 8) bilinear identity, there are rational (8, 8, 8) formulae too. For example the "doubling" process (see Theorem 2.1) gives us such a formula and as an example we derive a (4, 4, 4) identity here. By the (2, 2, 2) identity we have

$$\begin{pmatrix} Z_1 \\ Z_2 \end{pmatrix} = \begin{pmatrix} X_1 & X_2 \\ -X_2 & X_1 \end{pmatrix} \begin{pmatrix} Y_1 \\ Y_2 \end{pmatrix} = T^{(1)} \begin{pmatrix} Y_1 \\ Y_2 \end{pmatrix} \quad \text{and}$$

$$\begin{pmatrix} Z_3 \\ Z_4 \end{pmatrix} = \begin{pmatrix} X_3 & X_4 \\ -X_4 & X_3 \end{pmatrix} \begin{pmatrix} Y_3 \\ Y_4 \end{pmatrix} = T^{(2)} \begin{pmatrix} Y_3 \\ Y_4 \end{pmatrix}$$

in the notation of Theorem 2.1. Then by Theorem 2.1 we have

$$\mathbf{Z} = \begin{pmatrix} Z_1 \\ Z_2 \\ Z_3 \\ Z_4 \end{pmatrix} = \begin{pmatrix} T^{(1)} & T^{(2)} \\ T^{(2)} & X \end{pmatrix} \begin{pmatrix} Y_1 \\ Y_2 \\ Y_3 \\ Y_4 \end{pmatrix} = T\mathbf{Y},$$

say, where

$$X = -T^{(2)'^{-1}} T^{(1)'} T^{(2)}$$
$$= -\begin{pmatrix} X_3 & -X_4 \\ X_4 & X_3 \end{pmatrix}^{-1} \begin{pmatrix} X_1 & -X_2 \\ X_2 & X_1 \end{pmatrix} \begin{pmatrix} X_3 & X_4 \\ -X_4 & X_3 \end{pmatrix}.$$

Simplifying, we can get T explicitly as

$$T = \begin{pmatrix} X_1 & X_2 & X_3 & X_4 \\ -X_2 & X_1 & -X_4 & X_3 \\ X_3 & X_4 & \frac{X_1(X_3^2-X_4^2)+2X_2X_3X_4}{-(X_3^2+X_4^2)} & \frac{X_2(X_4^2-X_3^2)+2X_1X_3X_4}{-(X_3^2+X_4^2)} \\ -X_4 & X_3 & \frac{X_2(X_3^2-X_4^2)-2X_1X_3X_4}{-(X_3^2+X_4^2)} & \frac{X_1(X_3^2-X_4^2)+2X_2X_3X_4}{-(X_3^2+X_4^2)} \end{pmatrix}$$

and the four quantities Z_1, Z_2, Z_3, Z_4 can be written down immediately. It is a formidable exercise to verify directly that for these Z_1, Z_2, Z_3, Z_4 we have indeed

$$(X_1^2 + X_2^2 + X_3^2 + X_4^2)(Y_1^2 + Y_2^2 + Y_3^2 + Y_4^2) = Z_1^2 + Z_2^2 + Z_3^2 + Z_4^2$$

where

$$Z_1 = X_1Y_1 + X_2Y_2 + X_3Y_3 + X_4Y_4$$
$$Z_2 = -X_2Y_1 + X_1Y_2 - X_4Y_3 + X_3Y_4$$
$$Z_3 = X_3Y_1 + X_4Y_2 - \left(\frac{X_1(X_3^2 - X_4^2) + 2X_2X_3X_4}{X_3^2 + X_4^2}\right) Y_3 - \left(\frac{X_2(X_4^2 - X_3^2) + 2X_1X_3X_4}{X_3^2 + X_4^2}\right) Y_4$$
$$Z_4 = -X_4X_1 + X_3Y_2 - \left(\frac{X_2(X_3^2 - X_4^2) - 2X_1X_3X_4}{X_3^2 + X_4^2}\right) Y_3 - \left(\frac{X_1(X_3^2 - X_4^2) + 2X_2X_3X_4}{X_3^2 + X_4^2}\right) Y_4$$

Theorem 2.1 tells us precisely this, of course: but still one feels the urge to recheck one's formulae.

The beauty of this formula is, no doubt diffused on account of the existence of the much simpler bilinear $(4, 4, 4)$ formula.

We give one more identity, the $(12, 12, 26)$ one. Its beauty is that by restriction, it gives the $(10,10,16)$ identity. Writing $\mathbf{X} = (X_1, X_2, \ldots, X_{11}, X_{12})$ as a row vector, we have

$$(Z_1, Z_2, \ldots, Z_{26}) = (X_1, X_2, \ldots, X_{12})A(\mathbf{Y}),$$

where $A(\mathbf{Y})$ is the 12×26 matrix given opposite. Putting $X_{11} = X_{12} = Y_{11} = Y_{12} = 0$ in this we get

$$(Z_1, Z_2, \ldots, Z_{16}) = (X_1, \ldots, X_{10})B(\mathbf{Y})$$

where $B(\mathbf{Y})$ is the 10×16 top left hand corner minor of $A(\mathbf{Y})$. This gives

$$A(\underline{Y})=\begin{pmatrix}
Y_1 & Y_2 & Y_3 & Y_4 & Y_5 & Y_6 & Y_7 & Y_8 & Y_9 & Y_{10} & 0 & 0 & 0 & 0 & 0 & 0 & Y_{11} & Y_{12} & 0 & 0 & 0 & 0 & 0 & 0 & 0 & 0 \\
Y_2 & -Y_1 & Y_4 & -Y_3 & Y_6 & -Y_5 & -Y_8 & Y_7 & Y_{10} & -Y_9 & 0 & 0 & 0 & 0 & 0 & 0 & Y_{12} & -Y_{11} & 0 & 0 & 0 & 0 & 0 & 0 & 0 & 0 \\
Y_3 & -Y_4 & -Y_1 & Y_2 & Y_7 & Y_8 & -Y_5 & -Y_6 & 0 & 0 & Y_9 & Y_{10} & 0 & 0 & 0 & 0 & 0 & 0 & -Y_{11} & -Y_{12} & 0 & 0 & 0 & 0 & 0 & 0 \\
Y_4 & Y_3 & -Y_2 & -Y_1 & Y_8 & -Y_7 & Y_6 & -Y_5 & 0 & 0 & -Y_{10} & Y_9 & 0 & 0 & 0 & 0 & 0 & 0 & Y_{12} & -Y_{11} & 0 & 0 & 0 & 0 & 0 & 0 \\
Y_5 & -Y_6 & -Y_7 & -Y_8 & -Y_1 & Y_2 & Y_3 & Y_4 & 0 & 0 & 0 & 0 & Y_9 & Y_{10} & 0 & 0 & 0 & 0 & 0 & 0 & -Y_{11} & -Y_{12} & 0 & 0 & 0 & 0 \\
Y_6 & Y_5 & -Y_8 & Y_7 & -Y_2 & -Y_1 & -Y_4 & Y_3 & 0 & 0 & 0 & 0 & -Y_{10} & Y_9 & 0 & 0 & 0 & 0 & 0 & 0 & Y_{12} & -Y_{11} & 0 & 0 & 0 & 0 \\
Y_7 & Y_8 & Y_5 & -Y_6 & -Y_3 & Y_4 & -Y_1 & -Y_2 & 0 & 0 & 0 & 0 & 0 & 0 & Y_9 & -Y_{10} & 0 & 0 & 0 & 0 & 0 & 0 & Y_{11} & Y_{12} & 0 & 0 \\
Y_8 & -Y_7 & Y_6 & Y_5 & -Y_4 & Y_3 & Y_2 & -Y_1 & 0 & 0 & 0 & 0 & 0 & 0 & Y_{10} & Y_9 & 0 & 0 & 0 & 0 & 0 & 0 & Y_{12} & -Y_{11} & 0 & 0 \\
Y_9 & -Y_{10} & 0 & 0 & 0 & 0 & 0 & 0 & -Y_1 & Y_2 & -Y_3 & -Y_4 & -Y_5 & -Y_6 & -Y_7 & -Y_8 & 0 & 0 & 0 & 0 & 0 & 0 & 0 & 0 & -Y_{11} & -Y_{12} \\
Y_{10} & Y_9 & 0 & 0 & 0 & 0 & 0 & 0 & -Y_2 & -Y_1 & Y_4 & -Y_3 & Y_6 & -Y_5 & -Y_8 & Y_7 & 0 & 0 & 0 & 0 & 0 & 0 & 0 & 0 & Y_{12} & -Y_{11} \\
Y_{11} & -Y_{12} & 0 & 0 & 0 & 0 & 0 & 0 & 0 & 0 & 0 & 0 & 0 & 0 & 0 & 0 & -Y_1 & Y_2 & Y_3 & Y_4 & Y_5 & Y_6 & -Y_7 & Y_8 & Y_9 & Y_{10} \\
Y_{12} & Y_{11} & 0 & 0 & 0 & 0 & 0 & 0 & 0 & 0 & 0 & 0 & 0 & 0 & 0 & 0 & -Y_2 & -Y_1 & -Y_4 & Y_3 & -Y_6 & Y_5 & -Y_8 & -Y_7 & -Y_{10} & Y_9
\end{pmatrix}$$

the (10, 10, 16) bilinear identity mentioned before with some minor sign changes.

In all the previous examples of bilinear identities, we notice that the bilinear functions

$$Z_k = \sum\sum a_{ij} X_i Y_j$$

have all the a_{ij} integers, whereas they are supposed to be only in the ground field, in our case (Chapter 10), the real numbers. It is a remarkable fact that we can always choose the a_{ij} to be integers; indeed they can be chosen to be ± 1 (or zero of course in which case the term is no longer written). This is the content of the following.

Theorem 14.1 (Gabel (1974)). *Let $n \geq 1$ be an integer and let p be an integer such that $1 \leq p \leq \rho(n)$. Then there exist n bilinear forms Z_1, $Z_2, \ldots, Z_n$ in $X_1, \ldots, X_p$, $Y_1, \ldots, Y_n$ with integer coefficients such that*

$$\left(\sum_{i=1}^{p} X_i^2\right)\left(\sum_{j=1}^{n} Y_j^2\right) = \sum_{k=1}^{n} Z_k^2.$$

Further these bilinear functions can all be chosen with coefficients ± 1.

Remarks.

1. The two special properties of the ring $\mathbb{Z}$ viz.

(a) $2 \cdot m = 0 \Rightarrow m = 0$ $(m \in \mathbb{Z})$

(b) $a_1^2 + \ldots + a_s^2 = 0 \Rightarrow a_1 = a_2 = \ldots = a_s = 0$ $(a_j \in \mathbb{Z})$.

are used in the proof (see Lemma 1). If R is any ring in which (a) and (b) hold, then Gabel's theorem will be true in R.

2. Gabel calls the theorem the Integral Version of the Hurwitz-Radon-Eckmann theorem. Eckmann [E1] gave a group-theoretic proof of the Hurwitz-Radon theorem in 1943.

We now prove a series of lemmas; the theorem will follow from these.

Lemma 1. *Let p and n be integers satisfying*

$$1 \leq p \leq n$$

The following five statements are equivalent.

(1) *There exist p $n \times n$ matrices $A_1, \ldots, A_p$ over $\mathbb{Z}$, such that*
(a) $A_i A_i^t = I_n$ $(i = 1, \ldots, p)$
(b) $A_i A_j^t + A_j A_i^t = 0$ $(i \neq j)$,

(2) *There exist n $p \times n$ matrices $B_1, \ldots, B_n$ over $\mathbb{Z}$, such that*
(a) $B_i B_i^t = I_p$ $(i = 1, \ldots, n)$
(b) $B_i B_j^t + B_j B_i^t = 0$ $(i \neq j)$,

(3) *There exist $p-1$ $n \times n$ matrices $C_1, \ldots, C_{p-1}$ over $\mathbb{Z}$, such that*
(a) $C_i C_i^t = I_n$ $(i = 1, \ldots, p-1)$
(b) $C_i^2 = -I_n$ $(i = 1, \ldots, p-1)$
(c) $C_i C_j + C_j C_i = 0$ $(i \neq j)$,

(4) *There exists a $p \times n$ matrix T over $\mathbb{Z}[X_1, \ldots, X_n]$ such that*
(a) *the top row of T is $X_1, X_2, \ldots, X_n$*
(b) $TT^t = \left(\sum_1^n X_i^2\right) I_p$

(5) *There exist n bilinear forms $Z_1, Z_2, \ldots, Z_n$ with coefficients in $\mathbb{Z}$ in the variables*

$$X_1, X_2, \ldots, X_p, Y_1, Y_2, \ldots, Y_n$$

such that

$$(X_1^2 + \ldots + X_p^2)(Y_1^2 + \ldots + Y_n^2) = Z_1^2 + \ldots + Z_n^2.$$

Thus to prove (5) it is enough to prove any of (1), (2), (3) and (4).

Proof. (1)⇒(3). Put $C_i = A_{i+1} A_1^t$ $(1 \leq i \leq p-1)$. We check the conditions (a), (b), (c) of (3):

(a)

$$\begin{aligned} C_i C_i^t &= A_{i+1} A_1^t A_1 A_{i+1}^t \\ &= A_{i+1} I_n A_{i+1}^t \quad \text{(by (1a))} \\ &= I_n \quad \text{(again by (1a))}. \end{aligned}$$

(b)

$$\begin{aligned} C_i^2 &= A_{i+1} A_1^t A_{i+1} A_1^t \\ &= -A_1 A_{i+1}^t A_{i+1} A_1^t \quad \text{(by (1b))} \\ &= -A_1 I_n A_1^t \\ &= -I_n. \end{aligned}$$

(c)

$$\begin{aligned} C_i C_j + C_j C_i &= A_{i+1} A_1^t A_{j+1} A_1^t + A_{j+1} A_1^t A_{i+1} A_1^t \\ &= -A_{i+1} A_1^t A_1 A_{j+1}^t - A_{j+1} A_1^t A_1 A_{i+1}^t \\ &= -(A_{i+1} A_{j+1}^t + A_{j+1} A_{i+1}^t) \\ &= 0. \end{aligned}$$

(5)⇒(1). We have done this in Chapter 10 in the transpose notation. In the present notation we are given $(X_1^2 + \ldots + X_p^2)(Y_1^2 + \ldots + Y_n^2) = Z_1^2 + \ldots + Z_n^2$.

Write $(Z_1, \ldots, Z_n) = (Y_1, \ldots, Y_n)A$ where A is an $n \times n$ matrix with entries linear forms in $X_1, \ldots, X_p$ with coefficients in $\mathbb{Z}$. Then

$$(X_1^2 + \ldots + X_p^2)(Y_1, \ldots, Y_n) \begin{pmatrix} Y_1 \\ \vdots \\ Y_n \end{pmatrix} = (Z_1, \ldots, Z_n) \begin{pmatrix} Z_1 \\ \vdots \\ Z_n \end{pmatrix}$$

i.e.

$$\left(\sum_1^p X_j^2\right)(Y_1,\ldots,Y_n)\begin{pmatrix}Y_1\\ \vdots\\ Y_n\end{pmatrix} - (Y_1,\ldots,Y_n)AA'\begin{pmatrix}Y_1\\ \vdots\\ Y_n\end{pmatrix} = 0$$

i.e.

$$(Y_1,\ldots,Y_n)\left\{\left(\sum_1^p X_j^2\right)I_n - AA'\right\}\begin{pmatrix}Y_1\\ \vdots\\ Y_n\end{pmatrix} = 0.$$

Since this is true for all $Y_1,\ldots,Y_n$, we have

$$AA' = \left(\sum_1^p X_j^2\right)I_n.$$

Now write $A = A_1X_1 + \ldots + A_pX_p$ (the A_j are integral matrices). Then the above gives

$$(A_1X_1 + \ldots + A_pX_p)(A_1'X_1 + \ldots + A_p'X_p) = \left(\sum X_j^2\right)I_n$$

i.e.

(a) $A_iA_i' = I_n$ $(i = 1,\ldots,p)$

(b) $A_iA_j' + A_jA_i' = 0$ $(i \neq j)$; which gives (1).

(1)⇒(5). Starting with the equations (1a, b), define $A = A_1X_1+\ldots+A_pX_p$. Then by (a), (b) we have

$$AA' = \left(\sum_1^p X_j^2\right)I_n$$

and working backwards we arrive at the required identity (5).

(2)⇒(1). Given the $p\times n$ matrices $B_1,\ldots,B_n$ satisfying (2a, b), define the p matrices $A_1,\ldots,A_p$ by

$$A_1 = \begin{pmatrix}\text{1st row of } B_1\\ \text{1st row of } B_2\\ \cdots\cdots\cdots\\ \text{1st row of } B_n\end{pmatrix},\ldots, A_p = \begin{pmatrix}p\text{th row of } B_1\\ p\text{th row of } B_2\\ \cdots\cdots\cdots\\ p\text{th row of } B_n\end{pmatrix}.$$

Thus A_i is the matrix whose jth row is the ith row of B_j. These are indeed $n\times n$ matrices. We show (1a, b):

(a)

$$\begin{aligned}(A_iA_i^t)_{kl} &= (k\text{th row of } A_i)(l\text{th column of } A_i^t)\\ &= (k\text{th row of } A_i)(l\text{th row of } A_i)\\ &= (i\text{th row of } B_k)(i\text{th row of } B_l)\\ &= (B_kB_l')_{ii}\\ &= (B_kB_k')_{ii} \quad (\text{if } k = l)\\ &= (I_p)_{ii} \quad (\text{by (2a)})\\ &= 1\end{aligned}$$

as required. If $k \neq l$, then $(B_k B_l')_{ii} = (B_l B_k')_{ii}$ and so

$$\begin{aligned} 2((B_k B_l')_{ii}) &= (B_k B_l' + B_l B_k')_{ii} \\ &= 0 \end{aligned}$$

by (2b). It follows that $(B_k B_l^t)_{ii} = 0$ if $k \neq l$; which proves (1a).

The proof of (1b) goes as follows:

$$\begin{aligned} (A_i A_j^t + A_j A_i^t)_{kl} &= (k\text{th row of } A_i)(l\text{th row of } A_j) \\ &\quad + (k\text{th row of } A_j)(l\text{th row of } A_i) \\ &= (i\text{th row of } B_k)(j\text{th row of } B_l) \\ &\quad + (j\text{th row of } B_k)(i\text{th row of } B_l) \\ &= (B_k B_l')_{ij} + (B_k B_l')_{ji} \\ &= (B_k B_l' + B_l B_k')_{ij} \\ &= 0 \quad (\text{by b) of (2)}). \end{aligned}$$

(3)⇒(4). We are given $p-1$ $n \times n$ matrices $C_1, C_2, \ldots, C_{p-1}$ satisfying (3a,b,c). Now construct the $A_1, \ldots, A_p$ of (1) as described in the proof of the implication (3)⇒(1) and construct the $B_1, \ldots, B_n$ of (2) by letting

$$B_1 = \begin{pmatrix} \text{1st row of } A_1 \\ \text{1st row of } A_2 \\ \cdots\cdots\cdots\cdots \\ \text{1st row of } A_p \end{pmatrix}, \quad B_2 = \begin{pmatrix} \text{2nd row of } A_1 \\ \text{2nd row of } A_2 \\ \cdots\cdots\cdots\cdots \\ \text{2nd row of } A_p \end{pmatrix}, \ldots$$

Then these B's satisfy

(a) $B_i B_i^t = I_p$ $(1 \leq i \leq n)$

(b) $B_i B_j^t + B_j B_i^t = 0$ $(i \neq j)$

(c) The top row of $B_i = \varepsilon_i$ where

$$\varepsilon_1 = (1, 0, \ldots, 0), \varepsilon_2 = (0, 1, 0, \ldots, 0), \ldots, \varepsilon_n = (0, 0, \ldots, 0, 1).$$

So far this is easy to visualize. Now set

$$T = X_1 B_1 + \ldots + X_n B_n.$$

Then 'clearly' $TT^t = (\sum_1^n X_i^2) I_p$ as required by (4). This 'clear' part is best understood by an example: let $p = 2$, $n = 3$,

$$c_1 = \begin{pmatrix} r_1 & r_2 & r_3 \\ r_4 & r_5 & r_6 \\ r_7 & r_8 & r_9 \end{pmatrix};$$

then

$$A_1 = \begin{pmatrix} 1 & 0 & 0 \\ 0 & 1 & 0 \\ 0 & 0 & 1 \end{pmatrix}, \quad A_2 = C_1$$

and this gives the three 2×3 matrices:

$$B_1 = \begin{pmatrix} 1 & 0 & 0 \\ r_1 & r_2 & r_3 \end{pmatrix}, \; B_2 = \begin{pmatrix} 0 & 1 & 0 \\ r_4 & r_5 & r_6 \end{pmatrix}, \; B_3 = \begin{pmatrix} 0 & 0 & 1 \\ r_7 & r_8 & r_9 \end{pmatrix}$$

under the implication (3)⇒(2). These satisfy $B_iB_i^t = I_2$ $(i = 1, 2, 3)$ and $B_iB_j^t + B_jB_i^t = 0$ $(i \neq j)$. Then

$$T = X_1B_1 + X_2B_2 + X_3B_3$$
$$= \begin{pmatrix} X_1 & X_2 & X_3 \\ X_1r_1+X_2r_4+X_3r_7, & X_1r_2+X_2r_5+X_3r_8, & X_1r_3+X_2r_6+X_3r_9 \end{pmatrix}$$

Hence TT^t = multiply out the product and we have to show in (4), that this $= (X_1^2 + X_2^2 + X_3^2)I_2$.

Now the relations (2a,b) give the following:

$$r_1^2 + r_2^2 + r_3^2 = 1, \qquad r_1 = r_5 = r_9 = 0$$
$$r_4^2 + r_5^2 + r_6^2 = 1, \qquad r_7^2 + r_8^2 + r_9^2 = 1,$$

and

$$r_2 + r_4 = r_3 + r_7 = r_6 + r_8 = 0$$

$$r_1r_4 + r_2r_5 + r_3r_6 = r_1r_7 + r_2r_8 + r_3r_9 = r_4r_7 + r_5r_8 + r_6r_9 = 0$$

Plugging in these values in TT^t we find it is equal to $(X_1^2 + X_2^2 + X_3^2)I_2$ as required.

Finally to show that (4) implies each of (1), (2), (3), (5) it is enough to show, for example, that

(4)⇒(2). Here we are given a $p \times n$ matrix $T = (f_{ij})$, where $f_{ij} \in \mathbf{Z}[X_1, \ldots, X_n]$, satisfying $TT^t = (\sum_1^n X_i^2)I_p$. Write T fully and use $TT^t = \sum X_i^2 \cdot I_p$ to give

$$\sum_{j=1}^{n} f_{ij}^2 = X_1^2 + \ldots + X_n^2 \qquad (i = 1, 2, \ldots, p).$$

Here if any of the f_{ij} had degree greater than 1 in any X_k, then this would lead to an equation of the type

$$\alpha_1^2 + \ldots + \alpha_n^2 = 0 \qquad (\alpha_j \in \mathbf{Z}).$$

It follows that each f_{ij} is a linear function in the $X_1, \ldots, X_n$. So we may write $T = X_1B_1 + \ldots + X_nB_n$, where B_i's are $p \times n$ matrices with integer coefficients. Then

$$\left(\sum X_i^2\right) I_p = TT^t = (X_1B_1 + \ldots + X_nB_n)(X_1B_1^t + \ldots + X_nB_n^t)$$

$$= \sum_{i=1}^{n} X_i^2 B_iB_i^t + \sum_{i<j} X_iX_j(B_iB_j^t + B_jB_i^t).$$

It follows that $B_iB_i^t = I_p$ $(i = 1, \ldots, n)$ and $B_iB_j^t + B_jB_i^t = 0$ $(i \neq j)$, which proves (2). This completes the proof of Lemma 1. □

Lemma 2. *If each of the $n \times n$ matrices $C_1, \ldots, C_{p-1}$ in (3) of Lemma 1* *have their entries* 1, -1, 0, *then all the entries of the matrix* T, *in (4) of Lemma 1 will be linear forms in* $X_1, \ldots, X_n$ *in which each* X_j *occurs with a coefficient* 1 *or* -1.

Conversely if T *has its entries that are linear forms in* $X_1, \ldots, X_n$ *in which each* X_j *occurs with a coefficient* ± 1, *then under the implication (4)⇒(3) of Lemma 1, the matrices* $C_1, \ldots, C_{p-1}$ *have all their entries* 1, -1, 0.

Proof. Under the implications (3)⇒(1)⇒(2) of Lemma 1, we see that all the $B_1, \ldots, B_n$ of (2) have entries 1, -1, 0. From (2) we go to (4) by setting $T = X_1B_1 + \ldots + X_nB_n$, so T satisfies what is stated.

The converse is similar: from (4) go to (3), via (2) and (1). □

An immediate deduction is the following.

Lemma 3. *There exist eight* 16×16 *matrices* $M_1, \ldots, M_8$ *with integer entries, indeed with entries* 1, -1, 0, *such that*

(a) $M_i^2 = -I_{16}$ $(1 \le i \le 8)$.
(b) $M_iM_j + M_jM_i = 0$ $(i \ne j)$
(c) $M_iM_i^t = I_{16}$ $(1 \le i \le 8)$.

Proof. Consider the 9×16 matrix T over $\mathbb{Z}[X_1, \ldots, X_{16}]$ given on page 204. Check that $TT^t = (\sum_1^{16} X_i^2)I_{16}$. Thus we have (4) of Lemma 1, hence (3) of Lemma 1, i.e. there exist eight 16×16 matrices $M_1, \ldots, M_8$, with integer coefficients satisfying (a), (b), (c) above. Furthermore, since T has all its entries linear forms with coefficients ± 1, by Lemma 2, converse, the $M_1, \ldots, M_8$ have all their entries 1, -1, 0.

We now need a compact procedure to build up large matrices from small ones and so we introduce the so-called Kronecker product $A \times B$ of $A = (a_{ij})$, an $m \times n$ matrix, with $B = (b_{ij})$, a $p \times q$ matrix (all entries in $\mathbb{Z}$). If

$$A = \begin{pmatrix} a_{11} & a_{12} & \cdots & a_{1n} \\ a_{21} & a_{22} & \cdots & a_{2n} \\ \cdots & \cdots & \cdots & \cdots \\ a_{m1} & a_{m2} & \cdots & a_{mn} \end{pmatrix}$$

define

$$A \times B = \begin{pmatrix} a_{11}B & a_{12}B & \ldots & a_{1n}B \\ a_{21}B & a_{22}B & \ldots & a_{2n}B \\ \cdots & \cdots & \cdots & \cdots \\ a_{m1}B & a_{m2}B & \ldots & a_{mn}B \end{pmatrix}$$

an $mp \times nq$ matrix in blocks of $p \times q$. The following are verified by brute force:

$$T = \begin{pmatrix}
X_1 & X_2 & X_3 & X_4 & X_5 & X_6 & X_7 & X_8 & X_9 & X_{10} & X_{11} & X_{12} & X_{13} & X_{14} & X_{15} & X_{16} \\
-X_2 & X_1 & -X_4 & X_3 & -X_6 & X_5 & X_8 & -X_7 & X_{10} & -X_9 & X_{12} & -X_{11} & X_{14} & -X_{13} & -X_{16} & X_{15} \\
-X_3 & X_4 & X_1 & -X_2 & -X_7 & -X_8 & X_5 & X_6 & X_{11} & -X_{12} & -X_9 & X_{10} & X_{15} & X_{16} & -X_{13} & -X_{14} \\
-X_4 & -X_3 & X_2 & X_1 & -X_8 & X_7 & -X_6 & X_5 & X_{12} & X_{11} & -X_{10} & -X_9 & X_{16} & -X_{15} & X_{14} & -X_{13} \\
-X_5 & X_6 & X_7 & X_8 & X_1 & -X_2 & -X_3 & -X_4 & X_{13} & -X_{14} & -X_{15} & -X_{16} & -X_9 & X_{10} & X_{11} & X_{12} \\
-X_6 & -X_5 & X_8 & -X_7 & X_2 & X_1 & X_4 & -X_3 & X_{14} & X_{13} & -X_{16} & X_{15} & -X_{16} & -X_9 & -X_{12} & X_{11} \\
-X_7 & -X_8 & -X_5 & X_6 & X_3 & -X_4 & X_1 & X_2 & X_{15} & X_{16} & X_{13} & -X_{14} & -X_{11} & X_{12} & -X_9 & -X_{10} \\
-X_8 & X_7 & -X_6 & -X_5 & X_4 & X_3 & -X_2 & X_1 & X_{16} & -X_{15} & X_{14} & X_{13} & -X_{12} & -X_{11} & X_{10} & -X_9 \\
-X_9 & -X_{10} & -X_{11} & -X_{12} & -X_{13} & -X_{14} & -X_{15} & -X_{16} & X_1 & X_2 & X_3 & X_4 & X_5 & X_6 & X_7 & X_8
\end{pmatrix}$$

Lemma 4.

(i) $(A_1 \times B_1) \cdot (A_2 \times B_2) = A_1A_2 \times B_1B_2$
(ii) *If* $r, s \in \mathbf{Z}$, *then* $rA \times sB = rs(A \times B)$
(iii) *If* A, B *are orthogonal, then so is* $A \times B$.

We leave the proof as an exercise to the reader.

Now note the relation

$$\rho(n) + 8 = \rho(16n).$$

This coupled with the following result is what makes things work.

Lemma 5. *Suppose there exist* $n \times n$ *matrices* $C_1, \ldots, C_{p-1}$ *over* $\mathbf{Z}$ *such that*

(a) $C_iC_i^t = I_n \ (1 \le i \le p-1)$
(b) $C_i^2 = -I_n \ (1 \le i \le p-1)$
(c) $C_iC_j + C_jC_i = 0 \ (i \ne j)$.

Then there exist $16n \times 16n$ *matrices* $D_1, D_2, \ldots, D_{(p+8)-1}$ *over* $\mathbf{Z}$ *such that*

(a)′ $D_iD_i^t = I_{16n} \ (1 \le i \le (p+8)-1)$
(b)′ $D_i^2 = -I_{16n} \ (1 \le i \le (p+8)-1)$
(c)′ $D_iD_j + D_jD_i = 0 \ (i \ne j)$.

Furthermore, if $C_1, \ldots, C_{p-1}$ *have entries* $1, -1, 0$, *then the* $D_1, D_2, \ldots, D_{(p+8)-1}$ *can also be so chosen.*

Proof. Let $M_1, \ldots, M_8$ be as in Lemma 3 and put $M = M_1 \cdot M_2 \ldots M_8$. Now let

$$D_i = \begin{cases} C_i \times M & \text{if } i \le p-1 \\ I_n \times M_{i-p+1} & \text{if } i > p-1 \end{cases}$$

Using Lemma 4, it is easy to verify (a)′, (b)′, (c)′ above. That the D's have entries $1, -1, 0$ is clear from their construction from the C_i, the M_i, M and I_n. □

Lemma 6. *Suppose* n *is an integer of the form* 2^a $(a \ge 1)$, *then there exist* $\rho(n) - 1$ $n \times n$ *matrices* $A_1, A_2, \ldots, A_{\rho(n)-1}$ *with entries* $1, -1, 0$ *such that*

(a) $A_iA_i^t = I_n \ (1 \le i \le \rho(n)-1)$
(b) $A_i^2 = -I_n \ (1 \le i \le \rho(n)-1)$
(c) $A_iA_j + A_jA_i = 0 \ (i \ne j)$.

Proof. Use induction on a. For $a = 1, 2, 3$, the $2^a \times 2^a$ matrices in the upper left hand corner of the matrix T exhibited in the proof of Lemma 3 all satisfy (4) of Lemma 1. Since (4)$\Rightarrow$(3) we get respectively $\rho(2)-1$, $\rho(4)-1$, $\rho(8)-1$ matrices with entries $1, -1, 0$ (see Lemma 2) satisfying (a), (b), (c) above. Further $\rho(2^4) = 9$ and the matrix T of Lemma 3 is a matrix satisfying (4) of Lemma 1 and so again by (4)$\Rightarrow$(3) of Lemma 1 we get $A_1, \ldots, A_8$ $(8 = \rho(2^4) - 1)$ with entries $1, -1, 0$ satisfying (a), (b), (c) above. Thus Lemma 6 is true for $a = 1, 2, 3, 4$. It remains to prove the induction step from a to $a+4$. So suppose there exist $\rho(2^a) - 1$ $2^a \times 2^a$ matrices with entries $-1, 1, 0$ satisfying (a), (b), (c) above. By Lemma 5, there exist $(\rho(2^a)+8)-1$ $16 \cdot 2^a \times 16 \cdot 2^a$ matrices with entries $1, -1, 0$ also satisfying (a), (b), (c) above (n replaced by $16 \cdot 2^a$). But $16 \cdot 2^a = 2^{a+4}$ and $\rho(2^a) + 8 = \rho(2^{a+4})$. □

We have now to prove Lemma 6 without any restriction on n. To achieve this we first prove the following.

Lemma 7. *Let T be a $p \times n$ matrix over $\mathbf{Z}[X_1, \ldots, X_n]$ such that*

(a) *the top row of T is $X_1, \ldots, X_n$*

(b) $TT^t = (\sum_1^n X_i^2) I_p$.

Then for each integer $s \geq 1$, there exists a $p \times sn$ matrix $T(s)$ over $\mathbf{Z}[X_1, \ldots, X_n, \ldots, X_{sn}]$ such that

(a) *the top row of $T(s)$ is $X_1, X_2, \ldots, X_n, \ldots, X_{sn}$.*

(b) $T(s) \cdot T(s)^t = (\sum_{i=1}^{sn} X_i^2) I_p$.

Furthermore, if every entry of T is a form with coefficients ± 1 then $T(s)$ can also be so chosen.

Proof. Let T_i be the matrix obtained from $T = T_1$ by replacing each variable X_j occuring in T by the variable $X_{j+(i-1)n}$ and define

$$T(s) = (T_1, T_2, \ldots, T_s).$$

It is now almost obvious that $T(s)$ satisfies the required conditions. As we have always done, we verify it for the case $s = 2$:

$$T(2) = (T_1, T_2) = \begin{pmatrix} X_1, \ldots, X_n, & X_{n+1}, \ldots, X_{2n} \\ f_{21}, \ldots, f_{2n}, & f_{2,n+1}, \ldots, f_{2,2n} \\ \cdots\cdots\cdots & \cdots\cdots\cdots \\ f_{p1}, \ldots, f_{pn}, & f_{p,n+1}, \ldots, f_{p,2n} \end{pmatrix}$$

The top row is visibly as required. Further,

$$T(2)T(2)^t = \left(\sum_{i=1}^{2n} X_i^2 \right) I_p :$$

just write out and check. □

Lemma 8. *Let $n \geq 1$ be an integer and let $p = \rho(n)$. Then there exists a $p \times n$ matrix T over $\mathbf{Z}[X_1, \ldots, X_n]$ such that*

(a) *the top row of T is $X_1, \ldots, X_n$*

(b) $TT^t = (\sum_{i=1}^n X_i^2)I_p$.

Moreover T can be chosen so that every entry of T is a linear form in $X_1, \ldots, X_n$ with coefficients ± 1. (This is just Lemma 6 without any restriction on n).

Proof. Write $n = 2^a \cdot s$ for s odd. If $a = 0$, set

$$T = (X_1, X_2, \ldots, X_n).$$

Then T satisfies everything required by the lemma. So suppose $a \geq 1$. By Lemma 6, there exist $\rho(2^n) - 1$ $2^a \times 2^a$ matrices $A_1, A_2, \ldots, A_{\rho(2^n)-1}$, with entries 1, -1, 0 such that (a), (b), (c) of Lemma 6 are satisfied. Hence by (3)$\Rightarrow$(4) of Lemma 1, there exists a $\rho(2^a) \times 2^a$ matrix U with top row $(X_1, \ldots, X_{2^a})$ such that

$$UU^t = \sum_{i=1}^{2^a} X_i^2 I_{\rho(2^a)}.$$

Now use Lemma 7 to get the required T as

$$T = (U_1, U_2, \ldots, U_s) \qquad (U_1 = U)$$

and note that $\rho(2^a \cdot s) = \rho(2^a)$.

This completes the proof of Lemma 8. □

Proof of Theorem 14.1. For $p = \rho(n)$, the implication (4)$\Rightarrow$(5) of Lemma 1 gives the result, the last line of the enunciation of the theorem coming from Lemma 2.

If $p < \rho(n)$, we simply put $X_{p+1}, \ldots, X_{\rho(n)}$ equal to 0 in the identity

$$\left(\sum_{i=1}^{\rho(n)} X_i^2\right)\left(\sum_{j=1}^{n} Y_j^2\right) = \sum_{k=1}^{n} Z_k^2$$

to get what is required. □

Remark. The $(9, 16, 16)$ bilinear formula given on the second page of this chapter can be derived using Lemma 1 and the 9×16 matrix T of Lemma 3 as follows: write $T = X_1 B_1 + \ldots + X_{16} B_{16}$. Actually these X's should not be mixed up with the X's of the bilinear identity $(9, 16, 16)$; so let us call them $\xi_1, \ldots, \xi_{16}$ respectively so that

$$T(\underline{\xi}) = \xi_1 B_1 + \ldots + \xi_{16} B_{16}$$

These sixteen 9×16 matrices $B_1 \ldots, B_{16}$ are the B's of Lemma 1, (2).

From the proof of (2)⇒(1) of Lemma 1, construct the nine 16×16 matrices $A_1, \ldots, A_9$. This is a simple procedure, though occupying space, but the matrices $A_1, \ldots, A_9$ turn out to have most entries 0; indeed each has only sixteen non-zero entries (± 1 necessarily). Now use the proof (1)⇒ (5) of Lemma 1 and define $A = X_1 A_1 + \ldots + X_9 A_9$. Then

$$\underline{Z} = A\underline{Y}$$

($\underline{Y}$, $\underline{Z}$ are column vectors of order 16 and A is the required 16×16 matrix).

For more information and results on matrices with entries in $\{0, 1, -1\}$, coming from these Hurwitz-Radon ideas, see the book by Geramita and Seberry [G5].

Exercises

We have seen in Chapter 10 (Step 1, equation (10.3)) that the identity

$$(X_1^2 + \ldots + X_r^2)(Y_1^2 + \ldots + Y_n^2) = Z_1^2 + \ldots + Z_n^2$$

is equivalent to the existence of a family

$$\mathfrak{F} = (A_1, A_2, \ldots, A_{r-1})$$

of $n \times n$ matrices satisfying

(1) $A_j = -A_j^t$ $(j = 1, \ldots, r-1)$
(2) $A_i A_j = -A_j A_i$ $(i \neq j)$
(3) $A_j^2 = -I_n$ $(j = 1, \ldots, r-1)$.

We call such a family a Hurwitz-Radon (HR) family. The theorem of Hurwitz-Radon is then the following.

(a) The maximum r for which there is an HR family of $n \times n$ matrices is $\rho(n) - 1$, and conversely

(b) For $r = \rho(n) - 1$ there *is* such an HR family.

We give below a series of results (see [G4]) in the form of exercises to prove (b) of the above result. It will turn out that all matrices involved have as entries 0, 1, -1 only so that we get, in fact, the integral version of the Hurwitz-Radon theorem (b).

Let

$$A = \begin{pmatrix} 0 & 1 \\ -1 & 0 \end{pmatrix}, P = \begin{pmatrix} 0 & 1 \\ 1 & 0 \end{pmatrix}, Q = \begin{pmatrix} 1 & 0 \\ 0 & -1 \end{pmatrix}.$$

1. Prove that

(i) $\{A\}$ is an HR family of $\rho(2) - 1$ integral matrices of order 2.
(ii) $\{A \otimes I_2, P \otimes A, Q \otimes A\}$ is an HR-family of $\rho(4) - 1$ integral matrices of order 4.

(iii) $\{I_2 \otimes A \otimes I_2, I_2 \otimes P \otimes A, Q \otimes Q \otimes A, P \otimes Q \otimes A, A \otimes P \otimes Q, A \otimes P \otimes P, A \otimes Q \otimes I_2\}$ is an HR-family of $\rho(8) - 1$ integral matrices of order 8.

2. Suppose $\{M_1, \ldots, M_s\}$ is an HR family of integral matrices of order n. Show that

(i) $\{A \otimes I_n, Q \otimes M_i\ (i = 1, 2, \ldots, s)\}$ is an HR-family of $s + 1$ integral matrices of order $2n$.

(ii) If $\{L_1, \ldots, L_m\}$ is an HR-family of integral matrices of order k, show that

$$\{P \otimes I_k \otimes M_i (1 \leq i \leq s), Q \otimes L_j \otimes I_n (1 \leq j \leq m), A \otimes I_{nk}\}$$

is an HR-family of $s + m + 1$ integral matrices of order $2nk$.

(Hint: (i) and (iii) follow from $(A \otimes B)^t = A^t \otimes B^t$ while (ii) follows since the product of any two distinct members is skew-symmetric).

3. Show that Exercise 1 and Exercise 2, (i) give HR-families for $n = 2$, 4, 8, 16 and that Exercise 2, (ii) gives the transition from n to $16n$ with $k = 8$, $m = 7$.

Show also that the transition from $n = 2^a$ to $2^a b$ (b odd) is given by $\otimes I_b$.

Remark. This gives the integral version of the HR theorem part (b) over any integral domain D in which $2 \neq 0$.

4. From Theorem 1 of K.Y. Lam's paper [L1], work out the bilinear maps (ii), (v), (vi), (vii), (viii) and determine which of these are norm-preserving; when this is the case, write down the corresponding bilinear identities.

15

Artin-Schreier theory of formally real fields

A basic algebraic property of real numbers is that the only relations of the form $\sum a_i^2 = 0$ ($a_i \in \mathbb{R}$) are the trivial ones viz. $0^2 + \ldots + 0^2 = 0$. This observation led Artin and Schreier to call any field having this property *formally real*. We see that the definition of formally real fields inherently involves squares and indeed throughout the development of this theory, squares feature systematically into it. This is one of the reasons for including this topic in a book on squares.

Further, the theory of formally real fields led Artin (1927) to the solution of Hilbert's 17th problem which shows the additional connecting link between squares and the Artin-Schreier theory.

Note also that the classical application of Artin-Schreier theory is to the problem of determining which elements of a field are representable as sums of squares of elements of the field. For algebraic number fields (i.e. finite extension fields of the rationals $\mathbb{Q}$) the answer is the famous Hilbert-Landau theorem (Theorem 15.11 in this chapter).

We thus see that a chapter on Artin-Schreier theory fits well into the theme of squares. We proceed with the introduction of the main ingredients that appear in the development of this theory, namely order, real closure etc. There are many excellent books on this topic. Amongst the standard ones is [J1].

Definition 15.1. A field K is said to be *ordered* if a relation $>$ is defined on K that satisfies

(1) If $a, b \in K$, then either $a > b$ or $a = b$ or $b > a$;

(2) If $a > b$, $c \in K$, $c > 0$, then $ac > bc$;
(3) If $a > b$, $c \in K$, then $a + c > b + c$.

Equivalently we have

Definition 15.1′. K is ordered if a subset $P \subset K$ can be found such that

(1) $0 \notin P$;
(2) $P + P \subset P$, $P \cdot P \subset P$;
(3) $K = P \cup \{0\} \cup -P$. (disjoint union).

Clearly Definitions 15.1 and 15.1′ are equivalent, for say $a > b$ iff $a - b \in P$; conversely let $P = \{a \in K \mid a > 0\}$. The following are almost trivial consequences of the definition:

Corollary 15.1.

(a) *$P \neq \emptyset$ if $|K| \geq 2$.*
(b) *$P \cap -P = \emptyset$ for if $\alpha (\neq 0) \in P \cap -P$, then $\alpha \in P$, $\alpha \in -P$, i.e. $\alpha \in P$, $-\alpha \in P$ so $0 = \alpha + (-\alpha) \in P$ which is false as required.*
(c) *$(-P) + (-P) \subset -P$, $(-P)(-P) \subset P$.*
(d) *$\alpha^2 \in P$ for all $\alpha \in K^*$, for if $\alpha \in P$ then $\alpha^2 \in P$ by (2). If $\alpha \in -P$, then $-\alpha \in P$, so $(-\alpha)(-\alpha) \in P$ by (2) again, i.e. $\alpha^2 \in P$.*
(e) *All non-zero sums of squares $\in P$. This follows form (1) and (2) of Definition 15.1′.*
(f) *If P, P' are two orders of K such that $P \subset P'$ then $P = P'$. This follows from the definition.*

Definition 15.2. K is said to be *formally real* if -1 (or equivalently 0) cannot be expressed as a sum of non-zero squares in K.

Let $S(K)$ (or simply S when K is fixed) denote the set of all elements of K that are sums of squares of elements of K (including 0).

Corollary 15.2.

(a) *K ordered $\Rightarrow$ K formally real, for S^* (the non-zero elements of S) $\subset P$ by Corollary 15.1 (e), so no sum of squares can vanish.*
(b) char$K = p \Rightarrow K$ *not formally real, for then $1 + 1 + \ldots + 1$ (p times) $= 0$.*
(c) *S is closed under addition, multiplication and division (by non-zeros); closure under the first two are trivial. As for division, we have*

$$\sum a_i^2 / \sum b_j^2 = \sum a_i^2 \cdot \sum b_j^2 / (\sum b_j^2)^2$$
$$= \frac{\textit{a sum of squares}}{\textit{one square}} \in S.$$

(d) *If K is not formally real then $S = K$ if char $K \neq 2$ for then $-1 = a_1^2 + \dots a_r^2$ so if $\alpha \in K$, then*

$$\alpha = \left(\frac{\alpha+1}{2}\right)^2 - \left(\frac{\alpha-1}{2}\right)^2$$
$$= \left(\frac{\alpha+1}{2}\right)^2 + (a_1^2 + \dots + a_r^2)\left(\frac{\alpha-1}{2}\right)^2$$

which is a sum of $r+1$ squares and so in S. Hence $K \subseteq S$, and $K = S$.

If char $K = 2$, then $S = K$ need not hold, for example $K = \mathsf{F}_2(t)$.

We now prove the following important

Theorem 15.1. *K can be ordered iff K is formally real.*

First we have the following.

Lemma 1. *Let G be a subgroup of K^* such that*

(i) $G + G \subset G$

(ii) *G contains all non-zero squares of elements of K (i.e. $G \supset K^{*^2}$). Let $-a(\neq 0) \notin G$ and let*

$$\Gamma = G + aG = \{x + ay \mid x, y \in G\}.$$

Then Γ is a subgroup of K^*, closed under $+$ and $\supset G$.

Proof. Let x, y, x', y' be elements of G; then

$$(x+ay)\cdot\frac{1}{(x'+ay')} = (x+ay)(x'+ay')/(x'+ay')^2$$
$$= (xx' + a^2yy')/(x'+ay')^2 + a(xy' + x'y)/(x'+ay')^2$$
$$\in G$$

since x, x', y, $y\prime$ and all squares are in G. Also

$$(x+ay) + (x'+ay') = (x+x') + a(y+y') \in \Gamma.$$

Note also that $0 \notin \Gamma$ for if $0 = x + ay$ (x, $y \in G$) then $-a = x/y \in G$ - a contradiction.

Finally to show that $\Gamma \supset G$ we proceed as follows: since Γ is a subgroup of K^* so $1 \in \Gamma$, say $1 = x + ay$ (x, $y \in G$). Then if $g \in G$, we get $g = gx + gay \in \Gamma$ as required. □

Proof of Theorem 15.1. Let $\mathcal{F}$ be the set of all subgroups of K^* closed under addition and containing all squares. Then $S^* \in \mathcal{F}$, so $\mathcal{F}$ is non-empty.

By Zorn's Lemma, there exists a maximal element, G say, of $\mathcal{F}$. We show that G defines an order in K. We must show $K = G \cup 0 \cup -G$ (disjoint). For if not, then there is an $a \in K^*$ such that $a \notin G$, $-a \notin G$. By the Lemma, $\Gamma = G + aG \in \mathcal{F}$, so by the maximality of G, $\Gamma = G$. However, $1 \in G$, so $1 + a \cdot 1 \in \Gamma$, i.e. $1 + a \in \Gamma$, but $a \notin \Gamma$ (since $a \notin G$) so $a(1+a) \notin \Gamma$, i.e. $a^2 + 1 \cdot a \notin \Gamma$; but $a^2 \in G = \Gamma$ so $a^2 + 1 \cdot a \in \Gamma$; a contradiction. Thus $K = G \cup 0 \cup -G$. □

Example 1. $\mathbf{Q}$ has only one order viz. $P = \{\alpha \mid \alpha > 0\}$ ($\alpha > 0$ in the usual sense).

Proof. P is clearly an order. If P' is any other order, then since all squares are in P' and each positive rational (positive in the P-sense) is a sum of four rational squares, so each positive rational is in P', i.e. $P \subset P'$; but then by Corollary 15.1, (f), $P = P'$; so P is unique.

Example 2. $\mathbf{Q}(\sqrt{2})$ has the following two orders:

$$P_1 = \{a + b\sqrt{2} \mid a, b \in \mathbf{Q}, a + b\sqrt{2} > 0 \quad \text{in the usual sense}\}$$
$$P_2 = \{a + b\sqrt{2} \mid a, b \in \mathbf{Q}, a - b\sqrt{2} > 0 \quad \text{in the usual sense}\}.$$

and no others.

Proof. It is easy to check that these are certainly orders. We now show that there are no more.

Let σ be the automorphism $\sqrt{2} \to -\sqrt{2}$. First note that for any order P, σP is also an order:

$$\sigma P = \{\sigma\alpha \mid \alpha \in P\}.$$

(i) $0 \in \sigma P \Rightarrow \sigma^{-1} 0 \in P$, i.e. $0 \in P$ which is false.

(ii) Let $\sigma\alpha, \sigma\beta \in \sigma P$. Then $\sigma\alpha \pm \sigma\beta = \sigma(\alpha \pm \beta) \in \sigma P$ since $\alpha \pm \beta \in P$ as $\alpha, \beta \in P$.

(iii) Let $\alpha \in Q(\sqrt{2})^*$ say $\alpha = a + b\sqrt{2} \neq 0$. If $\alpha \notin \sigma P$ then $\sigma^{-1}\alpha \notin P$ so $\sigma^{-1}\alpha \in -P$, i.e. $\alpha \in \sigma(-P) = -\sigma P$.

□

Thus our orders P_1, P_2 are related by $P_2 = \sigma P_1$. Suppose now P is any order of $\mathbf{Q}(\sqrt{2})$. Then σP is also an order and by Example 1, both P and σP contain all the positive rationals.

Now if $\sqrt{2} \in P$, well and good, otherwise $-\sqrt{2} \in P$, so $\sigma(-\sqrt{2}) \in \sigma P$, i.e. $\sqrt{2} \in \sigma P$, and without loss of generality, $\sqrt{2} \in P$ (otherwise consider σP). We shall show that $P_1 \subset P$; then by Corollary 15.1, (f), $P_1 = P$. So let $a + b\sqrt{2} \in P_1$, i.e. $a + b\sqrt{2} > 0$, in the usual sense.

Case 1. $a,\ b > 0$. Then $a,\ b \in P$ (being rational) and $\sqrt{2} \in P$, so $a + b\sqrt{2} \in P$, since both $P + P \subset P$ and $P \cdot P \subset P$.

Case 2. $a < 0,\ b > 0$. Suppose $a + b\sqrt{2} \notin P$. Then $-a - b\sqrt{2} \in P$. But $-a+b\sqrt{2} \in P$, since $-a,\ b,\ \sqrt{2} \in P$, as they are all positive; so $a^2 - 2b^2 \in P$. Here $a + b\sqrt{2} > 0$, in the usual sense, as it is in P_1, i.e. $b\sqrt{2} > -a > 0$ or $b\sqrt{2} > |a|$. In other words $2b^2 > a^2$. Thus $0 > a^2 - 2b^2 \in P$ while $1/(2b^2 - a^2) \in P$ since it is a positive rational. Hence their product is $-1 \in P$. But $1 \in P$, so $-1 \in -P$. Thus $-1 \in P \cap -P$ which is false.

Case 3. $a > 0,\ b < 0$. Suppose $a + b\sqrt{2} \notin P$. Then $-a - b\sqrt{2} \in P$, but $a - b\sqrt{2} \in P$ since $a,\ -b,\ \sqrt{2} \in P$. So $0 > 2b^2 - a^2 \in P$ since $a + b\sqrt{2} > 0$, as it is in P_1; i.e. $a > -b\sqrt{2} > |b|\sqrt{2}$ because $b < 0$, so $a^2 > 2b^2$; while $1/(a^2 - 2b^2) \in P$, since it is a positive rational. So their product is $-1 \in P$ and as in Case 2, we get a contradiction.

Finally note that the case $a < 0,\ b < 0$ does not arise since then $a+b\sqrt{2} < 0$.

Thus in all cases $P_1 \subset P$ so $P_1 = P$. □

Definition 15.3. An element $\alpha \in K$ (ordered field) is said to be ***totally positive*** if α is positive in every ordering of K, i.e. $\alpha \in P$ for all order subsets P of K.

Remark. If K has no order, every α is taken as (vacuously) totally positive.

Thus for example, $3 + \sqrt{2}$ is totally positive in $\mathbb{Q}(\sqrt{2})$ since $3 + \sqrt{2} > 0$ and $3 - \sqrt{2} > 0$; where as $1 + \sqrt{2}$ is not totally positive since although $1 + \sqrt{2} > 0$, $1 - \sqrt{2} \not> 0$.

Totally positive elements of an ordered field are characterized by the following beautiful property.

Theorem 15.2. *Let K be an ordered field. An element $\alpha \in K$ is totally positive iff α is a sum of squares in K.*

First note the following: let L/K be an extension of fields. Suppose L is ordered by the subset $P \subset L$. Then the subset $P \cap K \subset K$ orders K (check). This order is said to be induced on K by the order P of L while the order P of L is called an extension of the order $P \cap K$ of K. It may be possible to extend an order in K to several orders in L, e.g. $\mathbb{Q}(\sqrt{2})/\mathbb{Q}$. Note also that if $\alpha \in K$ is positive as an element of P in L, then α is positive as an element of $P \cap K$ in K.

Proof of Theorem 15.2. First let α be a sum of squares. Since each order

P of K contains all sums of squares, so $\alpha \in P$ for all P, i.e. α is totally positive.

Conversely let α be totally positive. We must show that α is a sum of squares. Let $L = K(\sqrt{-a})$. If L is formally real (i.e. ordered, say by a subset P) then $-\alpha = (\sqrt{-\alpha})^2 > 0$, being a square in L, and so $-\alpha > 0$ in K also with respect to the restricted order $P \cap K$ of K, i.e. α is not totally positive - a contradiction. It follows that L is not formally real and so we have a relation

$$-1 = \sum_j \beta_j^2 \qquad (\beta_j \in L) \tag{15.1}$$

If $\sqrt{-\alpha} \in K$, then $L = K$ so $\beta_j \in K$ and we have

$$\alpha = \left(\frac{\alpha+1}{2}\right)^2 - \left(\frac{\alpha-1}{2}\right)^2 = \left(\frac{\alpha+1}{2}\right)^2 + \left(\frac{\alpha-1}{2}\right)^2 \sum \beta_j^2$$

which is a sum of squares in K as required.

If $\sqrt{-\alpha} \notin K$, then L is a quadratic extension of K and (15.1) gives

$$-1 = \sum_j (a_j + b_j\sqrt{-\alpha})^2 \qquad (a_j, b_j \in K),$$

i.e.

$$-1 = \sum a_j^2 - \alpha \sum b_j^2 + 2\sqrt{-\alpha} \sum_j a_j b_j.$$

But 1, $\sqrt{-\alpha}$ are linearly independent $/K$ and so

$$1 + \sum a_j^2 = \alpha \sum b_j^2, \qquad \sum a_j b_j = 0.$$

Hence

$$\alpha = \frac{1 + \sum a_j^2}{\sum b_j^2} = \frac{(1 + \sum a_j^2)(\sum b_j^2)}{(\sum b_j^2)^2}$$

which is a sum of squares in K. □

Basic Lemma. *Suppose K is formally real, but that the quadratic extension $K(\sqrt{a})$ is not; then $-a \in S(K)$.*

Proof. In $K(\sqrt{a})$ we have an equation

$$-1 = \sum_{j=1}^{n} (a_j + b_j\sqrt{a})^2 \qquad (a_j, b_j \in K).$$

Then as above $-1 = \sum a_j^2 + a \sum b_j^2$. Here $\sum b_j^2 \neq 0$ since K is formally real, hence

$$-a = \frac{1 + \sum a_j^2}{\sum b_j^2} = \frac{(1 + \sum a_j^2)(\sum b_j^2)}{(\sum b_j^2)^2}$$

which is a sum of squares in $K \in S(K)$. □

Definition 15.4. A field K is called *real closed* if:

(1) K is formally real;
(2) no proper algebraic extension of K is formally real.

The field $\mathbf{R}$ of real numbers is, of course, the most basic example of a real closed field. Are there other examples? The ubiquity of real closed fields is demonstrated by the following.

Remark. *Let K be formally real and let Ω be an algebraic closure of K. Then Ω contains a real closed field R containing K.*

Proof. Let $\mathcal{F}$ be the collection of all formally real subfields of Ω containing K. Now $\mathcal{F}$ is non-$\emptyset$ since $K \in \mathcal{F}$. By Zorn's lemma, $\mathcal{F}$ has a maximal element; call it R. If R is not real closed, it has a proper algebraic extension R' which is formally real and $R' \subset \Omega$, since Ω is algebraically closed. This contradicts the maximality of R in $\mathcal{F}$. Hence R is real closed. □

Our object now is the following.

Theorem 15.3. *If K is real closed then K has a unique order.*

Proof. It is enough to prove that for each non-zero α in K, exactly one of the following holds:

$$\begin{aligned} &\text{either} \quad \alpha \quad \text{is a square in } K \\ &\text{or} \quad -\alpha \quad \text{is a square in } K. \end{aligned} \tag{15.2}$$

This is because once (15.2) is proved and if P is any order of K, then

$$P \supset \{\alpha^2 | \alpha \in K^*\} = S^*(K)$$

and if $\alpha \in P$ and α is not a square, then by (15.2), $-\alpha$ is, i.e. $-\alpha \in P$ so $\alpha \in -P$ which is a contradiction. Thus $\alpha \in P \Rightarrow \alpha$ is a square. It follows that $P = S^*(K)$. So P is unique.

Now we prove (15.2). If α is a square in K well and good; otherwise let $L = K(\sqrt{\alpha})$. L is a proper algebraic extension of K, it is not formally real, i.e. there exist β_i, $\gamma_i \in K$ (not all zero) such that

$$0 = \sum (\beta_i + \gamma_i \sqrt{\alpha})^2$$

giving $\sum \beta_i^2 + \alpha \sum \gamma_i^2 = 0$, $\sum \beta_i \gamma_i = 0$. Now K is formally real so $\sum \gamma_i^2 \neq 0$. Hence

$$-\alpha = \frac{\sum \beta_i^2}{\sum \gamma_i^2} = \frac{\sum \beta_i^2 \sum \gamma_i^2}{(\sum \gamma_j^2)^2} = \quad \text{a sum of squares in } K.$$

Thus what we have proved is the following.

If α is not a square in K, then $-\alpha$ is a sum of squares in K. (15.3)

If now α is a sum of squares in K, then

$$-1 = \frac{-\alpha}{\alpha} = \sum \text{ squares}/\sum \text{ squares } = \sum \text{squares}$$

which contradicts the formal reality of K. Thus α is *not* a sum of squares in K. So what we have proved is that if α is not a square then α is not a sum of squares, i.e. that

$$\textit{if } \alpha \textit{ is a sum of squares then } \alpha \textit{ is a square.} \qquad (15.4)$$

(15.3) and (15.4) give that if α is not a square then $-\alpha$ is. □

Remarks.

1. Theorem 15.3 exhibits the structure of real closed fields viz. if K is real closed, then
 (i) for any $x \in K^*$, either $x \in {K^*}^2$ or $-x \in {K^*}^2$;
 (ii) K has a unique order in which $P = {K^*}^2$.
2. If σ is an automorphism of K, then σ preserves order. For $a > b \Rightarrow a - b \in P \Rightarrow a - b = c^2 \Rightarrow \sigma(a-b) = \sigma(c^2) \Rightarrow \sigma a - \sigma b = (\sigma c)^2 \in P \Rightarrow \sigma a > \sigma b$.

The following two properties of real closed fields R are important and useful.

(1) *Positive elements of R have square roots in R, where positive refers to the unique order $P = \{\alpha^2 | \alpha \in R^*\}$ of R.*

(2) *Every polynomial $f(X)$ of odd degree, with coefficients in R, has a root in R.*

Proof. (1) Let $x \in P$, so $x = \alpha^2$ $(\alpha \in R^*)$ i.e. $\sqrt{x} = \alpha \in R$.
(2) Use induction on the degree n of $f(X)$. The result is clear for $n = 1$. If $f(X)$ is reducible over R, one of its factors is of odd degree at most n and so has a root β in R by the induction hypothesis and this β is the required root of $f(X)$ in R.

So suppose $f(X)$ is irreducible over R. We shall get a contradiction as follows: let $L = R(\theta)$, θ a root of $f(X)$. Since L is an algebraic extension of R it is not formally real. So we have a relation

$$-1 = \sum \phi_i^2(\theta) \quad (\deg \phi_i(X) \leq n-1).$$

Then

$$1 + \sum \phi_i^2(X) = f(X) \cdot g(X) \qquad (15.5)$$

Comparing degrees, we get $\deg f + \deg g \leq 2(n-1)$ and even. It follows that $\deg g \leq n-2$ and odd. Hence by the induction hypothesis, $g(X)$ has a root β in R. Putting $X = \beta$ in (15.5) gives $1 + \sum \phi_i^2(\beta) = 0$ which contradicts the formal reality of R, noting that not all $\phi_j(\beta)$ are zero; for otherwise we get $-1 = 0$. □

Remark. A well known result in quadratic forms known as Springer's theorem states that if a form $f(X_1, \dots, X_n)$ is anisotropic over a field K and L is an extension of K of odd degree, then f is anisotropic over L. In Theorem 3.7 we gave a proof of this result for the diagonal form $f = X_1^2 + \dots + X_n^2$. The general case is no more difficult.

Using this we see that if K is formally real, so is every odd degree extension L of K. Furthermore, a proof of Property (2) above can be easily deduced as follows; the result is trivial if $n = 1$ so let $n \geq 3$. We claim that $f(X)$ must be reducible in $R[X]$, for otherwise $R(X)/(f(X))$ is a proper odd degree extension field of R and so is formally real as proved just above, contradicting the fact that R is real closed. So let $f = f_1 f_2$ be a proper factorization of f where, say $\deg f_1$ is odd and less than $\deg f$. By the induction hypothesis, f_1 has a root in R, hence so has f. □

Theorem 15.4. *Let K be an ordered field with respect to a fixed order in K, such that*

(1) *Positive elements have square roots in K.*
(2) *Any polynomial of odd degree $\in R[X]$ has a root in K.*

Then $\sqrt{-1} \notin K$ and $K(\sqrt{-1})$ is algebraically closed.

Proof. If $-1 = \alpha^2$ ($\alpha \in K$) then K is not formally real so not ordered. But it is given to be ordered so $\sqrt{-1} \notin K$.

Let $f(X) \in K(\sqrt{-1})[X]$ be any polynomial with coefficients in $K(\sqrt{-1})$. We have to prove that $f(X)$ has a root in $K(\sqrt{-1})$. Let $\sqrt{-1} = i$. Without loss of generality we may suppose $f(X) \in K[X]$; for consider the polynomial $f(X) \cdot \overline{f}(X) \in K[X]$ (where the 'bar' carries each coefficient of f to its conjugate, i.e. i to $-i$). Suppose $f(X) \cdot \overline{f}(X)$ is shown to have a root β in $K[X]$. If β is a root of $f(X)$, well and good; if β is a root of $\overline{f}(X)$, then $\overline{\beta}$ will be a root of $f(X)$. So $f(X)$ always has a root in $K(i)$ as required.

Next we show that if $\beta \neq 0$ every element $\alpha + i\beta$ of $K(i)$ has a square root $\xi + i\eta$ in $K(i)$: for we simply want to solve $(\xi + i\eta)^2 = \alpha + i\beta$, i.e. $\xi^2 - \eta^2 = \alpha$, $2\xi\eta = \beta$, or $\xi^2 - \beta^2/4\xi^2 = \alpha$, on eliminating η, i.e. $4\lambda^2 - 4\alpha\lambda - \beta^2 = 0$ with $\lambda = \xi^2$. The two solutions of this are $\{\alpha \pm \sqrt{(\alpha^2 + \beta^2)}\}/2$. Now by hypothesis (1) on K, since $\alpha^2 + \beta^2$ is 'positive', $\sqrt{\alpha^2 + \beta^2} \in K$ so both $\{\alpha \pm \sqrt{(\alpha^2 + \beta^2)}\}/2 \in K$. The 'positive' root is again positive, for clearly we have $\sqrt{\alpha^2 + \beta^2} > |\alpha|$, so $\sqrt{(\alpha^2 + \beta^2)} - |\alpha| > 0$, hence $\sqrt{(\alpha^2 + \beta^2)} + \alpha > 0$ certainly. Then

$$\xi^2 = \{\sqrt{(\alpha^2 + \beta^2)} + \alpha\}/2 > 0$$

and hence by (1), $\xi = \sqrt{\{\sqrt{(\alpha^2+\beta^2)}+\alpha\}/2} \in K$ and then $\eta = \beta/2\xi \in K$ also, as required.

Thus we have the following to note:

(a) that there is no extension field L of $K(i)$ of degree 2;

(b) that K has no extension of odd degree (by Hypothesis (2) in the theorem).

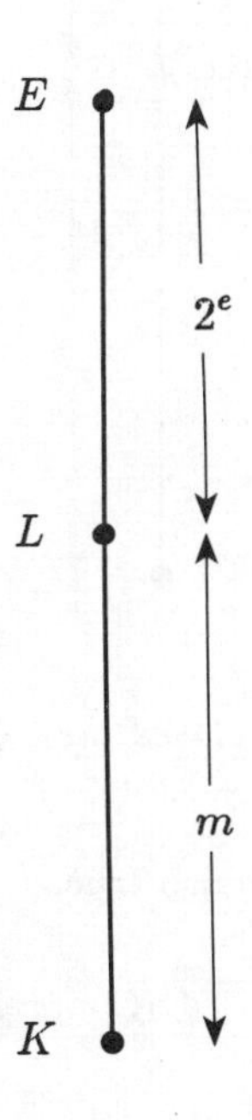

Let now $f(X) \in K[X]$ and we shall show that $f(X)$ has a root in $K(i)$. Let $E \supset K(i)$ be a splitting field of $(X^2+1)\cdot f(X)/K$. Let $G = \text{Gal}(E/K)$ and let $o(G) = 2^e \cdot m$, where m is odd. By Sylow's theorem, there exists a subgroup H of G of order 2^e. Let L be the fixed field of H so that the degrees are as shown.

By (b) above, $L = K$, i.e. $m = 1$, so $o(G) = 2^e = o(H)$. Now if $e > 1$, then $o(G(L/K(i))) = 2^{e-1}$. Let H be a subgroup of $G(L/K(i))$ such that $o(H) = 2^{e-2}$. The fixed field of H is a quadratic extension of $K(i)$. But such extensions do not exist. Hence $e = 1$ and the theorem is proved. □

In Theorem 15.4, if K is real closed, then both the hypotheses are satisfied as we have already verified before the theorem. Hence we have the following important corollary.

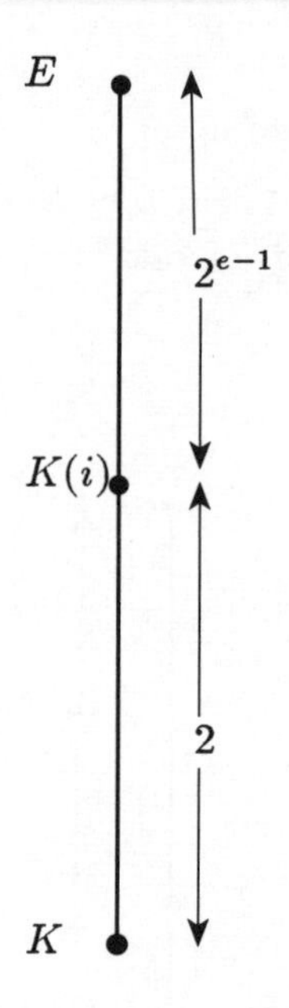

Theorem 15.5. *If K is real closed then $\sqrt{-1} \notin K$ and $K(\sqrt{-1})$ is algebraically closed.*

The converse of this result is also true.

Theorem 15.6. *Suppose $\sqrt{-1} \notin K$ and $K(\sqrt{-1})$ is algebraically closed, then K is real closed.*

Proof. First note that if $f(X) \in K[X]$ is irreducible then $\deg f = 1$ or 2 for $K(\theta) \subset K(\sqrt{-1})$ for any root θ of $f(X)$, since $K(\sqrt{-1})$ is algebraically closed, so θ is of degree 1 or 2 as claimed.

We now show that if α, $\beta \in K$, $(\beta \neq 0)$, then $\alpha^2 + \beta^2$ is a square in K: Consider the polynomial $g(X) = (X^2 - \alpha)^2 + \beta^2 \in K[X]$. This equals

$$(X - (\alpha + i\beta)^{1/2})(X + (\alpha + i\beta)^{1/2})(X - (\alpha - i\beta)^{1/2})(X + (\alpha - i\beta)^{1/2}).$$

Since $\beta \neq 0$ none of $\pm\alpha \pm i\beta \in K$, so $g(X)$ splits into two quadratic factors got by pairing the four linear factors above, two at a time. The linear factor pairing with $(X - (\alpha + i\beta)^{1/2})$ cannot be $(X + (\alpha + i\beta)^{1/2})$ since their product $= X^2 - (\alpha + i\beta) \notin K[X]$. So $(X - (\alpha + i\beta)^{1/2})$ must pair with one of $(X \pm (\alpha - i\beta)^{1/2})$, i.e. one of $(X - (\alpha + i\beta)^{1/2})(X \pm (\alpha - i\beta)^{1/2}) \in K[X]$. In either event $(\alpha^2 + \beta^2)^{1/2} \in K$, i.e. $\alpha^2 + \beta^2$ is a square in K as required. Thus a sum of two squares in K is a square in K. So by induction any sum of squares is a square. But -1 is not a square in K so it is not a sum of squares in K, i.e. K is formally real. Now any irreducible $f(X) \in K[X]$ is of degree 1 or 2, so a proper algebraic extension of K (E say) is of degree 2 and $\subset K(\sqrt{-1})$, since $K(\sqrt{-1})$ is algebraically closed. So $E = K(\sqrt{-1})$ which is not formally real. Thus K is real closed. □

We have now two characterizations of real closed fields:

(1) *K is real closed iff $\sqrt{-1} \notin K$ and $K(\sqrt{-1})$ is algebraically closed.*

(2) *K (supposed ordered) is real closed iff positive elements of K have a square root in K and any polynomial if odd degree $\in K[X]$ has a root in K.*

Alternatively the condition "positive elements of K have a square root in K" can be written $|K^*/K^{*^2}| = 2$ since for real closed fields K, denoting by P the set of positives, we have $P = K^{*^2}$.

Theorem 15.7. *Let R be a real closed extension field of K. Let $A = \{\alpha \in R | \alpha$ is algebraic $/K\}$. Then A is real closed.*

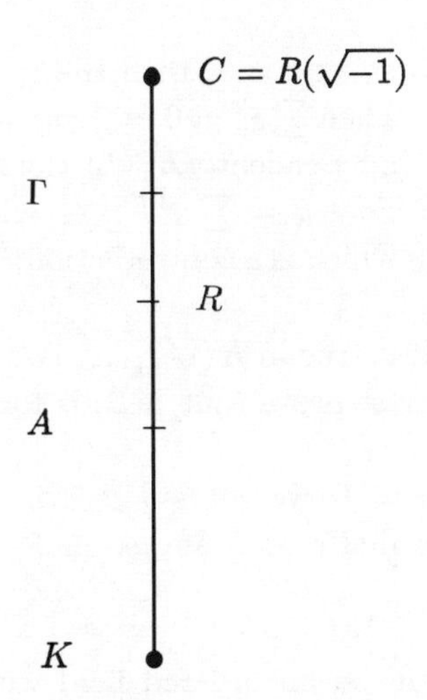

Proof. Let $C = R(\sqrt{-1})$ so that C is algebraically closed. Let

$$\Gamma = \{\beta \in C | \beta \text{ is algebraic } /K\}.$$

We claim that Γ is an algebraically closed field. For let $f(X) \in K[X]$ be any polynomial. $f(X) \in C[X]$ too, and so has a root ρ in C since C is algebraically closed. Now ρ is algebraic over Γ and Γ is algebraic over K, so ρ is algebraic over K, i.e. $\rho \in \Gamma$.

What do the elements of Γ look like?

They are elements of C and so of the form $\alpha + \sqrt{-1} \cdot \beta$ $(\alpha, \beta \in R)$. Since $\alpha + \sqrt{-1} \cdot \beta \in \Gamma$, so does $\alpha - \sqrt{-1}\beta$, clearly . Hence

$$\alpha = \frac{1}{2}\{\alpha + \sqrt{-1}\beta + \alpha - \sqrt{-1}\beta\} \in \Gamma,$$

so

$$\beta = (\alpha + \sqrt{-1}\beta - \alpha)\sqrt{-1} \in \Gamma,$$

i.e. α, β are algebraic over K, but $\alpha, \beta \in R$, so $\alpha, \beta \in A$. Thus elements of Γ are of the form $\alpha + \sqrt{-1}\beta$ where $\alpha, \beta \in A$, i.e. $\Gamma = A(\sqrt{-1})$ and since $\sqrt{-1} \notin R$ so $\sqrt{-1} \notin A$. This implies A is real closed. □

Theorem 15.8. *Let K be ordered and let Ω be an algebraic closure of K. Let E be the subfield of Ω obtained by adjoining to K, the square roots of all the positive elements of K; then E is formally real.*

Proof. Suppose we have a relation $\sum \xi_i^2 = 0$ in E. The ξ_i all belong to a finite extension of K, say $K(\sqrt{\beta_1}, \ldots, \sqrt{\beta_r})$ (β_j positive elements of K). So it is enough to prove that every field of the form $K(\sqrt{\beta_1}, \ldots, \sqrt{\beta_r})$, $\beta_j > 0$ in K, is formally real, i.e. that $\sum \xi_i^2 = 0$, $\xi_i \in K(\sqrt{\beta_1}, \ldots, \sqrt{\beta_r}) \Rightarrow \xi_i = 0$ for all i.

Use induction on r. First let $r = 1$, then the $\xi_i \in K(\sqrt{\beta})$, $\beta > 0$. Write $\xi_i = \eta_i + \beta\zeta_i$ $(\eta_i, \zeta_i \in K)$. Then $\sum \xi_i^2 = 0 \Rightarrow \sum \eta_i^2 + \beta \sum \zeta_i^2 = 0$, $\sum \eta_i \zeta_i = 0$. Since 1, $\sqrt{\beta}$ are linearly independent $/K$. In the first relation, $\sum \zeta_i^2 \neq 0$, since K is formally real. So $-\beta = \sum \eta_i^2 / \sum \zeta_i^2$ which is a sum of squares and so positive, i.e. $\beta < 0$ which is a contradiction. Thus $K(\sqrt{\beta})$ is formally real.

Now suppose the result is true in $K(\sqrt{\beta_1}, \ldots, \sqrt{\beta_{r-1}}) = L$. So suppose L is formally real and we must prove that $L(\beta)$ is formally real, where $\beta > 0$ (in K).

Now as for the case $r = 1$, we get $-\beta > 0$ in L (if we assume to the contrary that L is not formally real) and so in K, by the induced order – which is a contradiction. □

Definition 15.5. Let K be an ordered field with positive elements P. An extension field R of K is called a *real closure* of K (relative to P) if it satisfies the following three conditions:

(i) R is real closed;
(ii) R is algebraic over K (not necessarily finite);
(iii) The (unique) ordering of R is an extension of that of K (i.e. $P = R^{*^2} \cap K$).

We have the following existence and uniqueness result.

Theorem 15.9.

(i) *Every ordered field K possesses a real closure relative to the given ordering.*

(ii) *If K_1, K_2 are ordered fields and $\mathbf{R}_1$, $\mathbf{R}_1$ their real closures relative to the respective orderings, then any order isomorphism $f : K_1 \to K_2$ extends uniquely to an isomorphism $\tilde{f} : \mathbf{R}_1 \to \mathbf{R}_2$ (which is automatically an order isomorphism).*

Remarks.

1. This result implies that for each distinct ordering of K there exists a unique (up to order isomorphism) real closed field algebraic over K, for, if $\mathbf{R}_1$, $\mathbf{R}_2$ are two real closures of K, take $f : K \to K$ to be the identity isomorphism, then $\tilde{f} : \mathbf{R}_1 \to \mathbf{R}_2$ is an order isomorphism, i.e. $\mathbf{R}_1$ is order isomorphic to $\mathbf{R}_2$.

2. We present here only the proof of the existence. The uniqueness requires, classically, a technical theorem of Sturm and as this chapter is merely an introduction to the Artin-Schreier theory, we only refer the reader to Jacobson's *Algebra* Vol. III (or *Basic Algebra*, Vol. II). Recently, however, Knebusch [K2] has found an elegant proof of this uniqueness result which avoids Sturm's theorem.

Proof of the existence of real closures. Let Ω be an algebraic closure of K and let E be the subfield of Ω got by adjoining to K, the square roots of all the positive elements of K. By Theorem 15.8, E is formally real and of course Ω is an algebraic closure of E also. Hence there exists a real closed field $R : E \subset R \subset \Omega$ (see the Remark after Definition 15.4). This R is the required real closure of K for conditions (i) and (ii) are clearly satisfied. Indeed Theorem 15.8 and E need not even be brought into the picture for this.

As for condition (iii), let $\beta \in K$, $\beta > 0$ in K. Then $\sqrt{\beta} \in E$ (definition of E), i.e. $\beta = \rho^2$ $(\rho \in E)$. But $E \subset R$ so with $\rho \in R$, $\beta = \rho^2 > 0$ in R. Thus positive elements of E are positive in R. This proves (iii). □

We can apply these results to the field $\mathbf{Q}$ of rational numbers and prove the following beautiful theorem.

Theorem 15.10. *Let K be an algebraic number field with r real conjugates. Then K can be ordered in precisely r distinct ways.*

Proof. Let $K = \mathbf{Q}(\theta)$ and let $\operatorname{irr}(\theta, \mathbf{Q}) = f(X)$ be of degree n having r real zeros $\theta = \theta_1, \theta_2, \ldots, \theta_r$ and $n - r = 2s$ complex zeros. Let σ_i $(1 \le i \le r)$ be the isomorphisms of $K/\mathbb{Q}$ into $A/\mathbb{Q}$ (A being the field of all real algebraic numbers) given by $\sigma_i(\theta) = \theta_i$; indeed the elements of K are polynomials $g(\theta)$ of degree at most $n - 1$, so that σ_i carries $g(\theta)$ to the element $g(\theta_i)$ of $K_i = \mathbb{Q}(\theta_i)$. The ordering of K_i $(\subset A)$ imposed by the unique ordering of A provides an ordering $>_i$ of K, viz. for $\rho \in K$, say $\rho > 0$ iff $\sigma_i(\rho) >_i 0$ in K_i. Thus for each $i = 1, 2, \ldots, r$, we get an ordering $>_i$ of K. We now show

(a) any order of K coincides with one of these $>_i$;
(b) $>_1, >_2, \ldots, >_r$ are all distinct.

So suppose $>$ is an ordering of K and let $\mathcal{A}$ be a real closure of K relative to it. But K is algebraic over $\mathbb{Q}$, so $\mathcal{A}$ is a real closure of $\mathbb{Q}$ (check that it satisfies the three conditions of Definition 15.5). Hence by the uniqueness, we have an order isomorphism $\sigma : \mathcal{A}/\mathbb{Q} \to A/\mathbb{Q}$. The restriction of σ to K coincides with one of the σ_i and so the given order of K is the same as $>_i$. This proves (a).

Finally suppose $>_i$ and $>_j$ give the same ordering of K. Then we have an order preserving isomorphism $f : \mathbb{Q}(\theta_i) \to \mathbb{Q}(\theta_j)$ taking $\theta_i \to \theta_j$. But A is a real closure of $\mathbb{Q}(\theta_i)$ as well as of $\mathbb{Q}(\theta_j)$, so by Theorem 15.9, there exists an automorphism

$$\tilde{f} : A/\mathbb{Q} \to A/\mathbb{Q} \quad \text{taking} \quad \theta_i \to \theta_j.$$

However, since A is a real closure of $\mathbb{Q}$, we see by Theorem 15.9 that the identity is the only automorphism of $A/\mathbb{Q}$. It follows that $\theta_i = \theta_j$, which proves (b). □

Combining Theorems 15.10 and 15.2 we get the following.

Theorem 15.11 (Hilbert-Landau). *Let K be an algebraic number field with r real conjugates and let $\sigma_1, \ldots, \sigma_r$ be the r different isomorphisms of $K/\mathbb{Q}$ into the field A of real algebraic numbers. Then a non-zero element ρ of K is a sum of squares in K if and only if $\sigma_j(\rho) > 0$ for all $j = 1, 2, \ldots, r$.*

At the 1900 Paris International Congress of Mathematicians, Hilbert gave a list of 23 unsolved problems, the 17th of which is the following.

Let $f(X_1, \ldots, X_n)$ be a rational function (i.e. a polynomial divided by a polynomial) in the n independent indeterminates $X_1, \ldots, X_n$, with rational coefficients:

$$f(X_1, \ldots, X_n) \in \mathbb{Q}(X_1, \ldots, X_n);$$

if f is positive semi-definite, i.e. $f(a_1, \ldots, a_n) \ge 0$ for all real $a_1, \ldots, a_n$ for which $f(a_1, \ldots, a_n)$ is defined, then when is it a sum of squares in $\mathbb{Q}(X_1, \ldots, X_n)$?

In 1927, making essential use of Artin-Schreier theory and Sturm's theorem, Artin gave an affirmative answer to Hilbert's 'conjecture', by proving the following stronger result.

Theorem 15.12. *Let K be a subfield of the field $\mathbb{R}$ of real numbers having a unique ordering and let $f(X_1, \ldots, X_n) \in K(X_1, \ldots, X_n)$ be such that for all $a_i \in K$ for which $f(a_1, \ldots, a_n)$ is defined, we have $f(a_1, \ldots, a_n) \geq 0$ (we do not demand this for all $a_i \in \mathbb{R}$). Then $f(X_1, \ldots, X_n)$ is a sum of squares in $K(X_1, \ldots, X_n)$.*

Remarks.

1. Such fields K are not at all rare, for example, in addition to $\mathbb{Q}$ or A, K could be any algebraic number field having only one real conjugate.
2. There is no mention in Artin's result of how many squares are needed for such a representation. In Chapters 4 and 5 we described Pfister's solution to this problem where he shows that 2^n squares suffice.
3. Artin's original proof of Theorem 15.12 uses uniqueness of real closures and Sturm's theorem. Since these latter results are not covered here, there is no point giving his proof of Theorem 15.12. The reader is referred to [A6] or [J1].
4. Recently Knebusch [K2] has given an elegant proof of Artin's result avoiding the use of Sturm's theorem.

Exercises

1. (i) Show that if K is formally real then so is $K(X)$, where X is transcendental over K.

(ii) Prove that $K(X_1, \ldots, X_n)$ is formally real, where $X_1, \ldots, X_n$ are independent transcendentals over K.

2. Let $\mathbb{Q}$ be the field of rationals and let $K = \mathbb{Q}(X)$. Show that K has a non-countable number of distinct orderings.

3. (Cohn) An ordered field K is called archimedean if given $\alpha > 0$, $\beta > 0$ there exists an integer n such that $n\alpha > \beta$ (equivalently given $\alpha > 0$ there exists an integer n such that $n > \alpha$).

Let K be an ordered field with P as the set of positive elements. Show that $K(X)$ can be ordered as follows. For any element $r(X)$ of $K(X)$, write $r(X) = \alpha \cdot X^n \cdot f(X)/g(X)$, with $\alpha \in K$, $f(X)$, $g(X) \in K[X]$, and having constant term 1 and we say $r(X) > 0$ iff $\alpha > 0$ (i.e. $\alpha \in P$).

Show that $K(X)$ is not archimedean.

4.* Prove that any archimedean ordered field is order isomorphic to a subfield of $\mathbf{R}$.

5. (Cohn) Let K be ordered and let X, Y be independent indeterminates over K. Order $K(X)$ as in Exercise 3 and repeat the process for $K(X,Y) = K(X)(Y)$. Show that every element of $K(X,Y)$ is majorized by an element of $K(Y)$, but that there exists no element of $K(Y)$ between X and X^2.

6. Let K be formally real and let $K(\theta)$ be an algebraic extension of K. Show that $K(\theta)$ is formally real in the following cases:

(i) $\theta = \sqrt{a}$, $a \in K$, $a > 0$;
(ii) $\operatorname{irr}(\theta, K)$ is of odd degree.

7. (Kaplansky–Kneser). Let K be a field with $\operatorname{char} K \neq 2$ which is not formally real. Suppose $|K^*/K^{*^2}| = n$ is finite. Show that any non-singular quadratic form in n variables is universal.

8. (Artin–Schreier) (i) Let L/K be an extension of odd degree. Show that any ordering of K extends to an ordering of L.

(ii) Let $L = K(\alpha)$ where $[L : K]$ is even, and let P be an ordering of K. Suppose $N_{L/K}(\alpha) < 0$ with respect to P (of course $N_{L/K}(\alpha) \in K$). Show that P can be extended to an ordering of L.

(iii) Show that an ordering P of K can be extended to $K(\sqrt{a})$ iff $a > 0$ with respect to P.

Remark. (i) is indeed Springer's theorem; see Theorem 3.7.

16

Squares and sums of squares in fields and their extension fields

For any given field K, we would like the answers to the following questions:

(1) What proportion of elements of K are squares in K?

(2) What proportion of elements of K are sums of squares in K?

(3) Is there a bound $P = P(K)$, depending only on K, such that an element of K which is a sum of squares in K is already a sum of P squares in K?

(4) If L/K is an extension of fields, what are the answers to the above questions in L in relation to their answers in K?

Definition 16.1. If such a P as in (3) exists, it is called the ***pythagorean number*** of K. It is also sometimes called the ***reduced height*** of K.

For example $P(\mathbf{Q}) = 4$ by Lagrange's theorem and $P(\mathbf{R}) = 1$. We use the notation $G_m(K)$ to denote the set of elements of K^* which can be written as a sum of m squares in K. Thus $G_m(K) = V_{X_1^2+\dots+X_m^2}(K)$ in the notation of Chapter 2. Furthermore we have

$$S(K) - \{0\} = \bigcup_{m=1}^{\infty} G_m(K)$$

(using the notation of Chapter 15) and we write this as $G_\infty(K)$.

There is a host of fields giving a wide variety of results and in this chapter, we shall take up some of those which can be handled in an elementary fashion. We would like to calculate $G_\infty(K)$ in the following important cases: (a) global fields, (b) fields of transcendence degree n over real closed fields, i.e. finite extensions of $R(X_1, \dots, X_n)$, R real closed.

It is interesting to note that to solve the problem in case (a) we need to use the Hasse-Minkowski theorem and so we postpone it to Chapter 18 where a definition of global fields is given. For the solution of the problem in case (b) we need to use the theorem of Tsen-Lang. Indeed, in Chapter 5 we have already done this. As we saw there, we had to make considerable use of Pfister forms.

In this chapter, we shall be looking at fields K that can be handled in an elementary way.

In most cases it is neater to consider the multiplicative group K^* of K. Thus to answer the first question we need to know the order of the quotient K^*/K^{*^2}. Elements of this quotient are called *square classes*. Each element of the square class $K^{*^2}\alpha$ is a sum of n squares in K^* iff α is a sum of n squares in K^*.

We begin with finite fields and prove the following.

Theorem 16.1. *In F_q ($q = p^\alpha$, $p \neq 2$) one half of the elements are squares. The remaining half can all be written as a sum of two squares. For $p = 2$, we have $\mathsf{F}_{2^\alpha}^{*^2} = \mathsf{F}_2^*\alpha$.*

Proof. In $\mathsf{F}_2^*\alpha$, the mapping $x \mapsto x^2$ is an automorphism: $(x+y)^2 = x^2+y^2$ since $2 = 0$, and $(xy)^2 = x^2y^2$. Further if $x^2 = y^2$, then $x = y$ (for $x = -y \Rightarrow x = y$).

Next let $q = p^\alpha$, $p \neq 2$. Let $\nu \in \mathsf{F}_q$ be any element. Let $A = \{\nu - x^2 \mid x \in \mathsf{F}_q\}$, $B = \{y^2 \mid y \in \mathsf{F}_q\}$, and let r be a generator of F_q^*. Then the even powers of r are just the squares of F_q^* as required. Further $|A| = |B| = (q-1)/2 + 1 = (q+1)/2$. Hence $A \cap B \neq \emptyset$, i.e. there exists x_0, $y_0 \in \mathsf{F}_q$ such that $\nu - x_0^2 = y_0^2$. □

Fields K which are not formally real are easy to deal with in this regard; for let the Stufe of K be s, so that $-1 = t_1^2 + \ldots + t_s^2$ ($t_j \in K$). Then for any $\alpha \in K$, we have

$$\alpha = \left(\frac{\alpha+1}{2}\right)^2 - \left(\frac{\alpha-1}{2}\right)^2 = \left(\frac{\alpha+1}{2}\right)^2 + (t_1^2 + \ldots + t_s^2)\left(\frac{\alpha-1}{2}\right)^2$$
$$= \text{ a sum of } s+1 \text{ squares in } K.$$

Thus we have the following.

Theorem 16.2. *Let K be a non-formally real field of Stufe s. Then each element of K is a sum of $s+1$ squares and indeed $s \leq P(K) \leq s+1$ (for -1 is a sum of s squares and no fewer, so $s \leq P(K)$).*

As a corollary we have the following.

Let K be a field of characteristic p. Then each element of K is a sum of three squares; for $K \supset \mathsf{F}_p$ and since $s(\mathsf{F}_p) \leq 2$ so $s(K) \leq 2$ and the result follows by Theorem 16.1.

Now since there are fields with arbitrarily large Stufe, so given N, there are fields K, not formally real, in which a sum of squares is not always a sum of at most N squares. But is there such a bound for formally real fields? The following example settles this question in the negative.

Let $\mathsf{R}_n = \mathsf{R}(X_1, X_2, \ldots, X_n)$, $\mathsf{R}_0 = \mathsf{R}$ and let $K = \bigcup_{n=0}^{\infty} \mathsf{R}_n = \mathsf{R}(X_1, X_2, \ldots)$, the field of rational functions with real coefficients in an infinite number of independent indeterminates $X_1, X_2, \ldots$. We claim that the polynomial $X_1^2 + \ldots + X_n^2$ is not a sum of $n-1$ squares in K. For suppose it is; then on the right hand side only finitely many X's appear, say, without loss of generality,

$$X_1^2 + \ldots + X_n^2 = \phi_1^2(X_1, \ldots, X_m) + \ldots + \phi_{n-1}^2(X_1, \ldots, X_m).$$

Then

$$\begin{aligned} X_1^2 + \ldots &+ X_n^2 + X_{n+1}^2 + \ldots + X_m^2 \\ &= \phi_1^2 + \ldots + \phi_{n-1}^2 + X_{n+1}^2 + \ldots + X_m^2 \end{aligned}$$

which is a sum of $m-1$ squares in $\mathsf{R}(X_1, \ldots, X_m)$, contradicting Corollary 4 of Chapter 2. □

What we have said above is the following.

If $s(K) = 2^n$ then $2^n \leq P(K)$ so that by choosing n large we can get fields K for which $P(K)$ is arbitrarily large, i.e. $P(K)$ can be made arbitrarily large for a suitable non-formally real K. The example $K = \bigcup_{n=0}^{\infty} \mathsf{R}_n$ shows that this is true even for formally real fields.

Let now K be any field, not formally real, with Stufe $s = s(K)$. We have the following.

Theorem 16.3. $P(K(X)) = s(K) + 1$.

Proof. We know that $K(X)$ is also not formally real and indeed $s(K(X)) = s(K)$ (Theorem 11.8(i)). Hence by Theorem 16.2, $s(K) \leq P(K(X)) \leq s(K) + 1$. If $P(K(X))$ were equal to $s(K)$, then each element of $K(X)$ would be a sum of $s(K)$ squares in $K(X)$ (note that each element of $K(X)$ is always a sum of squares, indeed $r(X) = \left(\frac{r+1}{2}\right)^2 - \left(\frac{r-1}{2}\right)^2$ and -1 is a sum of squares). In particular

$$X = f_1^2(X) + \ldots + f_s^2(X)$$

where by Cassels' Lemma (Chapter 2), we may suppose $f_j(X) \in K[X]$. Here not all $f_j(X)$ are constants. Now equate the coefficients of the highest power of X on the right side to zero (since this highest power is even on the right side and zero on the left side). We get

$$a_1^2 + \ldots + a_s^2 = 0, \qquad a_j \in K$$

giving $s(K) < s$, a contradiction. □

Definition 16.2.

(i) K is called *quadratically closed* if each $x \in K$ is a square in K;

(ii) K is called *pythagorean* if any sum of 2 squares in K is itself a square in K, or equivalently if any sum of squares in K is a square in K or indeed $1 + y^2$ is a square in K for each $y \in K$.

For such a K, denote by $S(K)$ the set of all elements of K, including 0, that are sums of squares in K. Then $S(K) = K^2 = \{x^2 \mid x \in K\}$.

Theorem 16.4.

(i) *K quadratically closed $\Rightarrow$ K pythagorean.*

(ii) *K pythagorean and not formally real $\Rightarrow$ K quadratically closed.*

(iii) *K real closed $\Rightarrow$ K pythagorean*

(iv) *K_i pythagorean $\subset$ a field $L \Rightarrow K = \bigcap_i K_i$ is pythagorean.*

Proof. (i) Trivial.

(ii) Since K is not formally real, it follows that $S(K) = K$, i.e. each element of K is a sum of squares in K and so a square in K since K is pythagorean; so K is quadratically closed.

(iii) If $x = 1 + y^2$ is not a square for some $y \in K$, then $K(\sqrt{x})$ is a proper algebraic extension of K and so is not formally real since K is real closed. So by the basic lemma (see Chapter 15), there exist $z_i \in K$ such that $-x = \sum z_i^2$, i.e. $-1 - y^2 = \sum z_i^2$ giving $-1 =$ a sum of squares in K which is a contradiction.

(iv) Let $x, y \in K$, so $x, y \in K_i$ for all i. Since K_i is pythagorean $x^2 + y^2 = z_i^2$ $(z_i \in K_i)$. For any pair of indices i, j we have $z_i = \pm z_j$. (since all $z_i = \pm\sqrt{x^2 + y^2}$). Let i_0 be any fixed index and let j be an arbitrary index; then $\pm z_j \in K_j$, i.e. $z_{i_0} \in K_j$ (for all j) so $z_{i_0} \in \bigcap_j K_j = K$ and $z_{i_0}^2 = x^2 + y^2$. □

Given a field K there is the notion of a smallest pythagorean field K_p containing K - this is inspired by the last part of Theorem 16.4; namely, let $\Omega = \overline{K}$ be the algebraic closure of K and let

$$K_p = \bigcap (\text{all pythagorean subfields } K_i \text{ of } \Omega \text{ which contain } K) .$$

This intersection is not vacuous since Ω itself is in it (it is pythagorean because it is algebraically closed, so quadratically closed).

We have the following.

Definition 16.3. K_p is called the *pythagorean closure* (hull) of K.

It is clearly pythagorean by Theorem 16.4 (iv), and manifestly the smallest pythagorean field containing K.

Theorem 16.5. *If K is formally real, so is K_p.*

Proof. Let Ω be the algebraic closure of K. By the remark after Definition 15.4, there exists a real closed field $\Delta : K \subset \Delta \subset \Omega$ as shown.

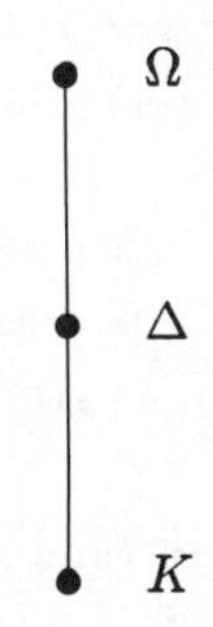

By Theorem 16.4(iv), Δ is pythagorean. Hence $K_p \subset \Delta$. It follows that K_p is formally real. □

An Example (Ribenboim [R5]). Let P be the smallest pythagorean field containing the rationals $\mathbb{Q}$, $\mathbb{A}$ the field of all algebraic numbers, and $\mathbb{R}$ the field of real numbers. We wish to look at P more closely. We have the following.

Theorem 16.6. *$P \subseteq \mathbb{A} \cap \mathbb{R}$ and is an infinite Galois extension of $\mathbb{Q}$.*

Proof. Since $\mathbb{A}$, $\mathbb{R}$ are pythagorean, so is $\mathbb{A} \cap \mathbb{R}$; hence $P \subset \mathbb{A} \cap \mathbb{R}$. Now let

$$\mathbb{Q}_0 = \mathbb{Q}, \mathbb{Q}_1 = \mathbb{Q}(\sqrt{a^2+b^2}) \quad (\text{all } a,\, b \in \mathbb{Q})$$
$$\mathbb{Q}_2 = \mathbb{Q}_1(\sqrt{a^2+b^2}) \quad (\text{all } a,\, b \in \mathbb{Q}_1), \ldots$$
$$\mathbb{Q}_n = \mathbb{Q}_{n-1}(\sqrt{a^2+b^2}) \quad (\text{all } a,\, b \in \mathbb{Q}_{n-1}), \ldots$$

Since any positive square free integer m is a sum of two squares iff $p|m \Rightarrow p \equiv 1(4)$, it follows that

$$\mathbb{Q}_1 = \mathbb{Q}(\sqrt{2}, \sqrt{p}) \quad (\text{all } p \equiv 1(4))$$

so that $[\mathbb{Q}_1 : \mathbb{Q}] = \infty$ and a fortiori $[P : \mathbb{Q}] = \infty$.

Next let σ be an isomorphism of P (into $\mathbb{A}$). To prove P is Galois, we have to show that all conjugates of P equal P, i.e. that $\sigma P = P$. Now $\sigma^{-1}P \supset \mathbb{Q}$ is pythagorean (just check), so by the minimality of P, $P \subset \sigma^{-1}P$, i.e. $\sigma P \subset P$ as required.

We now compare P with the field Φ of all numbers constructible by ruler and compass so that by definition $\Phi = \bigcup_{n=0}^{\infty} \Phi_n$, where $\Phi_0 = \mathbb{Q}$.

$$\Phi_1 = \mathbb{Q}(\sqrt{a})_{\text{all } a\in\mathbb{Q},\ a>0} \subset \mathbb{R},$$
$$\Phi_2 = \Phi_1(\sqrt{a})_{\text{all } a\in\Phi_1,\ a>0} \subset \mathbb{R},$$
$$\dots\dots\dots\dots\dots\dots\dots$$
$$\Phi_n = \Phi_{n-1}(\sqrt{a})_{\text{all } a\in\Phi_{n-1},\ a>0} \subset \mathbb{R}.$$
$$\dots\dots\dots\dots\dots\dots\dots$$

Gauss proved the following result regarding the elements x of Φ:

Theorem (Gauss). *$x \in \mathbb{R}$ is constructible iff there exists a Galois extension $K/\mathbb{Q}$ such that $x \in K$, $[K : \mathbb{Q}]$ is a power of 2.*

First note that $\mathbb{Q}_j \subseteq \Phi_j$ (all j) and so $P \subseteq \Phi$. Now we have the following.

Theorem 16.7. *Φ is pythagorean (and so $P \subseteq \Phi$ again).*

Proof. Let x, $y \in \Phi$ and let K, L be Galois extensions of $\mathbb{Q}$ of degree a power of 2 and such that $x \in K$, $y \in L$. Then $\mathbb{Q}(x^2+y^2) \subseteq \mathbb{Q}(x,y) \subseteq K \cdot L$ and KL is a Galois extension of $\mathbb{Q}$. But

$$[KL : \mathbb{Q}] = [KL : K \cap L][K \cap L : \mathbb{Q}]$$

and

$$\mathrm{Gal}(KL/K \cap L) \simeq \mathrm{Gal}(K/K \cap L) \times \mathrm{Gal}(L/K \cap L)$$

It follows that $[KL : \mathbb{Q}]$ is a power of 2.

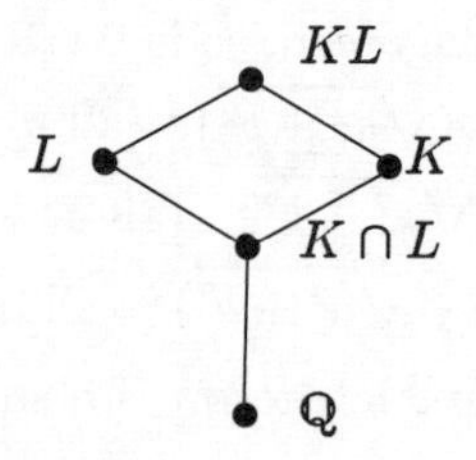

Now the conjugates of $\sqrt{x^2+y^2}$ over $\mathbb{Q}$ are $\pm\sqrt{(\sigma(x))^2+(\sigma(y))^2}$ where σ is an arbitrary automorphism of $KL/\mathbb{Q}$. Let M be the smallest Galois extension over $\mathbb{Q}$ containing KL and $\sqrt{x^2+y^2}$. Let $\sigma_1, \dots, \sigma_m$ be the automorphisms of KL such that

$$\sqrt{(\sigma_1(x))^2 + (\sigma_1(y))^2} \notin KL$$

$$\sqrt{(\sigma_2(x))^2 + (\sigma_2(y))^2} \notin KL\left(\sqrt{(\sigma_1(x))^2 + (\sigma_1(y))^2}\right)$$

$$\cdots\cdots\cdots\cdots\cdots\cdots$$

$$\sqrt{(\sigma_m(x))^2 + (\sigma_m(y))^2} \notin KL\left(\sqrt{(\sigma_i(x))^2 + (\sigma_i(y))^2}\right),$$

for all $i = 1, 2, \ldots, m-1$, and $M = KL\left(\sqrt{(\sigma_i(x))^2 + (\sigma_i(y))^2}\right)$, $i = 1, 2, \ldots, m$.

Now for extension fields $K \subseteq N_1, N_2 \subset K^*$, if N_1/K and N_2/K are Galois extensions then $N_1N_2/N_1 \cap N_2$ is Galois and

$$\mathrm{Gal}(N_1N_2/N_1 \cap N_2) \simeq G(N_1/N_1 \cap N_2) \times G(N_2/N_1 \cap N_2).$$

Repeated application of this to M gives

$$\mathrm{Gal}(M/KL) \simeq (\mathbf{Z}/(2))^m.$$

Thus $[M : \mathbb{Q}]$ is a power of 2 and by Gauss' theorem $\sqrt{(x^2 + y^2)}$ is constructible and so in Φ. It follows that Φ is pythagorean as required.

Remarks.

(i) It is easy to show that $P \neq \Phi$; indeed $\sqrt{(\sqrt{2} - 1)} \in \Phi$ but $\notin P$.

(ii) If K is a subfield of P such that $[K : \mathbb{Q}] < \infty$, then $[K : \mathbb{Q}]$ is a power of 2.

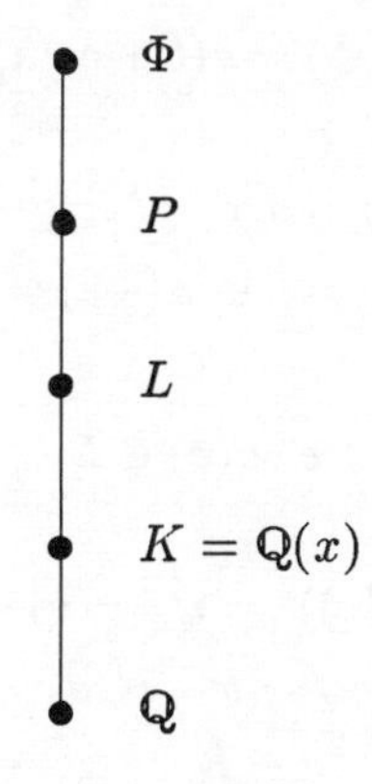

Proof. As K is a finite extension of $\mathbb{Q}$, it is simple, say $K = \mathbb{Q}(x)$. Since $x \in P \subset \Phi$, there exists a Galois extension $L/\mathbb{Q}$ containing x. But $[K : \mathbb{Q}]$ is a subfield of Φ, and so is a power of 2. □

Theorem 16.8 (Diller-Dress). *Let K be a field such that $-1 \notin K^2$. The following statements are equivalent.*

(i) *K is pythagorean, i.e. $K^2 + K^2 = K^2$.*
(ii) *K has no cyclic extension of degree 4.*
(iii) *K has no cyclic extension of degree 2^ν ($\nu \geq 2$).*

Proof. (iii)⇒(ii) is trivial and (ii)⇒(iii) follows because if K has a cyclic extension of degree 2^{ν_0} ($\nu_0 \geq 2$), then since $4|2^{\nu_0}$, K has a cyclic extension of degree 4.

(i)⇒(ii). Suppose E/K is a cyclic extension of degree 4. Let $G = \mathrm{Gal}(E/K)$ and L an intermediate field $K \subset L \subset E$, $H = \mathrm{Gal}(E/L)$. This quadratic extension L of K exists since G is cyclic of order 4.

$$\begin{array}{rl} E & \bullet \\ & \mid\ 2 \\ L = K(\sqrt{a}) & \bullet \\ & \mid\ 2 \\ K & \bullet \end{array}$$

We have $L = K(\sqrt{a})$, $a \in K$, $a \notin K^2$. Then since E/L is of degree 2, we have $E = L(\sqrt{b + c\sqrt{a}})$, b, $c \in K$, $c \neq 0$, $b + c\sqrt{a} \notin L^2$. Let $e = \sqrt{b + c\sqrt{a}}$ and let σ be a generator of G. Then $H = \{\varepsilon, \sigma^2\}$ and clearly we have

$$\sigma^2(e) = -e. \tag{16.1}$$

Furthermore

$$(\sigma(e))^2 = \sigma(e^2) = \sigma(b + c\sqrt{a}) = b - c\sqrt{a} \tag{16.2}$$

so

$$\begin{aligned} \sigma^2(e \cdot \sigma(e)) &= \sigma^2 e \cdot \sigma^3 e = -e \cdot \sigma(\sigma^2 e) \\ &= -e \cdot \sigma(-e) = e \cdot \sigma(e). \end{aligned}$$

Thus

$$e \cdot \sigma(e) \in L \tag{16.3}$$

Now, since $a, b, c \in K$,

$$\begin{aligned} (e \cdot \sigma(e))^2 &= e^2(\sigma(e))^2 = e^2 \cdot \sigma(e) \cdot \sigma(e) = e^2 \cdot \sigma(e^2) \\ &= (b + c\sqrt{a})(b - c\sqrt{a}) \quad \text{(by (16.2))} \\ &= b^2 - c^2 a \in K \end{aligned} \tag{16.4}$$

But by (16.3), $(e \cdot \sigma(e))^2 \in L^2$ so $(e \cdot \sigma(e))^2 \in L^2$ and K, i.e. $\in L^2 \cap K = K^2 \cup aK^2$. To see this let $(x + y\sqrt{a})^2 \in L^2 \cap K$ (x, $y \in K$). This equals $x^2 + ay^2 + 2xy\sqrt{a} \in K$, so $xy = 0$. If $x = 0$ then $x + y\sqrt{a} = y\sqrt{a}$ so $(x + y\sqrt{a})^2 = ay^2 \in aK^2$. If $y = 0$ then $x + y\sqrt{a} = x$ so $(x + y\sqrt{a})^2 = x^2 \in K^2$.

Thus $L^2 \cap K \subset K^2 \cup aK^2$ while the converse is trivial.

If $(e \cdot \sigma(e))^2 \in K^2$ then $e \cdot \sigma(e) \in K$ so

$$\sigma(e \cdot \sigma(e)) = e \cdot \sigma(e)$$

i.e.

$$\sigma(e) \cdot \sigma^2(e) = e \cdot \sigma(e).$$

Cancelling $\sigma(e)$ ($\neq 0$) this gives $\sigma^2(e) = e$, i.e. $-e = e$ (by (16.1)) which is false. It follows that $(e \cdot \sigma(e))^2 \in aK^2$, i.e. $b^2 - c^2 a = ax^2$ ($x \in K$) by (16.4) or $a = b^2/(c^2 + x^2)$, where note that $c^2 + x^2 \neq 0$ since $\sqrt{-1} \notin K^2$ by hypothesis. Now we are supposing K to be pythagorean, so $c^2 + x^2 \in K^2$. Hence $a \in K^2$ which is again false. Thus no E can exist.

(ii)⇒(i). Suppose to the contrary that K is not pythagorean, i.e. that for some $a \in K$, $b = 1 + a^2 \notin K^2$. Let $L = K(\sqrt{b})$, $E = K(\sqrt{b + \sqrt{b}})$. Put $e = \sqrt{b + \sqrt{b}}$, $c = \sqrt{b - \sqrt{b}}$. The four conjugates of e are e, $-e$, c, $-c$ and these are distinct since $-1 \notin K^2$ and char$K \neq 2$. But now $c = a\sqrt{b}/\sqrt{b + \sqrt{b}}$ (use $1 + a^2 = b$) and being a conjugate of e, there exists an automorphism σ of E/K such that $\sigma(e) = c$, i.e. $\sigma\sqrt{b} = -\sqrt{b}$. Then

$$\sigma^2 e = \sigma c = \sigma(a\sqrt{b}/e) = -a\sqrt{b}/c = -e.$$

Hence $\sigma^2 \neq \varepsilon$, so since Gal$(E/K)$ is of order 4, it follows that σ is a generator of Gal(E/K). i.e. E is cyclic of degree $4/K$ - a contradiction. □

Now it is easy to see that if char$K \neq 2$ and $b \notin K^2$ then the field $E = K(\sqrt{b + \sqrt{b}})$ is a cyclic extension of K of degree 4, for the generator $x = \sqrt{b + \sqrt{b}}$ satisfies $x^2 = b + \sqrt{b}$, i.e. $(x^2 - b)^2 = b$, so E/K is of degree 4. The four automorphisms are

$$\sigma_1 : \sqrt{b + \sqrt{b}} \to \sqrt{b - \sqrt{b}}$$

$$\sigma_2 : \sqrt{b + \sqrt{b}} \to -\sqrt{b + \sqrt{b}}$$

$$\sigma_3 : \sqrt{b + \sqrt{b}} \to -\sqrt{b - \sqrt{b}}$$

$$\sigma_4 : \varepsilon,$$

since the four conjugates of the generator are the roots of the above quartic. We have the following

Theorem 16.9.

(i) *Let K be a non-pythagorean field and K_p its pythagorean closure. Then there is a field E, $K \subset E \subset K_p$, which is cyclic of degree 4 over K.*

(ii) *If E is pythagorean and $K \subset E$ is such that $[E : K] < \infty$, then K is pythagorean.*

Proof. (i) Since K is not pythagorean, char $K \neq 2$. Further, we can find

$a_1, a_2 \in K$ such that $a_1^2 + a_2^2 \notin K^2$. Here $a_2 \neq 0$. Let $b = \left(\frac{a_1}{a_2}\right)^2 + 1 \notin K^2$ and let $E = K(\sqrt{b+\sqrt{b}})$. Then E is the required cyclic extension of K of degree 4. Note also that

$$\sqrt{b+\sqrt{b}} = \frac{1}{2}\sqrt{\left(\left(\frac{a_1}{a_2}\right)^2 + \left(1+\sqrt{1+\left(\frac{a_1}{a_2}\right)^2}\right)^2\right)} \cdot \sqrt{1+1} \in K_p$$

so $E \subset K_p$.

(ii) Suppose not; then there exist fields L_1, L_2 such that

$$K \subseteq L_1 \subset L_2 \subseteq E.$$

Now, neither L_1 nor L_2 are pythagorean, and there is no field between L_1 and L_2. Yet (i) implies there is an F, $L_1 \subset F \subseteq L_2$ and F a cyclic extension of degree 4 over L_1; it follows that $F = L_2$. But there again exists a (quadratic) field between L_1 and $L_2 = F$. □

Corollary. *Let P be the pythagorean closure of $\mathbb{Q}$ as in the example (see Theorem 16.6) and let K be a subfield of P such that $[P : K] < \infty$. Then $K = P$.*

Let us now take a look at question 4 posed at the start of this chapter. We have the following striking

Theorem 16.10. *Suppose that L is an algebraic extension of K of odd degree. Then $|K^*/K^{*^2}| = |L^*/L^{*^2}|$.*

Proof. Consider the commutative triangle

$$\begin{array}{ccc} & L^* & \\ {}^{\varphi}\nearrow & & \searrow_{N} \\ K^* & \xrightarrow{M} & K^* \end{array}$$

where φ is the injection map, N the norm map and M is defined by $x \mapsto x^m$, where $m = [L : K]$.

This induces the commutative triangle

$$\begin{array}{ccc} & L^*/L^{*^2} & \\ {}^{\overline{\varphi}}\nearrow & & \searrow_{\overline{N}} \\ K^*/K^{*^2} & \xrightarrow{\overline{M}} & K^*/K^{*^2} \end{array}$$

where $\overline{\varphi}, \overline{N}, \overline{M}$ are the induced maps. Since m is odd, $\overline{M}$ is the identity map. Thus $\overline{\varphi}, \overline{N}$ define a bijection between L^*/L^{*^2} and K^*/K^{*^2} □

For even degree extensions, the situation is not quite so simple. We first have the following

Definition 16.4. A sequence of maps $A \xrightarrow{f} B \xrightarrow{g} C$ of groups is said to be *exact* (at B) if the image of A under f ($\subset B$) equals the kernel of g.

We have the following.

Theorem 16.11. *Let $L = K(\sqrt{d})$ be a quadratic extension of K. Then the following sequence is exact.*

$$1 \xrightarrow{\phi} \{K^{*^2}, dK^{*^2}\} \xrightarrow{\theta} K^*/K^{*^2} \xrightarrow{\varepsilon} L^*/L^{*^2} \xrightarrow{N} K^*/K^{*^2}$$

where ε is the map induced by inclusion and N is the homomorphism induced by the norm L/K and ϕ, θ are the obvious inclusion maps.

Proof. (i) The exactness at $\{K^{*^2}, dK^{*^2}\}$ is trivial: d is not a square in K, so $\ker\theta = K^{*^2} = \text{Im}\phi$.
(ii) ε is the map $\alpha K^{*^2} \xrightarrow{\varepsilon} \alpha L^{*^2}$; so the coset $\alpha K^{*^2} \in \ker\varepsilon$ iff α is a square in L^*. Let

$$\alpha = (a + b\sqrt{d})^2 \quad \text{(a square in } L^*) \tag{16.5}$$

We have to prove that $\alpha = de^2$ or f^2 ($e, f \in K^*$). Now since $\alpha \in K^*$, (16.5) gives $\alpha = a^2 + db^2$, $ab = 0$. If $a = 0$, then $\alpha = db^2$; if $b = 0$, then $\alpha = a^2$.
(iii) $L^*/L^{*^2} \xrightarrow{N} K^*/K^{*^2}$ is the map

$$\alpha L^{*^2} \xrightarrow{N} N_{L/K}(\alpha)K^{*^2}$$

where $\alpha \in L^*$, say $\alpha = a + b\sqrt{d}$, a, $b \in K$, so $\ker N = \{\alpha L^{*^2} | N_{L/K}(\alpha) \in K^{*^2}\}$, say $N\alpha = \alpha\overline{\alpha} = c^2$ ($c \in K^*$). Let $\beta = c - \overline{\alpha} = c - (a - b\sqrt{d})$. Then

$$\begin{aligned} \alpha\beta^2 &= \alpha(c^2 + \overline{\alpha}^2 - 2c\overline{\alpha}) = \alpha(N\alpha + \overline{\alpha}^2 - 2c\overline{\alpha}) \\ &= \alpha \cdot N\alpha + \alpha\overline{\alpha} \cdot \overline{\alpha} - 2c\alpha\overline{\alpha} = (\alpha + \overline{\alpha})N\alpha - 2cN\alpha \\ &\in K, \end{aligned}$$

say $\alpha\beta^2 = r \in K$.

Now to show $\ker N = \text{Im}\varepsilon$, we have to show that such an $\alpha \in K^* \cdot L^{*^2}$; then the element αL^{*^2} of the kernel is the image under ε of αK^{*^2} (since $\alpha \in K^*$).

If $\beta = 0$ then $\overline{\alpha} = c (\neq 0) \in K^*$, i.e. $\alpha = a + b\sqrt{d} \in K^*$, i.e. $b = 0$ so $\alpha \in K^* \subset K^*L^{*^2}$. If $\beta \neq 0$ then by the above calculations $\alpha = r(1/\beta)^2 \in K^*L^{*^2}$ as required. □

As a corollary we have the following.

Theorem 16.12. *Let $L = K(\sqrt{d})$ be as in Theorem 16.11. Then K^*/K^{*^2} is finite iff L^*/L^{*^2} is finite. More precisely, we have*

$$\frac{1}{2}|K^*/K^{*^2}| \leq |L^*/L^{*^2}| \leq \frac{1}{2}|K^*/K^{*^2}|^2.$$

Proof. First note that

$$\ker \varepsilon = 2 \tag{16.6}$$

Now look at the maps $K^*/K^{*^2} \xrightarrow{\varepsilon} L^*/L^{*^2} \xrightarrow{N} K^*/K^{*^2}$. We have

$$(K^*/K^{*^2})/\ker \varepsilon \simeq \mathrm{Im}\varepsilon \subseteq L^*/L^{*^2};$$

hence

$$\frac{|K^*/K^{*^2}|}{\ker \varepsilon} = |\mathrm{Im}\varepsilon| = |\ker N| \leq |L^*/L^{*^2}| \tag{16.7}$$

and similarly

$$\frac{|L^*/L^{*^2}|}{|\ker N|} = |\mathrm{Im}N| \leq |K^*/K^{*^2}| \tag{16.8}$$

Hence $|K^*/K^{*^2}| \leq 2|L^*/L^{*^2}|$, using (16.6) and (16.7). Thus $|K^*/K^{*^2}|$ is finite if $|L^*/L^{*^2}|$ is and the first inequality follows. Next by (16.8),

$$\begin{aligned} |L^*/L^{*^2}| \leq |\ker N||K^*/K^{*^2}| &= |\mathrm{Im}\varepsilon||K^*/K^{*^2}| \\ &= \frac{|K^*/K^{*^2}|}{\ker \varepsilon} \cdot |K^*/K^{*^2}| = \frac{1}{2}|K^*/K^{*^2}| \end{aligned}$$

Thus $|L^*/L^{*^2}|$ is finite if $|K^*/K^{*^2}|$ is and the second inequality follows.

□

For extension fields of arbitrary degree the following result is true.

Theorem 16.13. *Let L be a finite extension of K of degree $2^r \cdot m$ (m odd). Then*

$$|K^*/K^{*^2}| \leq 2^r|L^*/L^{*^2}|.$$

Proof. Consider the kernel of the map

$$\Theta : K^*/K^{*^2} \to L^*/L^{*^2}$$

induced by injecting K into L. Suppose that bK^{*^2} is in the kernel ($b \in K^*$). Then L contains $\sqrt{b}$. Now let $b_1, \ldots, b_t$ be such that the $b_jK^{*^2}$ are in the kernel and $b_1, \ldots, b_t$ are multiplicatively independent modulo K^{*^2}. Then L contains the field

$$K_1 = K(\sqrt{b_1}, \ldots, \sqrt{b_t})$$

of degree 2^t over K; but $K_1 \subset L$, so $t \leq r$.

Now let t be maximal, so

$$b_1K^{*^2}, \ldots, b_tK^{*^2}$$

generate the kernel of Θ. Then the kernel has order precisely 2^t and the result follows.

□

Better estimates are available if K is not formally real. See for example, [L2] p. 216, Exercise 6. Indeed we have

Theorem 16.14. *Let L be a finite extension of K of degree $2^r m$ (m odd) and suppose that K is not formally real. Then*

$$|K^*/K^{*^2}| \le |L^*/L^{*^2}|$$

Proof. To prove this we require to know when there is equality in the left hand inequality of Theorem 16.12. It turns out that if

$$\frac{1}{2}|K^*/K^{*^2}| = |L^*/L^{*^2}| \tag{16.9}$$

then K is pythagorean and $L = K(\sqrt{-1})$. To see this we note that (16.9) holds precisely when

$$\mathrm{Im}G = L^*/L^{*^2}.$$

By exactness, this is precisely when

$$\ker N = L^*,$$

i.e. when $\mathrm{Norm}\alpha \in K^{*^2}$ for all $\alpha \in L^*$.

Put $\alpha = u + v\sqrt{d}$ $(u, v \in K)$, so that $\mathrm{Norm}\alpha = u^2 - dv^2 \in K^{*^2}$. Hence $-d \in K^{*^2}$ $(u = 0, v = 1)$. Then $K^2 + K^2 \subset K^2$, which means K is pythagorean and $d \in (-1)K^{*^2}$ as required.

Theorem 16.14 may now be proved as follows:

We have a chain of fields

$$K = M_0 \subset M_1 \subset M_2 \subset \cdots \subset M_r \subset L$$

where each extension M_{j+1} is of degree 2 over M_j, and L does not contain any extension of M_r of degree 2 over M_r. By Theorem 16.13 we have

$$|M_r^*/M_r^{*^2}| \le |L^*/L^{*^2}|.$$

Now we are given that $K = M_0$ is not formally real. Hence none of the quadratic extensions M_{j+1} over M_j is of the exceptional type satisfying (16.9). Hence

$$|M_j^*/M_j^{*^2}| \le |M_{j+1}^*/M_{j+1}^{*^2}| \quad (0 \le j \le r)$$

On putting everything together, we get

$$|K^*/K^{*^2}| = |M_0^*/M_0^{*^2}| \le |L^*/L^{*^2}|$$

as required. □

Then there is the question of whether or not L^*/L^{*^2} is finite given that K^*/K^{*^2} is, where L/K is a general extension. By Theorem 16.12, if $[L : K] = 2$, then the answer is "yes". By an easy application of Galois theory, it can be shown that the answer is again "yes" if $[L : K] = 2^\nu$ and L is a Galois extension of K. However, in general the answer is "no" even if

$[L:K] = 2^\nu$. A counterexample, elementary but not that easy, is given in [L2], p. 218–22.

For $L \supset K$ as in Theorem 16.13, we have the following.

Corollary.
(i) $|L^*/L^{*^2}|$ *finite* $\Rightarrow$ $|K^*/K^{*^2}|$ *finite.*
(ii) L *quadratically closed* $\Rightarrow$ K *quadratically closed.* □

One may prove many more results regarding pythagorean fields. See for example [L2], p. 251.

Exercises

1. Let $s = 2^n$ be any power of 2 and let K be a field of Stufe s. Let $L = K(X)$. Prove that $P(L) = s + 1$ ($s(L)$ is of course equal to s). (Hint: $-1 + X^2$ cannot be written as a sum of s squares in L.)

2. Let E/F be a cyclic extension of degree 4. Suppose $F(\sqrt{a})$ is the unique field between E and F. Show that $a = x^2 + y^2$ (x, $y \in F$).

3. Let $E = F(\sqrt{x^2 + y^2})$ be a quadratic extension of F (where x, $y \in F$). Show that there exists a field $K \supset E$ such that K is cyclic of degree 4 over F.

4. Let K be formally real and suppose $|K^*/K^{*^2}| = 2$. Show that K is pythagorean.

5. Let K be formally real. Show that each ordering of K can be extended to an ordering of K_p, the pythagorean hull of K.

6. Let K be pythagorean. Show that $|K^*/K^{*^2}| = \infty$ if and only if K has infinitely many orderings.

7. Let K be formally real and pythagorean. Show that a quadratic form f over K is universal if and only if f is isotropic.

8. Show that the following conditions are equivalent:
(i) K is pythagorean;
(ii) if a, b are different square classes in K^*/K^{*^2}, then K has an ordering at which a, b have different signs;
(iii) if c is a square class ($\neq -1$), then there exists an ordering of K at which c is positive.

17

Pourchet's theorem and related results

The main result to be proved in this chapter is the following (see [P6]).

Theorem 17.1 (Pourchet 1971). *Let $f(X) \in \mathbb{Q}(X)$ be a non-zero positive definite function. Then there exist $f_1(X)$, $f_2(X)$, $f_3(X)$, $f_4(X)$, $f_5(X) \in \mathbb{Q}(X)$ such that*

$$f(X) = f_1^2(X) + f_2^2(X) + f_3^2(X) + f_4^2(X) + f_5^2(X).$$

We shall prove a series of lemmas leading towards a more general result. We make heavy use of the local-global principle; in particular the Hasse-Minkowski theorem.

Let K be a field with $\operatorname{char} K \neq 2$. For a, $b \in K^*$, we denote the quadratic form $X_1^2 + aX_2^2 + bX_3^2 + abX_4^2$ by $[a, b]$. For a polynomial $f(X) \in K[X]$, $l(f)$ denotes the *leading coefficient* of f.

Lemma 17.1. *$[a, b]$ represents 0 in K if and only if $aX_2^2 + bX_3^2 + abX_4^2$ represents 0 in K.*

Proof. The if part is trivial. Conversely suppose $[a, b]$ represents 0 in K:

$$\alpha_1^2 + a\alpha_2^2 + b\alpha_3^2 + ab\alpha_4^2 = 0.$$

If $\alpha_1 = 0$, we are through. So let $\alpha_1 \neq 0$. Then using the equation $\alpha_1^2 = -a\alpha_2^2 - b\alpha_3^2 - ab\alpha_4^2$ we can verify that

$$a\{b(a\alpha_2\alpha_4 - \alpha_1\alpha_3)\}^2 + b\{a(b\alpha_3\alpha_4 + \alpha_1\alpha_2)\}^2 + ab(a\alpha_2^2 + b\alpha_3^2)^2 = 0.$$

This completes the proof. □

Lemma 17.2. *Suppose $[a,b]$ does not represent 0 in K. Then it does not represent 0 in $K[X]$ and if $f(X) = f_1^2(X) + af_2^2(X) + bf_3^2(X) + abf_4^2(X)$, where the $f_j(X) \in K[X]$; then $\deg f = 2 \max \deg f_j$. Further if $f(X) \neq 0$, then $[a,b]$ represents $l(f)$ in K.*

Proof. If $[a,b]$ represents 0 in $K(X)$, then

$$0 = f_1^2 + af_2^2 + bf_3^2 + abf_4^2,$$

where, on clearing the denominators, we may suppose $f_j(X) \in K[X]$. Then the leading coefficient of the right side is 0 and this gives a representation of 0 by $[a,b]$ in K, which is a contradiction.

If all the $f_j = 0$, then $f = 0$ and the degrees on both sides equal $-\infty$. So suppose the contrary. Now the degrees will match unless there is cancellation of the highest powers of X on the right side. But that would imply that $[a,b]$ represents 0 in K giving a contradiction again.

Finally $l(f) = l(f_1^2 + af_2^2 + bf_3^2 + abf_4^2)$ so $l(f)$ is represented by $[a,b]$ in K. □

Lemma 17.3. *Let $f(X) \in K[X]$ be a non-zero polynomial. Then $[a,b]$ represents $f(X)$ in $K[X]$ if and only if*

(i) *$[a,b]$ represents $l(f)$ in K, and*
(ii) *$[a,b]$ represents 0 in $K[X]/(p(X))$ for each prime factor $p(X)$ of $f(X)$ in $K[X]$ of odd multiplicity.*

Proof. First suppose $[a,b]$ represents $f(X)$ in $K[X]$:

$$f = f_1^2 + af_2^2 + bf_3^2 + abf_4^2.$$

Then by Lemma 17.2, $[a,b]$ represents $l(f)$ in K.
Next let $\Delta(X) = \gcd(f_1, f_2, f_3, f_4)$ and write $f_j = \Delta g_j$ $(j = 1,2,3,4)$. Then

$$\begin{aligned} f &= \Delta^2(g_1^2 + ag_2^2 + bg_3^2 + abg_4^2) \\ &= \Delta^2 g, \end{aligned}$$

say. Hence if $p | f$ with an odd multiplicity, then $p | g$. But $p \not| $ all the g_j since the g_j are coprime. The equation $g = g_1^2 + ag_2^2 + bg_3^2 + abg_4^2$ $(p|g,\ p \not|\ \text{all } g_j)$ modulo p gives (ii) above.

Now the converse: if $[a,b]$ represents 0 in K, it represents 0 in $K(X)$ and so represents all elements of $K(X)$, in particular $[a,b]$ represents $f(X)$ in $K[X]$ as required.

So suppose $[a,b]$ does not represent 0 in K. We make use of the following identity:

$$\begin{aligned}(X_1^2 + aX_2^2 + bX_3^2 + abX_4^2)&(Y_1^2 + aY_2^2 + bY_3^2 + abY_4^2)\\ &= (X_1Y_1 + aX_2Y_2 + bX_3Y_3 + abX_4Y_4)^2\\ &\quad + a(-X_1Y_2 + Y_1X_2 - bX_3Y_4 + bY_3X_4)^2\\ &\quad + b(-X_1Y_3 + Y_1X_3 + aX_2Y_4 - aY_2X_4)^2\\ &\quad + ab(-X_1Y_4 + Y_1X_4 - bX_2Y_3 + bY_2X_3)^2\\ &= B_1^2 + aB_2^2 + bB_3^2 + abB_4^2,\end{aligned} \tag{*}$$

say. On account of this identity and since $l(f)$ is represented by $[a, b]$, it is enough to prove the result for monic irreducible $f(X)$.

Now $[a, b]$ represents 0 in $K[X]/(f(X))$ (non-trivially), i.e. there exist polynomials $\gamma_j(X)$ whose degree is less than that of f, not all divisible by f, such that

$$\gamma_1^2 + a\gamma_2^2 + b\gamma_3^2 + ab\gamma_4^2 = fg \qquad (g \neq 0) \tag{17.1}$$

Amongst such systems of polynomials γ_j, select one for which

$$g \neq 0 \text{ and } \max \deg \gamma_j \text{ is least}\ . \tag{17.2}$$

By Lemma 17.2, $\deg f + \deg g = 2 \max \deg \gamma_j < 2 \deg f$ so

$$\deg g < \deg f.$$

Now by the division algorithm, write

$$\gamma_j = gq_j + s_j \tag{17.3}$$

where either $s_j = 0$ or $\deg s_j < \deg g$. Substituting in (17.1) gives $(gq_1 + s_1)^2 + a(gq_2 + s_2)^2 + b(gq_3 + s_3)^2 + ab(gq_4 + s_4)^2 = f \cdot g$, i.e.

$$s_1^2 + as_2^2 + bs_3^2 + abs_4^2 = gh \text{ (say.)} \tag{17.4}$$

Again by Lemma 17.2, $\deg g + \deg h = 2 \max \deg s_j < 2 \deg g$ so $\deg h < \deg g$.

Now multiply (17.1) and (17.4) to get

$$fg \cdot gh = B_1^2 + aB_2^2 + bB_3^2 + abB_4^2. \tag{17.5}$$

The congruences $\gamma_j = s_j \pmod{g}$, which follow from (17.3) imply that $B_j \equiv 0 \pmod{g}$ $(j = 1, 2, 3, 4)$, say $B_j = g \cdot t_j$ where $t_j(X) \in K[X]$. Cancelling g^2 in (17.5) $(g \neq 0)$ gives

$$fh = t_1^2 + at_2^2 + bt_3^2 + abt_4^2 \tag{17.6}$$

where, by the definitions of t_j and B_j, $\deg g + \deg t_j = \deg B_j \leq \deg \gamma_j + \deg s_j$. Thus

$$\begin{aligned}\deg g + \max \deg t_j &\leq \max \deg \gamma_j + \max \deg s_j\\ &< \max \deg \gamma_j + \deg g,\end{aligned}$$

i.e.

$$\max \deg t_j < \max \deg \gamma_j \tag{17.7}$$

Hence by (17.6), $\deg f + \deg h = 2\max\deg t_j < 2\max\deg\gamma_j = \deg f + \deg g$, by (17.1) and Lemma (17.2). Now (17.7) contradicts (17.2), so $h(X) = 0$. (17.4) now implies $[a, b]$ represents 0 in $K[X]$, so by Lemma 17.2 $[a, b]$ represents 0 in K. Since we are supposing to the contrary, this representation must be the trivial one, i.e. $s_j(X) = 0$ for $j = 1, 2, 3, 4$. Hence by (17.3), $g|\gamma_j$ for all j so $g^2|\gamma_1^2 + a\gamma_2^2 + b\gamma_3^2 + ab\gamma_4^2$ i.e. $g|f$ (by (17.1)). But f is irreducible and $\deg g < \deg f$ which is proved below (17.2). Hence $g(X)$ is a constant, say α. Then (17.1) becomes $\gamma_1^2 + a\gamma_2^2 + b\gamma_3^2 + ab\gamma_4^2 = \alpha f \neq 0$. Now apply Lemma 17.2 and we see that $[a, b]$ represents $l(\alpha f) = \alpha l(f) = \alpha$ over K, since f is monic ; so $[a, b]$ also represents $1/\alpha$ in K: if $\alpha = \alpha_1^2 + a\alpha_2^2 + b\alpha_3^2 + ab\alpha_4^2$ then $1/\alpha = \alpha/\alpha^2 = (\alpha_1/\alpha)^2 + a(\alpha_2/\alpha)^2 + b(\alpha_3/\alpha)^2 + ab(\alpha_4/\alpha)^2$, and $[a, b]$ represents αf in $K[X]$. So by the identity, $[a, b]$ represents $f(X)$ in $K[X]$. □

Lemma 17.4. *Let $f(X) \in \mathbb{Q}[X]$ be a non-zero polynomial. Then $[a, b]$ represents $f(X)$ in $\mathbb{Q}[X]$ if and only if $[a, b]$ represents $f(X)$ in each $\mathbb{Q}_p[X]$ ($p = \infty$ included).*

Proof. The 'only if' part is trivial. To prove the converse we proceed as follows.

Let $q^e(X)\|f(X)$ in $\mathbb{Q}[X]$ where $q(X)$ is irreducible and e is odd. Now for any fixed prime p, factor

$$q(X) = q_{1p}(X)\dots q_{s_p p}(X)$$

in $\mathbb{Q}_p[X]$ into irreducible factors. These q_{jp} are all distinct, for if say $q_{1p}^2|q(X)$, then $q_{1p}|q'(X)$; but $(q(X), q'(X))=1$ in $\mathbb{Q}_p[X]$ since $(q(X), q'(X))$ $= 1$ in $\mathbb{Q}[X]$, as $q(X)$ is separable, with char$\mathbb{Q} = 0$.

So the factorization of $f(X)$ into irreducible factors in $\mathbb{Q}_p[X]$ is

$$f(X) = q_{1p}^e(X)\dots q_{s_p p}^e(X)\dots,$$

the $q_{jp}(X)$ having the same odd multiplicity e as $q(X)$. Hence since $[a, b]$ represents $f(X)$ in $\mathbb{Q}_p[X]$ we have, by Lemma 17.3,

(i) $[a, b]$ represents $l(f)$ in $\mathbb{Q}_p$;
(ii) $[a, b]$ represents 0 in $\mathbb{Q}_p[X]/(q_{jp}(X))$ for $1 \le j \le s_p$.

These fields $\mathbb{Q}_p[X]/(q_{jp}(X))$, $1 \le j \le s_p$, $p = \infty$, 2, 3, 5, ... , constitute the totality of all completions of the field $\mathbb{Q}[X]/(q(X))$. So by the Hasse-Minkowski theorem, $[a, b]$ represents 0 in $\mathbb{Q}[X]/(q(X))$ for each prime factor $q(X)$ of $f(X)$ in $\mathbb{Q}[X]$, of odd multiplicity.

Now $[a, b]$ represents $l(f)$ in $\mathbb{Q}_p$ for all p, by (i) above; so the form $[a, b] + (-l(f)X_5^2)$ represents 0 in $\mathbb{Q}_p$ for all p. Hence by the Hasse-Minkowski theorem, it represents 0 in $\mathbb{Q}$. In this representation, if $X_5 \neq 0$, then $[a, b]$ represents $l(f)$ in $\mathbb{Q}$, while if $X_5 = 0$, then $[a, b]$ represents 0 in $\mathbb{Q}$ non-trivially, so is universal (Theorem 11.4). Thus again $[a, b]$ represents $l(f)$ in $\mathbf{Q}$.

We have proved the following:

(i) $[a, b]$ represents $l(f)$ in $\mathbf{Q}$;

(ii) $[a, b]$ represents 0 in $\mathbf{Q}[X]/(q(X))$ for all irreducible factors $q(X)$ of $f(X)$ in $\mathbf{Q}[X]$ of odd multiplicity.

Hence by Lemma 17.3, $[a, b]$ represents $f(X)$ in $\mathbf{Q}[X]$. □

Lemma 17.5. *The form $[a, b]$ represents each element of $\mathbf{Q}_p$ unless $p = \infty$ and a, b are both positive, in which case it represents all positive elements of $\mathbf{R}$.*

Proof. Let $p \neq \infty$ and let $\alpha \in \mathbf{Q}_p$ be any element. Consider the form $[a, b] + (-\alpha)X_5^2$ over $\mathbf{Q}_p$. Since it is a form in five variables, it represents 0 non-trivially in $\mathbf{Q}_p$. In this representation if $X_5 \neq 0$, then on dividing by X_5 we get a representation of α in $\mathbf{Q}_p$ by $[a, b]$ as required. If $X_5 = 0$, then $[a, b]$ represents 0 non-trivially in $\mathbf{Q}_p$ and so is universal. So again $[a, b]$ represents α in $\mathbf{Q}_p$ as required.

For $p = \infty$, the statement is obvious. □

Lemma 17.6. *Let L/K be a separable extension of degree $[L : K]$ and suppose $[a, b]$ does not represent 0 in K. Then*

(i) *for $[L : K]$ odd, $[a, b]$ does not represent 0 in L,*

(ii) *for $[L : K]$ even and $K = \mathbf{Q}_p$, $[a, b]$ does represent 0 in L.*

Proof. (i) We use induction on $[L : K]$. If $[L : K] = 1$, i.e. $L = K$, the result is obvious. So let $[L : K] = n + 1$ where n is even and we suppose the result is true for extensions of degree less than $n + 1$. Since it is a separable extension, we have $L = K(\alpha)$, where $\operatorname{irr}(\alpha, K)$ has degree $n+1$ and elements of L are polynomials in α of degree at most n with coefficients in K.

Suppose to the contrary that $[a, b]$ represents 0 in L. Then by Lemma 17.1, $aX^2 + bY^2 + abZ^2$ represents 0 in L: $af_1^2(\alpha) + bf_2^2(\alpha) + abf_3^2(\alpha) = 0$, $\deg f_j \leq n$. Thus α satisfies the equation

$$af_1^2(X) + bf_2^2(X) + abf_3^2(X) = 0$$

and so $\operatorname{irr}(\alpha, K)$ divides it:

$$af_1^2(X) + bf_2^2(X) + abf_3^2(X) = \operatorname{irr}(\alpha, k) \cdot h(X) \qquad (17.8)$$

Here the degree of the left side, which is at most $2n$, is even since the coefficients of terms of highest degree in X cannot cancel out otherwise we would get a relation $aa_n^2 + bb_n^2 + abc_n^2 = 0$, where a_n, b_n, c_n are the leading coefficients of $f_1(X)$, $f_2(X)$, $f_3(X)$. That is, the form $aX^2 + bY^2 + abZ^2$

represents 0 in K and so $[a,b]$ represents 0 in K which is contrary to the hypothesis.

Hence $\deg h \leq n-1$ and odd. So $h(X)$ has an irreducible factor $p(X)$ of odd degree. Let β be a root of $p(X)$ and divide $f_1(X)$, $f_2(X)$, $f_3(X)$ by $p(X)$:

$$\left.\begin{aligned} f_1(X) &= q_1(X)p(X) + \gamma_1(X) \\ f_2(X) &= q_2(X)p(X) + \gamma_2(X) \\ f_3(X) &= q_3(X)p(X) + \gamma_3(X) \end{aligned}\right\} \tag{17.9}$$

where $\deg \gamma_j < \deg p$ for $j = 1,2,3$. Here not all $\gamma_j(X)$ can be zero, otherwise $p(X)|$ each $f_j(X)$ so $p^2|$ each f_j^2 i.e. $p|\mathrm{irr}(\alpha, K)\cdot h$ by (17.8). But $p \not| \mathrm{irr}(\alpha, K)$ since this would give $\mathrm{irr}(\alpha, K) = p$ (or $p =$ constant) and this is not possible because $\deg p < n-1 < n+1 = \deg(\mathrm{irr}(\alpha, K))$ and $p(X)$ is a proper polynomial (it has a root β). Hence $p^2|h$ and so we could cancel a factor $p^2(X)$ right through (17.8) and go on doing this till we arrive at a set of equations (17.9) with not all $\gamma_j(X) = 0$. So we may already suppose not all $\gamma_j(X)$ are zero in (17.9).

Now $[K(\beta) : K] < n+1$ and odd. So by the induction hypothesis $[a,b]$ does not represent 0 in $K(\beta)$. Now replace f_1, f_2, f_3 in (17.8) by the expressions (17.9) and put $X = \beta$:

$$a\gamma_1^2(\beta) + b\gamma_2^2(\beta) + ab\gamma_3^2(\beta) = 0,$$

where not all $\gamma_j(\beta) = 0$, since $\deg \gamma_j < \deg p$ and β is a root of the irreducible polynomial $p(X)$. Hence by Lemma 17.1, $[a,b]$ represents 0 in $K(\beta)$. This is a contradiction to what was said above.

Remark. Actually the above result is just a special case of Springer's theorem in its full generality as was Theorem 3.7. The general result says that any quadratic form is isotropic over L if it is isotropic over K if the extension L/K is a separable extension of odd degree. In Theorem 3.7 the quadratic form was a sum of squares while in (i) above it is simply $[a,b]$.

(ii) The field L must contain a quadratic extension of $\mathbb{Q}_p$ and for this there are only finitely many possibilities. Further there are only finitely many possibilities for a, b up to a square factor. Hence there are only finitely many cases to look at. One has to distinguish between -1 being a quadratic residue or a non-residue. These are easily cleared. □

The Local Theorem. *Let a, b, $c \in \mathbb{Q}_p^*$ ($p = \infty$ included). Let $f(X) \in \mathbb{Q}_p[X]$ be a non-zero polynomial, separable (i.e. square-free) and of degree $2n$. For $p = \infty$, suppose in addition:*

(α) *if a, b, $c > 0$, then $f(X)$ is positive definite;*

(β) *if a, $b > 0$, $c < 0$, then $f(X)$ is not negative definite and n is odd.*

Then there exist five polynomials $f_1(X), \ldots, f_5(X) \in \mathbb{Q}_p[X]$, *such that*
(i) $f = f_1^2 + af_2^2 + bf_3^2 + abf_4^2 + cf_5^2$,
(ii) $\deg(f - cf_5^2) = 2n$,
(iii) $\gcd(f_1, f_2, f_3, f_4, f_5) = 1$.

Proof. If $[a, b]$ represents 0 in $\mathbb{Q}_p$ (p finite or ∞) then it represents 0 in $\mathbb{Q}_p(X)$ so it represents all elements of $\mathbb{Q}_p(X)$; in particular it represents $f(X)$. On taking $f_5 = 0$, we get the result, noting that condition (ii) is then trivial, while (iii) follows since if $g|$ all f_j then $g^2|f$, which is false since f is separable. So in future suppose $[a, b]$ does not represent 0 in $\mathbb{Q}_p$.

First let p be finite.

Case 1. $f(X)$ *has all its prime factors of even degree.*

Let $p(X)$ be such a factor. The field $\mathbb{Q}_p[X]/(p(X))$ is an extension of $\mathbb{Q}_p$ of degree equal to the degree of $p(X)$, which is even. Hence by Lemma 17.6, $[a, b]$ represents 0 in $\mathbb{Q}_p[X]/(p(X))$. Also $[a, b]$ represents $l(f)$ in $\mathbb{Q}_p$ by Lemma 17.5. Hence by Lemma 17.3, $[a, b]$ represents $f(X)$ in $\mathbb{Q}_p[X]$ again, so our result follows with $f_5 = 0$ again.

Case 2. $f(X)$ *has a prime factor of odd degree.*

Call it $p(X)$, where without loss of generality, $\deg p \leq n$; for otherwise if $f = p \cdot g$ where $\deg g < n$ then g has a factor of odd degree less than n, which will do instead of $p(X)$ to begin with.

Lemma A. *There exists a polynomial* $g(X)$ *of degree* n, $g \nmid f$ *and polynomials* f_2, f_3, f_4, *not all zero and each of degree less than* n *such that*

$$f(X) \equiv af_2^2(X) + bf_3^2(X) + abf_4^2(X) \pmod{g(X)}$$

i.e. $f = af_2^2 + bf_3^2 + abf_4^2 + gh$, *so* $\deg h = n$ *since* $\deg f_2, f_3, f_4 < n$.

Proof. We have the following subcases.
1. n *even* (where $\deg f = 2n > 0$). Let $g(X)$ be any irreducible polynomial of $\deg n$ which $\nmid f(X)$ (there are an infinity of such polynomials). Let $L = \mathbb{Q}_p[X]/(g(X))$ so that $[L : \mathbb{Q}_p] = \deg g = n$ is even. By Lemma 17.6, $[a, b]$ represents 0 in L; so by Lemma 17.1, the form $aX^2 + bY^2 + abZ^2$ represents 0 in L and hence represents all elements of L, in particular $f(X) + (g(X))$, i.e. there exist polynomials f_2, f_3, $f_4 \in \mathbb{Q}_p[X]$ such that

$$a(f_2 + (g))^2 + b(f_3 + (g))^2 + ab(f_4 + (g))^2 = f + (g),$$

$af_2^2 + bf_3^2 + abf_4^2 \equiv f \pmod g$ as required. Note that the degrees of f_2, f_3, f_4 can always be made less than $n = \deg g$ by dividing by g and considering the remainders instead.

2. *n odd.* As we are in Case 2 of the theorem we may suppose $f(X)$ has a prime factor $p(X)$ of odd degree at most n. If this $p(X)$ has degree n, then as $f(X)$ is separable, there exists a $\xi \in \mathbb{Q}_p^*$, sufficiently near 0 in the topology of $\mathbb{Q}_p$, i.e. divisible by a high power of p, such that $f(X) - a\xi^2$ also has a prime factor of degree n; call it $g(X)$. (This follows from a problem given in Bourbaki: [B3], Chapter 6, Problem 12, p.465. However it is desirable to dispense with the rather deep machinery of Bourbaki and so we give an elementary and easier reference; see the Appendix to this chapter.)

Thus $f(X) - a\xi^2 \equiv 0(\bmod g(X))$ which solves the required congruence with $f_2 = \xi$, $f_3 = f_4 = 0$.

If $\deg p(X) < n$, we proceed as follows: select any irreducible polynomial $q(X)$ of degree $n - \deg p$ (which is even and positive) and which does not divide $f(X)$. As in Subcase 1 above, if $L = \mathbb{Q}_p[X]/(q(X))$ so that $[L : \mathbb{Q}_p] = \deg q$, which is even, then,

$$f(X) \equiv a\phi_1^2(X) + b\phi_2^2(X) + ab\phi_3^2(X) \qquad (\bmod q(X))$$

is solvable. Also $f(X) \equiv 0 \; (\bmod p(X))$ since $p|f$ i.e. $f(X) \equiv a0^2 + b0^2 + ab0^2$ $(\bmod p(X))$ is solvable. Letting $g(X) = p(X) \cdot q(X)$ we get the lemma.

□

Lemma B. *The local theorem holds with* $c = -1$. *More explicitly let* $f(X)$ *satisfy the hypothesis of the local theorem. Then*

$$f(X) = f_1^2(X) + af_2^2(X) + bf_3^2(X) + abf_4^2(X) - f_5^2(X)$$

where $\deg(f + f_5^2) = 2n$ *and* $\gcd(f_1, f_2, f_3, f_4) = 1$.

Proof. By Lemma A, for any $\lambda \in \mathbb{Q}_p^*$, we have

$$\begin{aligned} f &= \left(\frac{\lambda g + \lambda^{-1}h}{2}\right)^2 + af_2^2 + bf_3^2 + abf_4^2 - \left(\frac{\lambda g - \lambda^{-1}h}{2}\right)^2 \\ &= f_1^2 + af_2^2 + bf_3^2 + abf_4^2 - f_5^2 \end{aligned}$$

as required. However, we must check the other conditions: let

$$g(X) = \alpha_n X^n + \dots$$
$$h(X) = \beta_n X^n + \dots .$$

Then

$$f(X) = \alpha_n \beta_n X^{2n} + \dots,$$

so

$$f + f_5^2 = \alpha_n \beta_n X^{2n} + \left(\frac{\lambda\alpha_n - \lambda^{-1}\beta_n}{2}\right)^2 X^{2n} + \dots$$

and this has degree $2n$ if and only if $\lambda^2 + \beta_n/\alpha_n \neq 0$. Thus if we avoid just these two values of λ, then

$$\deg(f + f_5^2) = 2n.$$

Next, f_2, f_3, f_4 can have only finitely many irreducible common factors. Let $p(X)$ be such a factor; we shall show that if λ is suitably chosen then $p(X) \not| f_1(X)$. This will complete the proof of the last condition (iii).

Now $p \not|$ both g and h for then $p^2 | f$ which is false since f is separable. We see that $p | f_1$ if and only if $\lambda^2 g + h \equiv 0 \pmod{p(X)}$. We claim that this happens for at most one value of λ^2, i.e. for at most two values of λ:

$$\lambda_1^2 g + h \equiv 0(p), \quad \lambda_2^2 g + h \equiv 0(p)$$

implies

$$(\lambda_1^2 - \lambda_2^2)g \equiv 0(p), \quad (\lambda_1^2 - \lambda_2^2)h \equiv 0(p),$$

which implies $p|g$ and $p|h$ unless $\lambda_1^2 = \lambda_2^2$, as required.

So if λ avoids these two values, then $p \not| f_1$. Since $p(X)$ can have only finitely many possibilities, we need only select λ avoiding these corresponding values and the two earlier ones. Since $\mathbf{Q}_p$ is infinite, such a selection of λ is possible and all the requirements of Lemma B are satisfied. □

Proof of the local theorem for $p \neq \infty$. Let $\phi(X) = (-1/c)f(X)$ and apply Lemma B to $\phi(X)$. So there exist $\phi_1, \phi_2, \phi_3, \phi_4, \phi_5 \in \mathbf{Q}_p[X]$ such that

(1) $\phi = \phi_1^2 + a\phi_2^2 + b\phi_3^2 + ab\phi_4^2 - \phi_5^2$
(2) $\deg(\phi + \phi_5^2) = 2n$
(3) $\gcd(\phi_1, \phi_2, \phi_3, \phi_4) = 1$.

By Lemma 17.5, $[a, b]$ represents (in particular) $-c$:

$$-c = r_1^2 + ar_2^2 + br_3^3 + abr_4^2.$$

Then

$$\begin{aligned} f(X) &= -c\phi(X) \\ &= -c(\phi_1^2 + a\phi_2^2 + b\phi_3^2 + ab\phi_4^2 - \phi_5^2) \\ &= (r_1^2 + ar_2^2 + b_3^2 + abr_4^2)(\phi_1^2 + a\phi_2^3 + b\phi_3^2 + ab\phi_4^2) + c\phi_5^2 \\ &= (r_1\phi_1 + ar_2\phi_2 + \ldots)^2 + a(\ldots)^2 + b(\ldots)^2 + ab(\ldots)^2 + c\phi_5^2 \\ &= f_1^2 + af_2^2 + bf_3^2 + abf_4^2 + cf_5^2 \end{aligned}$$

where $f_5 = \phi_5$, $f_1 = r_1\phi_1 + r_2\phi_2 + \ldots$, etc.

We now have to verify (ii) and (iii) of the local theorem:

(ii) $\deg(f - cf_5^2) = \deg(-c\phi - c\phi_5^2) = 2n$ by 2 above.

(iii) $\gcd(f_1, f_2, f_2, f_4) = \gcd(r_1\phi_1 + ar_2\phi_2 + br_3\phi_3 + abr_4\phi_4, -r_1\phi_2 + r_2\phi_1 - br_3\phi_4 + br_2\phi_3, -r_1\phi_3 + r_3\phi_1 + ar_2\phi_4 - ar_1\phi_2, -r_1\phi_4 + r_4\phi_1 - r_2\phi_3 + r_3\phi_2) =$

$(\phi_1, \phi_2, \phi_3, \phi_4)$ since the connecting determinant, which is

$$\begin{vmatrix} r_1 & ar_2 & br_3 & abr_4 \\ r_2 & -r_1 & br_2 & -br_3 \\ r_3 & -ar_4 & -r_1 & ar_2 \\ r_4 & r_3 & -r_2 & -r_1 \end{vmatrix}$$

evaluates to

$$-(r_1^2 + ar_2^2 + br_3^2 + abr_4^2) = -c^2 \neq 0.$$

We have used the generalization to four terms of the result that $(g_1, g_2) = (ag_1 + bg_2, cg_1 + dg_2)$ if $ad - bc \neq 0$.

But $(\phi_1, \phi_2, \phi_3, \phi_4) = 1$ by (3) above, so $(f_1, f_2, f_3, f_4) = 1$. This completes the proof of the local theorem in the non-archimedean case.

Finally, let $p = \infty$, so that $\mathbb{Q}_p = \mathbb{R}$. Since $[a, b]$ does not represent 0 in $\mathbb{R}$ then clearly $a, b > 0$. We have again to consider cases.

Case 1. $f(X)$ has a factor of degree n in $\mathbb{R}[X]$ and $c < 0$:
Write $-c = r^2$. Now we want

$$f = f_1^2 + af_2^2 + bf_3^2 + abf_4^2 - (rf_5)^2$$

which is the case $c = -1$ and, as in Lemma B, we get the result.

Case 2: $c > 0$. Then f is positive definite by (α) and of course separable. So the irreducible factors of $f(X)$ in $\mathbb{R}[X]$ are all quadratic, which we see by factorizing $f(X)$ in $\mathbb{C}[X]$ as:

$$f(X) = \prod_{j=1}^{r}(X - \alpha_j) \prod_{j=1}^{s}(X - (\beta_j + ir_j)(X - (\beta_j - ir_j)).$$

If to the contrary $r > 0$, then take X rational satisfying $\alpha_{r-1} < x < \alpha_r$ if $r \geq 2$, $X < \alpha_r$ if $r = 1$, (without loss of generality we suppose $\alpha_1 < \ldots < \alpha_r$). Then for this value of X, $f(X) < 0$ since the complex factors are always positive.

Further the quadratic factors of f are all to the exponent 1 since f is separable. Denote any such factor by $p(X)$. Then $\mathbb{R}[X]/(p(X)) \simeq \mathbb{C}$ and so $[a, b]$ represents 0 in $\mathbb{R}[X]/(p(X))$. Also $[a, b]$ represents $l(f)$ by Lemma 17.5, since $l(f) > 0$, as f is positive definite. Hence by Lemma 17.3, $[a, b]$ represents $f(X)$ in $\mathbb{R}[X]$ and the result follows with $f_5 = 1$.

Case 3. $f(X)$ has no factor of degree n in $\mathbb{R}[X]$: Let the factoriztion of $f(X)$ in $\mathbb{R}[X]$ be

$$f(X) = l(f)(X - \alpha_1)\ldots(X - \alpha_r)(X^2 + \beta_1 X + \gamma_1)\ldots(X^2 + \beta_s X + \gamma_s).$$

Hence $2n = r + 2s$, so $2|r$ and we can always get a factor of $f(X)$ of degree n unless $r = 0$. So $r = 0$ giving $s = n$, which is odd.

Now if $l(f) < 0$, then f is negative definite, since each factor $X^2 + \beta X + \gamma = (X + \frac{\beta}{2})^2 + (\frac{-D}{4}) > 0$ because $D = \beta^2 - 4\gamma < 0$. So $c > 0$ for otherwise, by hypothesis, f is not negative definite (don't forget $a, b > 0$, n is odd). But then $a, b, c > 0$, so f is positive definite. It follows that $f = 0$, which is a contradiction. Hence $l(f) > 0$. So f is positive definite and the factorization of $f(X)$ is

$$f(X) = l(f)(X^2 + \beta_1 X + \gamma_1) \dots (X^2 + \beta_s X + \gamma_s),$$

where s is odd and all factors are distinct since f is separable. Then exactly as in Case (ii) above we get the theorem.

Note that once f is shown to be positive definite (as in the above or in Case (ii)) we could use the fact that such an f is a sum of two squares in $\mathsf{R}(X)$; see Theorem 4.1.

Lastly note that a, b cannot have opposite signs for then $[a, b]$ represents 0 in $\mathbf{R}$ which we have supposed is not the case.

This completes the proof of the local theorem. □

Remark. The hypothesis (α) and (β) in the theorem are necessary.

Proof. Clearly (i) implies (α). To see that (i) and (ii) implies (β) we proceed as follows: let $a, b > 0$, $c < 0$, and suppose to the contrary that f is negative definite, where $\deg f = 2n$, n odd. We have

$$f = f_1^2 + af_2^2 + bf_3^2 + abf_4 + cf_5^2$$

$$\deg(f - cf_5^2) = 2n.$$

Write $a = a_1^2$, $b = b_1^2$, $-c = c_1^2$ and put $g_1 = f_1$, $g_2 = a_1 f_2$, $g_3 = b_1 f_3$, $g_4 = a_1 b_1 f_4$, $g_5 = c_1 f_5$. Then

$$f = g_1^2 + g_2^2 + g_3^2 + g_4^2 - g_5^2$$

with $\deg(f + g_5^2) = 2n$, i.e. $f + g_5^2 = g_1^2 + g_2^2 + g_3^2 + g_4^2$; so by Lemma 9.2, $\max_{1 \leq j \leq 4} \deg g_j = n$. Without loss of generality let $\deg g_1 = n$. Since $\deg(f + g_5^2) = 2n$, $\deg g_5 \leq n$. Thus the highest term in $g_1 \pm g_5$ cannot cancel out for both signs, i.e. $\deg(g_1 \pm g_5) = n$ for at least one sign. Moreover, as n is odd, $g_1 \pm g_5$ has a real root θ (the sign chosen so that the degree is n). Now put $X = \theta$ in the relation $f = (g_1 + g_5)(g_1 - g_5) + g_2^2 + g_3^2 + g_4^2$ to give, since f is negative definite,

$$0 \leq g_2^2(\theta) + g_3^2(\theta) + g_4^2(\theta) = f(\theta) \leq 0.$$

It follows that $f(\theta) = 0$. But f is separable, so $f'(\theta) \neq 0$, $f(X)$ is continuous at $X = \theta$, $f(\theta) = 0$, and $f'(\theta) \neq 0$. Hence f changes sign at θ, contradicting the fact that $f(X)$ is negative definite. This completes the proof of the remark. □

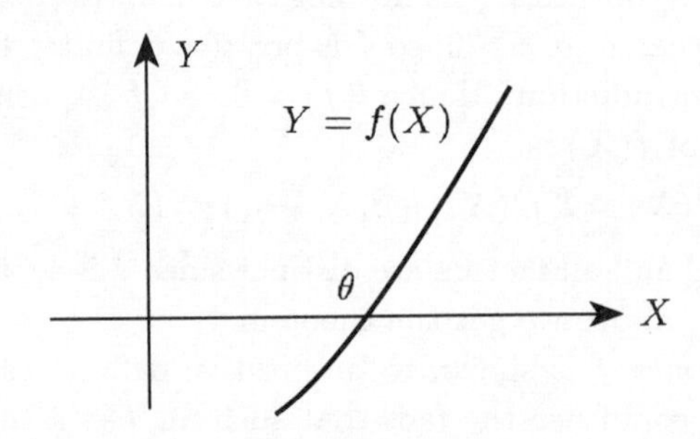

The Approximation Lemma. *Let $f(X)$ be as in the Local Theorem, i.e. $f = f_1^2 + af_2^2 + bf_3^2 + abf_4^2 + cf_5^2$, $\deg(f - cf_5^2) = 2n$, $\gcd(f_1, f_2, f_3, f_4) = 1$. Let $|\ |_p$ be the valuation defined by the topology of $\mathbb{Q}_p$. Then there exists $\eta > 0$, such that for all polynomials $g(X) \in \mathbb{Q}_p[X]$, with $\deg g \leq n$, $|g(X) - f_5(X)|_p < \eta$, the polynomial $f(X) - cg^2(X)$ is of degree $2n$ and is represented by $[a, b]$ in $\mathbb{Q}_p[X]$.*

Remark. If $f(X) = a_0 + a_1X + \ldots + a_nX^n$, then by $|f|_p$, we simply mean $\max_{0 \leq j \leq n} |a_j|_p$. If $|f - g|_p$ is small, we say f is *near* g.

The *moral of the Local Theorem* is that $[a, b]$ represents $f - cf_5^2$ while the *moral of the Approximation Lemma* is that given g near f_5 and that $[a, b]$ represents $f - cf_5^2$, we have $[a, b]$ represents $f - cg^2$.

For a proof of the Approximation Lemma see the Appendix to this chapter.

The Global Theorem. *Let a, b, $c \in \mathbb{Q}^*$. Let $f(X) \in \mathbb{Q}[X]$ be a non-zero polynomial of degree $2n$. Suppose*

(α) *whenever a, b, $c > 0$, f is positive definite;*

(β) *whenever a, $b > 0$, $c < 0$, f is not negative definite and n is odd.*

Then there exist f_1, f_2, f_3, f_4, $f_5 \in \mathbb{Q}[X]$ such that

(1) $f = f_1^2 + af_2^2 + bf_3^2 + abf_4^2 + cf_5^2$;

(2) $\deg(f - cf_5^2) = 2n$.

Proof. We may suppose f is separable for, writing $f = g^2 f_1$, it is enough to prove the result for f_1.

Let $\Omega = \{2, 3, 5, 7, \ldots, \infty\}$ be the set of all primes of $\mathbb{Q}$ and let $T = \{p \in \Omega | [a, b]$ does not represent 0 in $\mathbb{Q}_p\}$.

We may disregard the infinite prime as $[a, b]$ represents 0 if and only if

$aX^2 + bY^2 + abZ^2$ represents 0 (see Lemma 17.1); and for the case of three variables if the form represents 0 in all but one local $\mathbb{Q}_p$ then it represents 0 in that remaining local $\mathbb{Q}_p$ and so in $\mathbb{Q}$ by the Hasse-Minkowski theorem.

In T is $\emptyset$, then $[a, b]$ represents 0 in all $\mathbb{Q}_p$ and so in $\mathbb{Q}$ and so in $\mathbb{Q}(X)$. Hence $[a, b]$ represents all elements of $\mathbb{Q}(X)$, in particular $f(X)$; and the global theorem follows with $f_5 = 0$.

Let $T \neq \emptyset$. For $p \in T$, there exist $f_{1p}, f_{2p}, f_{3p}, f_{4p}, f_{5p} \in \mathbb{Q}_p[X]$ satisfying the conditions of the local theorem. Then for each $p \in T$, find an $\eta_p > 0$ using the Approximation Lemma, such that if $g(X) \in \mathbb{Q}_p[X]$, $\deg g \leq n$, $|g(X) - f_{5p}(X)|_p < \eta_p$, then

$$f - cg^2 \text{ is of degree } 2n \text{ and is represented by } [a, b] \text{ in } \mathbb{Q}_p[X]. \qquad (17.10)$$

By the weak approximation theorem on valuations, there exists a polynomial $f_5(X) \in \mathbb{Q}[X]$ such that

$$|f_5(X) - f_{5p}(X)|_p < \eta_p$$

for all $p \in T$. Then $f(X) - cf_5^2(X) \in \mathbb{Q}[X] \subset \mathbb{Q}_p[X]$ and is of degree $2n$ (since f_5 is close to f_{5p} for all $p \in T$, so the highest terms cannot cancel out as they do not cancel out in $f - cf_{5p}^2$), and is represented by $[a, b]$ in $\mathbb{Q}_p[X]$ by (17.10), for all $p \in T$. For $p \notin T$, $f - cf_5^2$ is represented by $[a, b]$ in $\mathbb{Q}_p[X]$ by the definition of T. Thus for each p, the polynomial $f(X) - cf_5^2(X)$, belonging to $\mathbb{Q}[X]$, is represented by $[a, b]$ in $\mathbb{Q}_p[X]$. Hence by Lemma 17.4, it is represented by $[a, b]$ in $\mathbb{Q}[X]$, i.e.

$$f = f_1^2 + af_2^2 + bf_3^2 + abf_4^2 + cf_5^2. \qquad \square$$

We are now in a position to prove our main theorem.

Theorem 17.1 (Pourchet). *Let $f(X) \in \mathbb{Q}[X]$ be a positive definite polynomial; then $f(X)$ is a sum of five squares of polynomials in $\mathbb{Q}[X]$.*

Proof. Take $a = b = c = 1$ in the global theorem. We see that f is of even degree, say $2n$. The condition $\deg(f - f_5^2) = 2n$ is a side condition, the main result being $f = f_1^2 + f_2^2 + f_3^2 + f_4^2 + f_5^2$, $f_i \in \mathbb{Q}[X]$. $\square$

Not all positive definite functions require their full quota of five squares for the representation. We wish to classify those polynomials which can be expressed as a sum of four squares. We have the following.

Theorem 17.2 (Pourchet). *Let $f(X) \in \mathbb{Q}[X] - \{0\}$. The following conditions are equivalent:*

(i) *$f(X)$ is a sum of 4 squares in $\mathbb{Q}[X]$;*

(ii) *$l(f) > 0$ and for all $p(X)|f(X)$, where $p(X)$ is irreducible and of odd* *multiplicity as a factor of $f(X)$, the Stufe of $\mathbb{Q}[X]/(p(X))$ is at most 2;*

(iii) *$f(X)$ is positive definite and in $\mathbb{Q}_2[X]$, every prime factor of f of odd multiplicity has even degree.*

Proof. The equivalence of (i) and (ii) follows from Lemma 17.3 with $a = b = 1$, noting that every rational is a sum of four rational squares.

Proof of (i) implies (iii). Clearly $f(X)$ is positive definite. If $p^e(X) \,\|\, f(X)$, where $p(X)$ is irreducible and e is odd, then by Lemma 17.3 again, $[a, b] = [1, 1]$ represents 0 in $\mathbb{Q}_2[X]/(p(X))$. Since the Stufe of $\mathbb{Q}_2 = 4$, $[1, 1]$ does not represent 0 in $\mathbb{Q}_2$ and the degree of $p(X)$ has to be even by Lemma 17.6 as required.

Proof of (iii) implies (i). We shall show that $f(X)$ is a sum of four squares in each $\mathbb{Q}_p[X]$ ($p = 2, 3, \dots, \infty$). Then by Lemma 17.4, $f(X)$ is a sum of four squares in $\mathbb{Q}[X]$. Now $f(X)$ is positive definite, so a sum of four squares in $\mathbb{Q}_\infty[X] = \mathbb{R}[X]$ (indeed a sum of two squares, by Theorem 4.1) and by Lemma 17.3, $f(X)$ is a sum of four squares in $\mathbb{Q}_2[X]$.

For $p \neq 2, \infty$, $[1, 1]$ represents 0 in $\mathbb{Q}_p$, hence in $\mathbb{Q}_p(X)$ and so is universal in $\mathbb{Q}_p(X)$. Thus in particular, $[1, 1]$ represents $f(X)$ in $\mathbb{Q}_p(X)$ hence in $\mathbb{Q}_p[X]$. □

For quadratic polynomials, we have a very simple condition for being expressible as a sum of four squares:

Theorem 17.3 (Pourchet). *Let $f(X) = aX^2 + bX + c \in \mathbb{Q}[X]$, $a \neq 0$. Then $f(X)$ is a sum of four squares in $\mathbb{Q}[X]$ if and only if $a > 0$ and $4ac - b^2$ is a sum of three squares in $\mathbb{Q}$.*

Proof. We have $f(X) = \frac{1}{a}\{(aX + \frac{b}{2})^2 + (4ac - b^2)/4\}$. First suppose $a > 0$ and $4ac - b^2$ is a sum of three squares in $\mathbb{Q}$. Then $1/a$, being a positive rational, is a sum of four squares in $\mathbb{Q}$, while the curly bracket above is visibly a sum of four squares, so by the Euler identity, $f(X)$ is a sum of four squares in $\mathbb{Q}[X]$. □

Conversely let $aX^2 + bX + c = f_1^2 + f_2^2 + f_3^2 + f_4^2$, where the $f_j(X) \in \mathbb{Q}[X]$. These f_j are necessarily linear for otherwise, by equating to zero the coefficients of the highest power of X, gives 0 as a sum of squares in $\mathbb{Q}$.

Thus

$$aX^2 + bX + c = \sum_1^4 (a_jX + b_j)^2$$
$$= \left(\sum a_j^2\right) X^2 + \left(2\sum a_jb_j\right) X + \sum b_j^2.$$

Then

$$4ac - b^2 = 4\left(\sum a_j^2\right)\left(\sum b_j^2\right) - 4\left(\sum a_jb_j\right)^2$$
$$= 4\left[\left(\sum a_jb_j\right)^2 + \quad \text{three more squares}\right] - 4\left(\sum a_jb_j\right)^2$$
$$= \quad \text{a sum of three squares.}$$

□

The following result characterizes polynomials that can be expressed as a sum of two squares.

Theorem 17.4 (Pourchet). *Let $a \in \mathbb{Q}^*$ and let $f(X) \in \mathbb{Q}[X] - \{0\}$. The form $X_1^2 + aX_2^2$ represents $f(X)$ in $\mathbb{Q}[X]$ if and only if:*

(i) *it represents $l(f)$ in $\mathbb{Q}$;*

(ii) *for each irreducible $p(X)|f(X)$, with an odd multiplicity, the field $\mathbb{Q}[X]/(p(X))$ (as an extension of $\mathbb{Q}$) $\supset \mathbb{Q}(\sqrt{-a})$.*

Proof. First suppose

$$f = f_1^2 + af_2^2 \tag{17.11}$$

In $\mathbb{Q}[X]/(p(X))$, this becomes

$$f(X) + (p(X)) = \{f_1^2(X) + (p(X))\} + \{af_2^2(X) + (p(X))\},$$

i.e. $0 = \overline{f}_1^2 + \overline{a}\overline{f}_2^2$ (since $p|f$), i.e. $-\overline{a} = \overline{f}_1^2/\overline{f}_2^2$ so $\sqrt{-\overline{a}} = \overline{f}_1/\overline{f}_2$ as required.

To prove (i) equate the leading coefficients on both sides of (17.11). If the highest degree terms are of the same degree in each of f, f_1^2, f_2^2, then we get $l(f) = (l(f_1))^2 + a(l(f_2))^2$.

If $\deg f = 2n$ and $\deg f_1^2 = 2n$, $\deg f_2^2 < 2n$ or $\deg f_1^2 < 2n$, $\deg f_2^2 = 2n$, then we get $l(f) = (l(f_1))^2$ or $a(l(f_2))^2$.

If $\deg f$ is odd, then $\deg f_1^2$, $\deg f_2^2$ exceeds $\deg f$ and we have a cancellation:

$$(l(f_1))^2 + a(l(f_2))^2 = 0.$$

So $X_1^2 + aX_2^2$ represents 0 in $\mathbb{Q}$, so in $\mathbb{Q}(X)$ and hence represents all elements of $\mathbb{Q}(X)$, in particular it represents $f(X)$.

To prove the converse, first let $f(X)$ be irreducible and monic. Suppose

$E = \mathbb{Q}[X]/(f(X)) \supset \mathbb{Q}(\sqrt{-a}) = F$. Then $-a$ is a square in E, i.e. there exists $g(X) \in \mathbb{Q}[X]$ such that

$$a \equiv g^2(X) \pmod{f(X)},$$

i.e. $a + g^2 \equiv o(f)$.

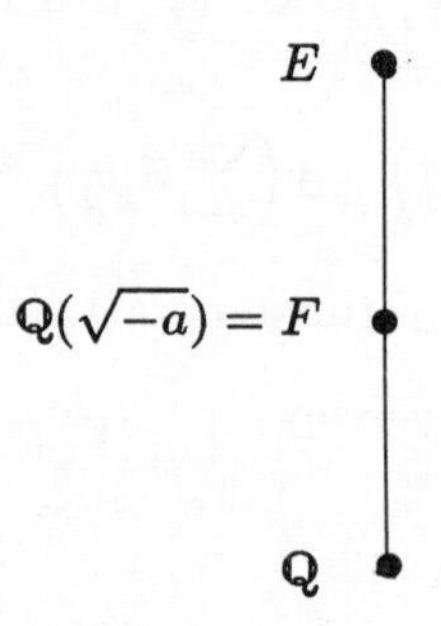

Now $E = \mathbb{Q}(\theta)$, θ a root of $f(X)$. Then

$$\begin{aligned} f(X) &= \prod (X - \theta^{(j)}) = N_{E/\mathbb{Q}}(X - \theta) \\ &= N_{F/\mathbb{Q}}(N_{E/F}(X - \theta)). \end{aligned}$$

We have $N_{E/F}(X - \theta) = g(X) + \sqrt{-a}h(X)$; for any extension E/F, if $\sigma_1, \ldots, \sigma_r$ are the embeddings of E (over F) in any algebraic closure and if

$$f(X) = a_0 + a_1 X + \ldots + a_n X^n \in E[X],$$

then, by definition $N_{E/F}(f(X)) = \prod_1^r f^{\sigma_i}(X)$, where $f^{\sigma_i}(X) = \sigma_i(a_0) + \sigma_i(a_1)X + \ldots + \sigma_i(a_n)X^n$). So

$$\begin{aligned} f(X) &= N_{F/\mathbb{Q}}(g(X) + \sqrt{-a} \cdot h(X)) \\ &= (g + \sqrt{-a}h)(g - \sqrt{-a}h) \\ &= g^2 + ah^2. \end{aligned}$$

If $f(X)$ is general, then an argument similar to that in the proof of Lemma 17.3 gives the result, the identity required being

$$(X_1^2 + aX_2^2)(Y_1^2 + aY_2^2) = (X_1Y_1 + aX_2Y_2)^2 + a(X_1Y_2 - X_2Y_1)^2. \quad \square$$

Appendix for Chapter 17

First we deal with Subcase 2 (n odd) of Lemma A regarding the factorization of $f(x) - a\xi^2$.

Lemma (i). *Let $u(X), v(X) \in \mathbb{Q}_p[X]$ have precise degrees r, s and suppose that $u(X), v(X)$ have no common factors. Then there is a neighbourhood $\mathcal{M}$ of $u(X)v(X)$ in the space of polynomials in $Q_p[X]$ of degree $r + s$ such* *that every $H(X) \in \mathcal{M}$ factorizes*

$$H(X) = u_H(X) \cdot v_H(X)$$

where $u_H(X), v_H(X)$ have degrees r, s respectively and they are near $u(X)$, $v(X)$.

Proof. See the proof of Lemma 4.1 of Chapter 6, p. 105, of Cassels [C2], which assumes an additional condition but which can be easily adapted to give a proof of what we need. □

Now we give a proof of the Approximation Lemma as promised earlier. We need the following

Lemma (ii). *Let $f_1, \ldots, f_m \in \mathbf{Q}_p[X]$ with no common factors and $\max_j(\deg f_j) = n$. Then every $s \in \mathbf{Q}_p[X]$ of degree at most $2n$ can be put in the form*

$$s = \Sigma f_j h_j \qquad (*)$$

where $h_j \in \mathbf{Q}_p[X], \deg h_j \leq n$.

Furthermore there is an $e > 0$ depending only on $f_1, \ldots, f_m$ such that the h_j can be chosen with

$$|h_j| \leq e|s| \quad (1 \leq j \leq m),$$

where for $f(X) \in \mathbf{Q}_p[X]$, we denote by $|f|$ the maximum of the value of the coefficients: $|\Sigma a_j X^j| = \max |a_j|_p$.

Proof. Since the f_j have no common factor, there are certainly $h_j(X) \in \mathbf{Q}_p[X]$ satisfying $(*)$.

Without loss of generality put $\deg f_1 = n$. For $j > 1$ let $h_j = u_j f_j + v_j$ where $u_j, v_j \in \mathbf{Q}_p[X]$, and $\deg v_j < n$. We may replace the h_j by

$$h_1 + \Sigma u_j f_j, \quad v_j (j \neq 1)$$

and then $\deg h_j < n$ $(j \neq 1)$. Clearly $\deg s \leq 2n$ now implies $\deg h_1 \leq n$.

In particular, there are $h_j^{(r)} \in \mathbf{Q}_p[X]$ of degree at most n such that

$$\sum_j h_j^{(r)} f_j = X^r \quad (0 \leq r \leq 2n).$$

If now $s(X) = \sum_0^{2n} s_r X^r$ $(s_r \in \mathbf{Q}_p)$, we may take

$$h_j = \sum_r h_j^{(r)} s_r.$$

Then

$$\sum h_j f_j = s \text{ and } |h_j| \leq |s| \max_{j,r} |h_j^{(r)}| = e|s|,$$

say. □

The Approximation Lemma. *Let $c_j \in \mathbb{Q}_p (1 \le j \le m)$. Let $g \in \mathbb{Q}_p[X]$ be of degree $2n$ and square-free. Suppose g is representable as*

$$g = \sum c_j f_j^2,$$

where $f_j \in \mathbb{Q}_p[X]$, $\deg f_j \le n$. Then so is every g^* of degree $2n$ in some p-adic neighbourhood of g.

Proof. Without loss of generality put $|c_j|_p \le 1\ (1 \le j \le m)$. We note that $\Sigma c_j (f_j + h_j)^2 = \Sigma c_j f_j^2 + \Sigma 2 c_j f_j h_j + \Sigma c_j h_j^2$. Since g is square-free, the f_j have no common factor.

Now let e be as in Lemma (ii) above, but with $2c_j f_j$ instead of f_j. Suppose that

$$|g^* - g| \le 1/2e^2.$$

By Lemma (ii), we may choose h_j so that

$$\sum_j 2 c_j f_j h_j = g^* - g,$$

$$|h_j| \le e|g - g^*| \le 1/2e.$$

Then

$$|g^* - \sum c_j (f_j + h_j)^2 = |\sum c_j h_j^2| \le e^2 |g^* - g|^2 \le \frac{1}{2}|g^* - g|.$$

We now proceed by successive approximation. Suppose that we have already found $F_j \in \mathbb{Q}_p[X]$ of degree at most n such that

$$|F_j| \le 1/2e,$$

and $(1/2e^2) > |g^* - \sum_j c_j (f_j + F_j)^2| = \eta$ (say). Consider the following sum, where the h_j are to be determined: $\sum c_j (f_j + F_j + h_j)^2$; this is

$$\sum c_j (f_j + F_j)^2 + 2\sum c_j f_j h_j + 2\sum c_j F_j h_j + \sum c_j h_j^2.$$

We may choose the h_j so that $\sum 2 c_j f_j h_j = g^* - \sum c_j (f_j + F_j)^2$,

$$|h_j| \le e\eta.$$

Then

$$|g^* - \sum c_j (f_j + F_j + h_j)^2| \le \max\{\max_j |2 c_j F_j h_j|, \max_j |h_j|^2\}.$$

Here $|2 c_j F_j h_j| \le \frac{1}{2e} \cdot e\eta \le \eta/2$, $|h_j|^2 \le e^2 \eta^2 \le \eta/2$.

Hence we may replace F_j by $F_j + h_j$ and η by $\eta/2$. The F_j clearly converge to limits F_j^* and $\sum c_j (f_j + F_j^*)^2 = g^*$ as required. □

18

Examples of the Stufe and pythagoras number of fields using the Hasse-Minkowski theorem

We now make use of the Hasse-Minkowski theorem and discuss examples of the Stufe and pythagoras number of fields, so that this chapter is really a sequel to Chapters 3 and, partly, 16, except that they were free of the Hasse-Minkowski theorem.

We shall take up algebraic number fields first and give, to begin with, a quick survey of some preliminaries that we shall require.

Let K be an algebraic number field, i.e. a finite extension of $\mathbb{Q}$ with say $[K : \mathbb{Q}] = n$. For a rational prime p, we have the decomposition in K

$$(p) = \mathfrak{p}^{e_\mathfrak{p}} \ldots$$

(1) $e_\mathfrak{p}$ is called the *ramification index* of $\mathfrak{p}$ over p.

(2) If $N_{K/\mathbb{Q}}(\mathfrak{p}) = p^{f_\mathfrak{p}}$, then $f_\mathfrak{p}$ is called the residue class degree of $\mathfrak{p}$ over p.

We have the fundamental relations

(3) (i) $\sum_{\mathfrak{p}|p} e_\mathfrak{p} f_\mathfrak{p} = n$.
(ii) $[K_\mathfrak{p} : \mathbb{Q}_p] = e_\mathfrak{p} f_\mathfrak{p}$.

(4) If $\mathfrak{p}$ is an infinite prime of K (K can be $\mathbb{Q}$), i.e. $\mathfrak{p}$ is either one of the r real embeddings of K in $\mathbb{R}$ or one of the $2s$ complex embeddings of K in $\mathbb{C}$, so that $r + 2s = n$, then $K_\mathfrak{p} = \mathbb{R}$ if $\mathfrak{p}$ is real and $K_\mathfrak{p} = \mathbb{C}$ if $\mathfrak{p}$ is complex.

(5) If K is the cyclotomic field $\mathbb{Q}(e^{2\pi i/m})$ (m odd) and $\mathfrak{p}$ is a prime factor of p in K, then $e_\mathfrak{p} = 1$ and $f_\mathfrak{p}$ is the order of 2 modulo m, i.e. the least positive integer f such that $2^f \equiv 1 \pmod{m}$.

For a detailed discussion of all these results see [B2] or [O1].

Theorem 18.1.

$$s(\mathbb{Q}_p) = \begin{cases} 1 & \text{if } p \equiv 1\ (4). \\ 2 & \text{if } p \equiv 3\ (4). \\ 4 & \text{if } p = 2. \end{cases}$$

Proof. First let p be odd. The quadratic form $X_1^2 + X_2^2 + X_3^2$ represents 0 nontrivially in $\mathbb{Q}_p$; see [B2], page 50. It follows that $s(\mathbb{Q}_p) \leq 2$. Now if $p \equiv 1$ (4), the congruence $-1 \equiv X^2 \pmod{p}$ is solvable in $\mathbb{Z}$ and so by Hensel's Lemma, $-1 = X^2$ is solvable in $\mathbb{Q}_p$; hence $s(\mathbb{Q}_p) = 1$, if $p \equiv 1(4)$.

If $p \equiv 3(4)$, we know that $-1 \equiv X^2 \pmod{p}$ is *not* solvable in $\mathbb{Z}$, whereas $-1 = X^2 + Y^2 \pmod{p}$ *is*. So again by Hensel's Lemma, $s(\mathbb{Q}_p) = 2$, if $p \equiv 3(4)$.

Finally for $p = 2$, the form $X_1^2 + X_2^2 + X_3^2 + X_4^2$ does *not* represent 0 in $\mathbb{Q}_2$ whereas $X_1^2 + \ldots + X_5^2$ *does*; hence $s(\mathbb{Q}_2) = 4$ as required. □

One may similarly prove the following.

Theorem 18.1′. *Let $K_\mathfrak{p}$ be a $\mathfrak{p}$-adic field and p the rational prime lying below $\mathfrak{p}$. Then*

$$s(K_\mathfrak{p}) = \begin{cases} 1 & \text{if } p^{f_\mathfrak{p}} \equiv 1(4), \\ 2 & \text{if } p^{f_\mathfrak{p}} \equiv 3(4) \text{ or if } p = 2 \text{ and } [K_\mathfrak{p} : \mathbb{Q}_2] \text{ is even}, \\ 4 & \text{if } p = 2 \text{ and } [K_\mathfrak{p} : \mathbb{Q}_2] \text{ is odd}. \end{cases}$$

While considering cyclotomic fields, we shall give proofs of those parts of this result that are needed.

Suppose now $K = \mathbb{Q}(\alpha)$ is an algebraic number field of degree n with $\operatorname{irr}(\alpha, \mathbb{Q}) = p(X)$ say. The number of distinct orderings of K is equal to the number of real roots of $p(X) = 0$ (always distinct since $p(X)$ is irreducible and separable). An element β of K is said to be totally positive, written $\beta >> 0$, if $\beta > 0$ under all orderings of K.

In 1902, Hilbert conjectured that every totally positive β in K is a sum of four squares in K. The first published proof of this was given by Siegel [S7] in 1921. Using the Hasse-Minkowski theorem we can easily prove the following.

Theorem 18.2 (Hilbert-Siegel). *Let $\beta \in K$ be totally positive; then β is a sum of four squares in K.*

Proof. Consider the quadratic form

$$X_1^2 + X_2^2 + X_3^2 + X_4^2 - \beta X_5^2$$

over $K_\mathfrak{p}$. If $\mathfrak{p}$ is finite, this represents 0 in $K_\mathfrak{p}$, being a form in five variables.

If $\mathfrak{p}$ is infinite and complex, $K_\mathfrak{p} = \mathbb{C}$ and the form again represents 0 in $\mathbb{C}$ since $\mathbb{C}$ is algebraically closed. Finally if $\mathfrak{p}$ is infinite and real, then $K_\mathfrak{p} = \mathbb{R}$. Now $\beta >> 0$ and so $\beta > 0$ in $\mathbb{R}$, i.e. the form is indefinite and again it represents 0 in $\mathbb{R}$. Thus it represents 0 in all $K_\mathfrak{p}$ and so by the Hasse-Minkowski theorem it represents 0 in K. In this representation, if $X_5 \neq 0$, then $\beta = (X_1/X_5)^2 + \ldots + (X_4/X_5)^2$ as required. If $X_5 = 0$, then $X_1^2 + \ldots + X_4^2$ represents 0 in K nontrivially and so is universal. In particular it represents β in K as required. This completes the proof. □

If K is totally complex then (vacuously), each element of K is totally positive and so a sum of four squares in K. In particular

$$-1 = a_1^2 + a_2^2 + a_3^2 + a_4^2 \qquad (a_j \in K).$$

Hence we have the following.

Theorem 18.3. *Let K be a totally complex (i.e. not a formally real) algebraic number field. Then $s(K) \leq 4$.*

Remarks.

1. Theorem 18.2 says that if K is any algebraic number field, then $P(K) \leq 4$.
2. If K is any field which is not formally real, then we have already seen (Theorem 16.2) that

$$s(K) \leq P(K) \leq s(K) + 1$$

Here both extremes are possible:

(i) $K = \mathbb{F}_{2^\alpha}$, the finite field of 2^α elements; then $s(K) = 1$ and indeed each element of K is a square in K (see Theorem 16.1) so $P(K) = 1$, or $K = \mathbb{C}$, $s(\mathbb{C}) = P(\mathbb{C}) = 1$.

(ii) $K = \mathbb{F}_5$; then $P(K) = 2$, $s(K) = 1$.

Using the powerful Hasse-Minkowski theorem it is easy to settle the quesiton of the exact determination of $s(K)$ for any algebraic number field K. Indeed $s(K) = 1$, 2, 4 (or ∞ if K is formally real). Further $s(K) = 1$ iff $i = \sqrt{-1} \in K$. Thus the problem boils down to deciding whether $s(K) = 2$ or 4. We have the following.

Theorem 18.4. *Let K be a totally complex algebraic number field, not containing $\sqrt{-1}$. Then $s(K) = 2$ iff for each prime $\mathfrak{p}$ of K lying above 2, the local degree $[K_\mathfrak{p} : \mathbb{Q}_2]$ $(= e_\mathfrak{p} f_\mathfrak{p})$ is even; otherwise $s(K) = 4$.*

Remark. If $(2) = \mathfrak{p}^{e_\mathfrak{p}} \ldots$ is the ideal factorization of 2 in K and $N_{K/\mathbb{Q}}\mathfrak{p} = 2^{f_\mathfrak{p}}$, then $[K_\mathfrak{p} : \mathbb{Q}_2] = e_\mathfrak{p} f_\mathfrak{p}$ and the requirement for $s(K)$ to be 2 is that $2 \mid e_\mathfrak{p} f_\mathfrak{p}$ for each $\mathfrak{p}$.

Proof. We shall assume Theorem 18.1′ in the proof. By the Hasse-Minkowski theorem, $s(K) \leq 2$ iff $s(K_\mathfrak{p}) \leq 2$ for all completions $K_\mathfrak{p}$ of K at $\mathfrak{p}$. First let $\mathfrak{p}$ be infinite. Since K is totally complex, $\mathfrak{p}$ is never real and so $K_\mathfrak{p} = \mathbb{C}$ and $s(\mathbb{C}) = 1 \leq 2$. Next let $\mathfrak{p}$ be an ideal factor in K of an odd rational p. Since $s(\mathbb{Q}_p) \leq 2$, it follows that $s(K_\mathfrak{p}) \leq 2$. Finally let $\mathfrak{p}$ be an ideal factor of 2 in K. We have $[K_\mathfrak{p} : \mathbb{Q}_2] = e_\mathfrak{p} f_\mathfrak{p}$ and $s(\mathbb{Q}_2) = 4$. Now it is well known that

$$s(K_\mathfrak{p}) = 4 \text{ iff } [K_\mathfrak{p} : \mathbb{Q}_2] \text{ is odd} \tag{18.1}$$

(indeed '$\Leftarrow$' is Springer's theorem).

So $s(K_\mathfrak{p}) \leq 2$ iff $e_\mathfrak{p} f_\mathfrak{p}$ is even. Thus $s(K_\mathfrak{p}) \leq 2$ for all $\mathfrak{p}$ of K iff at least one of $e_\mathfrak{p}$, $f_\mathfrak{p}$ is even for each $\mathfrak{p}$ lying above 2. □

Remark. The proof of the implication in (18.1) is well known in fact for general quadratic forms. When K is the cyclotomic field, we shall give a simple proof of it later. Right now we deduce some easy corollaries of Theorem 18.4.

Corollary 1. *Let $K = \mathbb{Q}(\sqrt{-d})$, $1 < d$, a square-free integer. Then $s(K) = 2$ iff d is a sum of three squares in $\mathbb{Z}$.*

Proof. If $-d \equiv 1$ (8), then (2) splits as $\mathfrak{p}\mathfrak{p}'$ in K and so $e_\mathfrak{p}$, $e_{\mathfrak{p}'}$, $f_\mathfrak{p}$, $f_{\mathfrak{p}'}$ are all 1. Hence $s(K) = 4$. In all other cases, one or the other of the following occurs:

(2) remains prime: $(2) = \mathfrak{p}$. Then

$$e_\mathfrak{p} = 1, \qquad f_\mathfrak{p} = 2$$

(2) ramifies: $(2) = \mathfrak{p}^2$. Then $e_\mathfrak{p} = 2$, $f_\mathfrak{p} = 1$.

In both cases $s(K) = 2$.

Finally $-d \equiv 1$ (8) iff $d \equiv 7$ (8), iff d is not a sum of three squares in $\mathbb{Z}$. □

Remark. The precise statement for the splitting of (2) in $\mathbb{Q}(\sqrt{m})$ (m square free) is the following: let

$$D = \begin{cases} 4m & \text{if } m \equiv 2 \text{ or } 3 \bmod 4, \\ m & \text{if } m \equiv 1 \bmod 4. \end{cases}$$

Then $\mathbb{Q}(\sqrt{m}) = \mathbb{Q}(\sqrt{D})$ and D is the discriminant of K. We have:

(2) ramifies in K iff D is even i.e. $m \equiv 2, 3$ (4);

(2) splits in K iff $(\frac{D}{2}) = 1$, i.e. $D \equiv 1$ (8), or $m \equiv 1$ (8);

(2) remains prime in K iff $(\frac{D}{2}) = -1$, i.e. $D \equiv 5$ (8), or $m \equiv 5$ (8).

The above corollary gives us a highbrow proof of Theorem 3.2.

In the case of cyclotomic fields $K = \mathbb{Q}(\zeta)$, $\zeta = e^{2\pi i/p}$ for p prime, the result of Theorem 18.4 is especially expedient in calculating the Stufe, since here any even $\mathfrak{p}$ is unramified; i.e. $e_\mathfrak{p} = 1$; whereas $f_\mathfrak{p}$ is precisely the order of 2 (mod p) for each $\mathfrak{p}$. Hence the following.

Corollary 2. *Let $K = \mathbb{Q}(e^{2\pi/p})$ (p odd prime). Then $s(K) = 2$ iff order of 2 modulo p is even.*

Remark. The beauty of the statements of Corollaries 1 and 2 is that they are fully in global terms. The statement of Theorem 18.4 includes local terms.

We now want to give a purely elementary proof of (18.1) for the case of cyclotomic fields. The Stufe of the general cyclotomic field $\mathbb{Q}(e^{2\pi i/m})$ can then be determined as above. Indeed we have the following.

Theorem 18.5. *Let $K^{(m)} = \mathbb{Q}(e^{2\pi i/m}) = \mathbb{Q}(\zeta)$, where $m \geq 3$ and odd, and let f be the multiplicative order of 2 mod$m$. Then*

$$s(K^{(m)}) = \begin{cases} 2 & \text{if } f \text{ is even,} \\ 4 & \text{if } f \text{ is odd.} \end{cases}$$

Furthermore if $2 \mid |m$, then $K^{(m)} = K^{(m/2)}$ (and $m/2$ is odd) and if $4|m$ then $\sqrt{-1} \in K^{(m)}$ and so $s(K^{(m)}) = 1$.

We first prove (18.1) in the form of the following two lemmas:

Lemma 1. *Let K be any field and let L/K be a normal separable extension of odd degree. Suppose $s(K) = 4$; then $s(L) = 4$.*

Lemma 2. *Let $K_\mathfrak{p}^{(m)}/\mathbb{Q}_2$ be an extension of even degree, where $\mathfrak{p}|2$. Then $s(K_\mathfrak{p}^{(m)}) \leq 2$.*

Remarks.

1. Lemma 1 is Theorem 3.7 (Springer). We shall give a different proof for the special case.
2. See Lemma 17.6 with $a = b = 1$.

Proof of Lemma 1. Let $G = \text{Gal}(L/K)$ so that $o(G)$ is odd. Suppose, to the contrary that $s(L) \leq 2$, say

$$-1 = X^2 + Y^2 \qquad (X, Y \in L).$$

Then $\prod_{\sigma \in G}(-1)^\sigma = \prod_{\sigma \in G}(X^2 + Y^2)^\sigma$ i.e.

$$-1 = \prod_{\sigma \in G}(X^\sigma + \sqrt{-1}Y^\sigma)(X^\sigma - \sqrt{-1}Y^\sigma) \tag{18.2}$$

Now let $\prod_{\sigma\in G}(X^\sigma + \sqrt{-1}Y^\sigma) = U + \sqrt{-1}V (U, V \in L(\sqrt{-1}))$. Then for $\tau \in G$, we have

$$\tau(U + \sqrt{-1}V) = \tau \prod_{\sigma\in G}(X^\sigma + \sqrt{-1}Y^\sigma) = \prod_{\sigma\in G}(X^{\tau\sigma} + \sqrt{-1}Y^{\tau\sigma})$$
$$= U + \sqrt{-1}V,$$

since as σ runs through G, so does $\tau\sigma$. It follows that $U + \sqrt{-1}V \in K(\sqrt{-1})$, i.e. that $U, V \in K$. Hence

$$\prod_{\sigma\in G}(X^\sigma - \sqrt{-1}Y^\sigma) = U - \sqrt{-1}V.$$

(11.2) now gives

$$\begin{aligned} -1 &= (U + \sqrt{-1}V)(U - \sqrt{-1}V) \\ &= U^2 + V^2 \qquad (U, V \in K) \end{aligned}$$

i.e. $s(K) \leq 2$ – a contradiction. □

Proof of Lemma 2. Let $G = \mathrm{Gal}(K_\mathfrak{p}^{(m)}/\mathbb{Q}_2)$. If we can find some intermediate field L between $\mathbb{Q}_2$ and $K_\mathfrak{p}^{(m)}$, then we can complete the proof by induction as follows: since the degree $[K_\mathfrak{p}^{(m)} : \mathbb{Q}_2]$ is even, either the degree $[K_\mathfrak{p}^{(m)} : L]$ is even or the degree $[L : \mathbb{Q}_2]$ is even, or both.

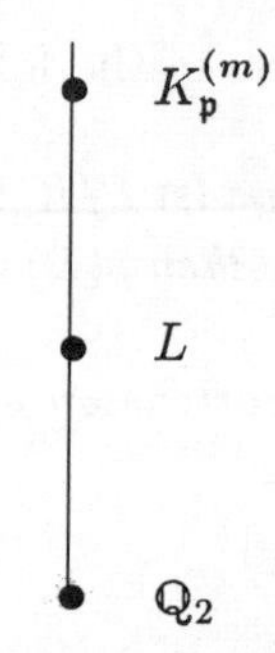

If $[L : \mathbb{Q}_2]$ is even, then by induction hypothesis, $s(L) \leq 2$, so a fortiori $s(K_\mathfrak{p}^{(m)}) \leq 2$. If $[K_\mathfrak{p}^{(m)} : L]$ is even but $[L : \mathbb{Q}_2]$ is odd, then by Lemma 1, $s(L) = 4$ and so again by the induction hypothesis, $s(K_\mathfrak{p}^{(m)}) \leq 2$. To find L, we proceed as follows: $2|o(G)$ so by Cauchy's theorem there exists an element of order 2 in G; call it a. Let $H = \{e, a\}$ and let L be the fixed field of H. Then L is an intermediate field as required, unless $H = G$, i.e. $L = \mathbb{Q}_2$. But then $K_\mathfrak{p}^{(m)}$ is a quadratic extension of $\mathbb{Q}_2$ and the possibilities for $K_\mathfrak{p}^{(m)}$ are the following: the coset representatives of $\mathbb{Q}_2^{*2}$ in $\mathbb{Q}_2^*$ are $\alpha = 1$, -1, -3, -5, -2, -6, -10, -14. The seven proper quadratic extensions of $\mathbb{Q}_2$ are $\mathbb{Q}_2(\sqrt{\alpha})$, $\alpha \neq 1$. We have $s(\mathbb{Q}_2(\sqrt{-1})) = 1$, while the Stufe of any other quadratic extension is at most 2, e.g. in $\mathbb{Q}_2(\sqrt{-3})$ we have

$$0 = (\sqrt{-3})^2 + 1^2 + 1^2 + 1^2,$$

hence $s(\mathbb{Q}_2(\sqrt{-3})) \leq 3$ so ≤ 2 and so on. □

Theorem 18.5 now follows exactly like Theorem 18.4.

Corollary 1. *Let $p \equiv 7$ (8) be a prime, then $s(K^{(p)}) = 4$.*

Proof. We have $(2/p) = (-1)^{(p^2-1)/8} = 1$; hence $2 \in G_p^{*^2}$. But $o(G_p^{*^2}) = (p-1)/2$. Thus the order of 2 (mod p) divides $(p-1)/2$ which is odd since $p \equiv 7$ (8). Thus the order of 2 mod p is odd so $s(K^{(p)}) = 4$. □

We thus arrive at the result of P. and S. Chawla (and of Shapiro and Leep) proved earlier by elementary methods.

Corollary 2. *Let $p \equiv 3$ or 5 (8) be a prime which divides m. Then $s(K^{(m)}) = 2$.*

Proof. It is enough to show that f, the order of 2 mod m, is even and for this it is enough to show that the order of 2 mod p is even. We have $(2/p) = (-1)^{(p^2-1)/8} = -1$, i.e. $2^{(p-1)/2} \equiv -1(p)$ so $f \nmid (p-1)/2$ i.e. $2f \nmid (p-1)$; but $2^{p-1} \equiv 1(p)$ i.e. $f \mid (p-1)$ and $2 \mid (p-1)$. It follows that $2 \mid f$. □

We thus arrive at the result of S. Chawla proved earlier in an elementary way.

If $m \equiv 1$ (8), one can have both the possibilities: $s(K^{(m)}) = 2$ or 4, for example:

(1) $m = 17$, $f = 8$, $s(K^{(m)}) = 2$.
(2) $m = 73$, $f = 9$, $s(K^{(m)}) = 4$.

For alternative proofs of these results, including Theorem 18.4, see [B1], [C17], [F1], [M1].

Let us now go back to the pythagoras number of fields. We shall use Pourchet's theorem for algebraic number fields K; although in Chapter 17 we only covered the case $K = \mathbb{Q}$, all the proofs go through just as easily for a general K. We have the following [H5].

Theorem 18.6 (Hsia-Johnson-Pourchet). *Let K be a formally real algebraic number field. Then $P(K) = 3$ or 4; it is 4 iff there exists a dyadic prime $\mathfrak{p}$ of K (i.e. $\mathfrak{p}|2$) such that $[K_{\mathfrak{p}} : \mathbb{Q}_2]$ is odd.*

Proof. First let $\mathfrak{p}$ be a dyadic prime of K at which the local degree $[K_\mathfrak{p} : \mathbb{Q}_2]$ is odd. Since $s(\mathbb{Q}_2) = 4$ it follows, by Springer's theorem, that $s(K_\mathfrak{p}) = 4$:

$$-1 = \alpha_1^2 + \ldots + \alpha_4^2 \qquad (\alpha_j \in K_\mathfrak{p}; 4 \text{ least}).$$

By the weak approximation theorem, there exists an element $a \in K$ which is positive at all the real primes of K and as near as we like to -1 at $\mathfrak{p}$. It follows that a needs precisely four squares for its representation as a sum of squares in K; so $P(K) \geq 4$. By Theorem 18.2, therefore, $P(K) = 4$.

Conversely suppose $[K_\mathfrak{p} : \mathbb{Q}_2]$ is even for all $\mathfrak{p}|2$. Since $s(\mathbb{Q}_2) = 4$, we see by (18.1) that $s(K_\mathfrak{p}) = 2$ for all $\mathfrak{p}|2$. For $\mathfrak{p} \not| 2$, if $\mathfrak{p}$ is finite, again $s(K_\mathfrak{p}) \leq 2$, while if $\mathfrak{p}$ is infinite complex, $K_\mathfrak{p} = \mathbb{C}$ so $s(K_\mathfrak{p}) = 1 \leq 2$. So any sum of squares in such a $K_\mathfrak{p}$ is a sum of at most three squares (in $K_\mathfrak{p}$).

Let now $a \in K$ be a sum of squares in K, so a is a sum of squares in $K_\mathfrak{p}$ for all $\mathfrak{p}$ and so a sum of at most three squares in $K_\mathfrak{p}$ for $\mathfrak{p}$ finite and infinite complex (as proved above); but if $\mathfrak{p}$ is infinite real, since a is a sum of squares in such a $K_\mathfrak{p} = \mathbb{R}$, we have $a > 0$ and thus in fact a is a single square in $\mathbb{R}$. Thus a is a sum of at most three squares in all $K_\mathfrak{p}$ (without exception), so by the Hasse-Minkowski theorem, it is a sum of at most three squares in K, hence $P(K) \leq 3$. Now it is easy to show that $P(K) \neq 1$ or 2; see Exercise 5.6 and Exercise 2 of this chapter and don't forget K is formally real. Hence $P(K) = 3$. □

As a corollary to Theorem 18.6, we calculate the pythagorean number of a real quadratic field.

Let $K = \mathbb{Q}(\sqrt{d})$, $d > 0$, a square-free integer. If $d \equiv 1\ (8)$, then $2 = \mathfrak{p}\mathfrak{p}'$ splits, so the local degrees at $\mathfrak{p}$ and $\mathfrak{p}'$ are each equal to 1: $[K_\mathfrak{p} : \mathbb{Q}_2] = [K_{\mathfrak{p}'} : \mathbb{Q}_2] = 1$. Hence $P(K) = 4$ by Theorem 18.6.

If $d \not\equiv 1\ (8)$, then either 2 remains prime in K or ramifies, so the local degree at the (unique) dyadic prime of K (2 or $\mathfrak{p}$) is 2. It follows that $P(K) = 3$. We have proved the

Corollary. *Let $d > 0$ be square-free. Then*

$$P(\mathbb{Q}(\sqrt{d})) = \begin{cases} 4 \text{ if } d \equiv 1\ (8), \\ 3 \text{ otherwise.} \end{cases}$$

We now go over to the function fields $K(X)$, where K is a formally real algebraic number field (remember Chapter 17 was entirely $\mathbb{Q}(X)$).

Theorem 18.7 (Hsia-Johnson-Pourchet). *Let K be a formally real algebraic number field. Then*

$$P(K(X)) = P(K) + 1.$$

Proof. We know that $P(K) = 3$ or 4. First let $P(K) = 4$ and let $a >> 0$ be a totally positive element of K requiring $P(K)$ to be the sum of four squares for its representation as a sum of squares in K:

$$a = a_1^2 + a_2^2 + a_3^2 + a_4^2.$$

The polynomial $X^2 + a \in K(X)$ is a sum of five squares in $K(X)$, and so is positive definite. We claim that it cannot be a sum of four squares in $K(X)$; for suppose $X^2 + a = f_1^2(X) + \ldots + f_4^2(X)$ where, by Cassels' lemma, we may suppose $f_j(X) \in K[X]$. Then by Corollary 3 of Chapter 2, a is a sum of three squares in K, contradicting the choice of a. Hence $X^2 + a$ requires its full quota of five squares for its representation as a sum of squares in $K(X)$. It follows that $P(K(X)) \geq 5$. But by Pourchet's theorem $P(K(X)) \leq 5$. So $P(K(X)) = 5$.

Next let $P(K) = 3$ and let $f(X)$ be a positive definite function in $K(X)$. We wish to show that $f(X)$ is a sum of four squares in $K(X)$. As usual, we may suppose without loss of generality that $f(X) \in K[X]$. We make use of the following result (cf. Lemma 17.4) proved in exactly the same way Lemma 17.4 was proved for $\mathbb{Q}$:

Lemma 3. *Let $f(X) \in K[X]$, K an algebraic number field, be a non-zero polynomial. Then $[a, b]$ represents $f(X)$ in $K[X]$ iff $[a, b]$ represents $f(X)$ in each $K_{\mathfrak{p}}[X]$ ($\mathfrak{p}$ finite or infinite).*

Taking $a = b = 1$, it is enough to prove that $f(X)$ is a sum of four squares in $K_{\mathfrak{p}}[X]$ for all $\mathfrak{p}$. Now if $\mathfrak{p}$ is infinite real or complex, so that $K_{\mathfrak{p}} = \mathbb{R}$ or $\mathbb{C}$, then $f(X)$ is a sum of at most two squares in $K_{\mathfrak{p}}[X]$ (see Theorem 4.1). Furthermore if $\mathfrak{p}$ is non-dyadic, $s(K_{\mathfrak{p}}) = 2$, so any element of $K_{\mathfrak{p}}(X)$, in particular $f(X)$ is a sum of three squares in $K_{\mathfrak{p}}(X)$.

Finally let $\mathfrak{p}$ be a dyadic prime. Since $P(K) = 3$, by Theorem 18.6, $[K_{\mathfrak{p}} : \mathbb{Q}_2]$ is even for all $\mathfrak{p}|2$. Thus for all $\mathfrak{p}|2$, $s(K_{\mathfrak{p}}) = 2$ (see (18.1)) so again for such $\mathfrak{p}$, $f(X)$ is a sum of three squares in $K_{\mathfrak{p}}(X)$.

Thus $f(X)$ is locally a sum of at most three squares and so a sum of four squares. Hence $f(X)$ is a sum of four squares in $K(X)$. It follows that $P(K(X)) = 4$ as required. □

As easy corollaries we prove the following.

Theorem 18.8.

(i) *Let K be a formally real algebraic number field with $[K : \mathbb{Q}]$ odd. Then $P(K(X)) = 5$.*

(ii) *Let $K = \mathbb{Q}(\sqrt{d})$, where d is square-free and positive, be a formally real quadratic extension of $\mathbb{Q}$. Then*

$$P(K(X)) = \begin{cases} 5 & \text{if } d \equiv 1 \ (8) \\ 4 & \text{otherwise.} \end{cases}$$

Proof. (i) The local degree formula gives

$$[K : \mathbb{Q}] = \sum_{\mathfrak{p}|2} [K_{\mathfrak{p}} : \mathbb{Q}_2].$$

Since the left side is odd, at least one term on the right is odd. It follows from Theorem 18.6 that $P(K) = 4$. Hence by Theorem 18.7, $P(K(X)) = 5$.
(ii) Is immediate from Theorem 18.7 and the Corollary to Theorem 18.6. □

If K *is a non-formally real algebraic number field* (*not containing* $\sqrt{-1}$), so that $s(K) = 2$ or 4, then

$$\begin{aligned} P(K) &= s(K) + 1 \quad \text{(Theorem 16.3).} \\ &= \begin{cases} 3 \text{ if } s(K) = 2, \\ 5 \text{ if } s(K) = 4, \end{cases} \\ &= \begin{cases} 3 \text{ if the local degree } [K_{\mathfrak{p}} : \mathbb{Q}_2] \text{ at all dyadic } \mathfrak{p} \text{ is even,} \\ 5 \text{ otherwise.} \end{cases} \end{aligned}$$

The pythagoras number of the $\mathfrak{p}$*-adic fields* can also be worked out using our results so far:

First let $\mathfrak{p}$ be non-dyadic. Now $K_{\mathfrak{p}}$ is never formally real (-1 is always a sum of squares in $K_{\mathfrak{p}}$). Let $s(K_{\mathfrak{p}})$ be the Stufe of $K_{\mathfrak{p}}$. Then by Theorem 16.3,

$$\begin{aligned} P(K_{\mathfrak{p}}(X)) &= s(K_{\mathfrak{p}}) + 1 \\ &= s(K_{\mathfrak{p}}(X)) + 1 \quad \text{(Theorem 11.8(i))} \\ &= \begin{cases} 2 \text{ if } s(K_{\mathfrak{p}}) = 1, \\ 3 \text{ if } s(K_{\mathfrak{p}}) = 2. \end{cases} \end{aligned}$$

Next if $\mathfrak{p}$ is a dyadic prime then we have the following table:

$s(K_{\mathfrak{p}}) = s(K_{\mathfrak{p}}(X))$	$P(K_{\mathfrak{p}})$	$P(K_{\mathfrak{p}})(X)$
1	2	2
2	3	3
4	4	5

Finally, it is possible to extend the complete discussion about sums of four squares in $\mathbb{Q}(X)$ to sums of four squares in $K(X)$, where K is a formally real algebraic number field. For details see [H5]. For details regarding sums of two squares in $K(X)$, see [H5]; in Chapter 17, we only covered the case $K = \mathbb{Q}$.

There are many interesting fields whose Stufe, pythagoras number, number of square classes (i.e. order of K^*/K^{*^2}) etc. need to be determined and a variety of fields and examples are available. The most common fields are the following:

1. Global fields. These are:

(a) Algebraic number fields, i.e. finite extensions of $\mathbb{Q}$,

(b) Function fields in one variable over F_q (finite field) i.e. finite extensions of $\mathsf{F}_q(X)$.

2. Local fields. These are fields F with a discrete non-archimedean valuation v such that F is complete with respect to v; discrete means $v : F \to \mathbb{Z}$, rather than $F \to \mathbb{R}$.

Of special interest among local fields F are the $\mathfrak{p}$-adic fields, i.e. completions of global fields at non-archimedean primes (finite primes). Their residue class fields $\bar{F}$ are finite. These turn out to be finite extensions of $\mathbb{Q}_p$ and the fields $\mathsf{F}_q((t))$.

3. Real closed fields R. Most results proved for
(1) algebraic number fields,
(2) $K_\mathfrak{p}$,
(3) the real numbers $\mathbb{R}$ and the fields $\mathbb{R}(X)$, $\mathbb{R}(X, Y)$ etc.
are valid for
(1) global fields,
(2) local fields,
(3) the fields R, $R(X)$, $R(X, Y)$ etc. where R is any real closed field,
respectively.

We shall end this chapter with an interesting example provided by the field

$$L = K((t)) = \{a_m t^m + a_{m+1} t^{m+1} + \ldots + \ldots \,|\, a_j \in K, m \in \mathbb{Z}\},$$

the set of all formal Laurent series, $m < 0$ allowed, under the usual series addition and multiplication. We have already made use of this field in Chapter 4 (Dubois' counterexample). We now give a more detailed account of some of its properties. Our aim is the following:

Theorem 18.9.

(i) *Given an ordering $>$ on K, there exist precisely two orderings on $L = K((t))$ extending $>$, one making t positive, the other making t negative.*

In particular L is formally real if K is.

(ii) *If K is formally real and pythagorean, so is L.*

(iii) *If K is real closed, then the field $K((t_1))\ldots((t_n))$ is pythagorean and has 2^{n+1} square classes (and 2^n orderings, by induction.)*

Before giving the proof we verify that L is a local field. Indeed the *valuation* v is given by

$$v(x) = v(a_m t^m + \ldots) = m \in \mathbb{Z}.$$

The three properties

(i) $v(x) = \infty$ iff $x = 0$
(ii) $v(xy) = v(x) + v(y)$
(iii) $v(x + y) = \min(v(x), v(y))$

are all easily checked, giving a *non-archimedean discrete valuation.* The *valuation ring* A of L is clearly $K[[t]]$, the ring of all power series in t:

$$A = K[[t]] = \{x \in L \mid v(x) \geq 0\}.$$

The *unique maximal ideal* $\mathfrak{p} = \{x \in L \mid v(x) \geq 1\}$. This is a *principal ideal* generated by any element π with $v(\pi) = 1$ e.g. $\pi = t$. π is determined up to a unit in A and is called a *local uniformizer* of A (or of L). The field $A/\mathfrak{p} = \bar{L}$ is the *residue class field* of L (relative to v). The mapping $a \to \bar{a} = a + \mathfrak{p}$ of $A \overset{\text{onto}}{\to} \bar{L}$ is called the *projection* of A onto $\bar{L}$. The *group of units* U of A is given by

$$\begin{aligned} U &= \{x \in A \mid x \notin \mathfrak{p}\} \\ &= \{x \in L^* \mid v(x) = 0\} \end{aligned}$$

Each $y \in L^*$ uniquely equals $u\pi^{v(y)}$ $(u \in U)$.

Now let $x = a_m t^m + \ldots$, $a_m \neq 0$. Say $x > 0$ iff $a_m > 0$ in K. It is easy to see that all the axioms of an ordering in L are satisfied. This order is unique in L (extending $>$ of K) in which $t > 0$ since the power series $1 + a_1 t + a_2 t^2 + \ldots$ is a square in L:

$$\begin{gathered} L \text{ is a local field and} \\ a_0 + a_1 t + \ldots \text{ is a square in } L \text{ iff } a_0 \text{ is a square in } A/\mathfrak{p}. \end{gathered} \tag{18.3}$$

Now t and $-t$ play the same role in K; indeed $t \to -t$ induces a K-antomorphism of L, so L also has a unique order extending $>$ of K in which $t < 0$. In this ordering

$$x = a_m t^m + \ldots > 0 \text{ iff } (-1)^m a_m > 0.$$

This proves (i).

To prove (ii), we proceed as follows: let

$$\begin{aligned} x &= a_m t^m + \ldots, \qquad a_m \neq 0 \\ y &= b_n t^n + \ldots, \qquad b_n \neq 0. \end{aligned}$$

Without loss of generality assume $m \leq n$. First let $m < n$; then by (18.3) $x^2 + y^2 = (a_m t^m)^2(1 + \ldots)$ is a square in L as required. Next let $m = n$ and write $a_m^2 + b_m^2 = 0$ which implies K has Stufe 1, whereas K is formally real.

Then $x^2 + y^2 = (c_m t^m)^2(1 + \ldots)$, which is a square in L as above. This proves (ii).

Finally we prove (iii). Note that amongst other things, (iii) shows the existence of fields with arbitrarily large number of square classes (2^n for any n).

Since K is real closed, it is pythagorean, so by induction, using (ii), we see that $K((t_1))\ldots((t_n))$ is pythagorean. That it has 2^n orderings follows by induction and (i), noting that K, being real and closed has a unique order. Finally it remains to show that $K((t_1))\cdots((t_n))$ has 2^{n+1} square classes. We use induction on n. Suppose a field F has N square classes. Then any non-zero element of $F((t))$ can be written as

$$at^{\alpha}(1 + b_1 t + b_2 t^2 + \ldots)$$

with $a \in F^*$, $b_1, b_2, \ldots, \in F$.

The power series is a square in $F((t))$, so if $a_1, \ldots, a_N$ are representatives of the square classes for F, then $a_1, \ldots, a_N, ta_1, \ldots, ta_N$ are representatives of the square classes of $F((t))$.

That completes the proof. □

Exercises

1. Let $L = K((t))$ and suppose K is formally real. Show that $P(L) = P(K)$. Hint: if $f \neq 0$ is a sum of squares in L, then show first that $f \equiv a$ (modulo L^{*^2}), where $a \in G_\infty(K)$.

If K is not formally real and the Stufe of K is s, deduce $P(L) = s + 1$.

2. Show that if K is a formally real algebraic number field then $P(K) \neq 2$.

Hint: Find a non-dyadic prime $\mathfrak{p}$ such that $-1 \notin K_{\mathfrak{p}}^{*^2}$. Show that if π is a uniformizer, then the quadratic form $X^2 + Y^2 + \pi Z^2$ is anisotropic over $K\mathfrak{p}$. So π, which is a SOS in $K_{\mathfrak{p}}$, is not a SOS of two squares (nor 1), so $P(K_{\mathfrak{p}}) > 2$. Now by the approximation theorem and the local square theorem, produce an $a \in K^*$ which is totally positive (i.e. is a SOS) but not a sum of two squares in K, so $P(K) > 2$.

Appendix 1 (for Chapter 10) Reduction of matrices to canonical forms

We have here six propositions and their corollaries, three for symmetric, and three for skew-symmetric, matrices over the reals, the complexes and the quaternions. The results over the reals and the complexes are standard material and we shall only state the propositions and their corollaries here. Propositions 5 and 6, are over the quaternions, and we give proofs on lines similar to those over the reals and the complexes.

Proposition 1. *Let S be a real symmetric $n \times n$ matrix; then there exists a real orthogonal matrix $\mathbf{O}$, such that $S = \mathbf{O}\Lambda\mathbf{O}'$, where $\Lambda = \mathrm{diag}(\lambda_1, \dots, \lambda_n)$.*

Corollary. *If in addition $S^2 = I_n$, then*

$$\Lambda = \begin{pmatrix} I_r & 0 \\ 0 & I_{n-r} \end{pmatrix}.$$

Proposition 2. *Let T be a real skew-symmetric $n \times n$ matrix and let n be even; then there exists a real orthogonal matrix $\mathbf{O}$ such that*

$$T = \mathbf{O}\begin{pmatrix} 0 & \Lambda \\ -\Lambda & 0 \end{pmatrix}\mathbf{O}', \qquad \textit{where} \qquad \Lambda = \begin{pmatrix} \lambda & & 0 \\ & \ddots & \\ 0 & & \lambda_{n/2} \end{pmatrix},$$

$\lambda_j \geq 0$ for all j.

Corollary. *If in addition $T^2 = -I_n$, then we can take $\Lambda = I_{n/2}$.*

Proposition 3. *Let* $\mathbf{S}$ *be a complex symmetric* $n \times n$ *matrix; then there exists a unitary matrix* $\mathbf{U}$ *such that* $\mathbf{S} = \mathbf{U}\Lambda\mathbf{U}'$, *where* $\Lambda = \begin{pmatrix} \lambda_1 & & 0 \\ & \ddots & \\ 0 & & \lambda_n \end{pmatrix}$, $\lambda_j \geq 0$ *for all* j.

Corollary. *If in addition,* $\mathbf{S}\bar{\mathbf{S}} = \mathbf{I}_n$, *then we can take* $\lambda_j = 1$ *for all* j.

Proposition 4. *Let* $\mathbf{T}$ *be a complex skew-symmetric* $n \times n$ *matrix,* n *even; then there exists a unitary matrix* $\mathbf{U}$ *such that* $\mathbf{T} = \mathbf{U}\begin{pmatrix} 0 & \Lambda \\ -\Lambda & 0 \end{pmatrix}\mathbf{U}'$, *where*

$$\Lambda = \begin{pmatrix} \lambda_1 & & 0 \\ & \ddots & \\ 0 & & \lambda_{n/2} \end{pmatrix}, \qquad \lambda_j \geq 0 \quad \textit{for all} \quad j$$

Corollary. *If in addition* $\mathbf{T}\bar{\mathbf{T}} = -\mathbf{I}_n$, *then* $\Lambda = \mathbf{I}_{n/2}$.

Proposition 5. *Let* $\mathbf{S}$ *be a quaternion matrix with* $\bar{\mathbf{S}}' = \mathbf{S}$ *(the bar indicates the conjugate quaternion). Then there exists an* $\mathbf{O}$ *such that* $\mathbf{O}\bar{\mathbf{O}}' = \mathbf{I}_n$, *and* $\mathbf{S} = \mathbf{O}\Lambda\bar{\mathbf{O}}'$, *where* $\Lambda = \begin{pmatrix} \lambda_1 & & 0 \\ & \ddots & \\ 0 & & \lambda_n \end{pmatrix}$.

Proof. Write $\mathbf{S} = \mathbf{S}_0 + \epsilon_1\mathbf{S}_1 + \epsilon_2\mathbf{S}_2 + \epsilon_3\mathbf{S}_3$. Then $\bar{\mathbf{S}}' = \mathbf{S}$ implies

$$\mathbf{S}_0' - \epsilon_1\mathbf{S}_1' - \epsilon_2\mathbf{S}_2' - \epsilon_3\mathbf{S}_3' = \mathbf{S}_0 + \epsilon_1\mathbf{S}_1 + \epsilon_2\mathbf{S}_2 + \epsilon_3\mathbf{S}_3,$$

i.e. $\mathbf{S}_0$ is symmetric while $\mathbf{S}_1$, $\mathbf{S}_2$, $\mathbf{S}_3$ are skew-symmetric real matrices.

Now try to solve the equation

$$\mathbf{S}\mathbf{x} = \lambda\mathbf{x} \qquad (\lambda \text{ real}).$$

Writing $\mathbf{x} = \mathbf{u}_0 + \epsilon_1\mathbf{u}_1 + \epsilon_2\mathbf{u}_2 + \epsilon_3\mathbf{u}_3$, this gives

$$(\mathbf{S}_0 + \epsilon_1\mathbf{S}_1 + \epsilon_2\mathbf{S}_2 + \epsilon_3\mathbf{S}_3)(\mathbf{u}_0 + \epsilon_1\mathbf{u}_1 + \epsilon_2\mathbf{u}_2 + \epsilon_3\mathbf{u}_3) = \lambda(\mathbf{u}_0 + \epsilon_1\mathbf{u}_1 + \epsilon_2\mathbf{u}_2 + \epsilon_3\mathbf{u}_3).$$

Equating components, we get four equations, which we write in matrix form:

$$\begin{pmatrix} \mathbf{S}_0 - \lambda\mathbf{I} & -\mathbf{S}_1 & -\mathbf{S}_2 & -\mathbf{S}_3 \\ \mathbf{S}_1 & \mathbf{S}_0 - \lambda\mathbf{I} & -\mathbf{S}_3 & \mathbf{S}_2 \\ \mathbf{S}_2 & \mathbf{S}_3 & \mathbf{S}_0 - \lambda\mathbf{I} & -\mathbf{S}_1 \\ \mathbf{S}_3 & -\mathbf{S}_2 & \mathbf{S}_1 & \mathbf{S}_0 - \lambda\mathbf{I} \end{pmatrix}\begin{pmatrix} \mathbf{u}_0 \\ \mathbf{u}_1 \\ \mathbf{u}_2 \\ \mathbf{u}_3 \end{pmatrix} = 0,$$

i.e. say $(T - \lambda I)\mathbf{u} = 0$. Here $\mathbf{T}$ is symmetric since $\mathbf{S}_1$, $\mathbf{S}_2$, $\mathbf{S}_3$ are skew-symmetric and $\mathbf{S}_0$ is symmetric. Thus all the characteristic roots of $\mathbf{T}$ are real. Select one, call it λ. Now normalize the vector $\mathbf{x} =$

$\mathbf{u}_0 + \epsilon_1\mathbf{u}_1 + \epsilon_2\mathbf{u}_2 + \epsilon_3\mathbf{u}_3$, i.e. make $\overline{\mathbf{x}}'\mathbf{x} = 1$, in other words

$$(\mathbf{u}_0' - \epsilon_1\mathbf{u}_1' - \epsilon_2\mathbf{u}_2' - \epsilon_3\mathbf{u}_3')(\mathbf{u}_0 + \epsilon_1\mathbf{u}_1 + \epsilon_2\mathbf{u}_2 + \epsilon_3\mathbf{u}_3) = 1$$

or

$$\mathbf{u}_0'\mathbf{u}_0 + \mathbf{u}_1'\mathbf{u}_1 + \mathbf{u}_2'\mathbf{u}_2 + \mathbf{u}_3'\mathbf{u}_3 = 1 \qquad \text{(i)}$$

$$\mathbf{u}_0'\mathbf{u}_1 - \mathbf{u}_1'\mathbf{u}_0 + \mathbf{u}_3'\mathbf{u}_2 - \mathbf{u}_2'\mathbf{u}_3 = 0 \qquad \text{(ii)}$$

$$\mathbf{u}_0'\mathbf{u}_2 - \mathbf{u}_2'\mathbf{u}_0 + \mathbf{u}_1'\mathbf{u}_3 - \mathbf{u}_3'\mathbf{u}_1 = 0 \qquad \text{(iii)}$$

$$\mathbf{u}_0'\mathbf{u}_3 - \mathbf{u}_3'\mathbf{u}_0 + \mathbf{u}_1'\mathbf{u}_2 - \mathbf{u}_2'\mathbf{u}_1 = 0 \qquad \text{(iv)}$$

Here (ii), (iii), (iv) turn out to be trivially true since $\mathbf{u}_0'\mathbf{u}_1 = \mathbf{u}_1'\mathbf{u}_0$ etc., being scalar, and so all terms just cancel out. However (i) is the normalization condition.

Now complete the (orthonormal) basis $\mathbf{x} = \mathbf{x}^{(1)}, \mathbf{x}^{(2)}, \ldots, \mathbf{x}^{(n)}$ of $\mathbb{C}^n/\mathbb{C}$ and let

$$\mathbf{X} = (\mathbf{x}^{(1)}, \mathbf{x}^{(2)}, \ldots, \mathbf{x}^{(n)}).$$

Then

$$\bar{\mathbf{X}}'\mathbf{S}\mathbf{X} = \begin{pmatrix} \lambda & 0 & 0 & \ldots \\ 0 & & & \\ 0 & & \mathbf{S}^* & \end{pmatrix}$$

where $\bar{\mathbf{S}}^{*\prime} = \mathbf{S}^*$ for $\bar{\mathbf{X}}'\mathbf{S}\mathbf{X} = (\bar{\mathbf{x}}^{(i)\prime}\mathbf{S}\mathbf{x}^{(j)})_{ij}$ and for $i = j = 1$, this is equal to $\bar{\mathbf{x}}^{(1)\prime}\mathbf{S}\mathbf{x}^{(1)} = \lambda\bar{\mathbf{x}}^{(1)\prime}\mathbf{x}^{(1)} = \lambda$. When $i = 1$, $(\bar{\mathbf{x}}^{(1)\prime}\mathbf{S})' = \mathbf{S}'\bar{\mathbf{x}}^{(1)} = \bar{\mathbf{S}}\bar{\mathbf{x}}^{(1)} = \overline{\mathbf{S}\mathbf{x}^{(1)}} = \lambda\bar{\mathbf{x}}^{(1)}$ (since $\mathbf{S}' = \bar{\mathbf{S}}$). So $\bar{\mathbf{x}}^{(1)\prime}\mathbf{S} = \lambda\bar{\mathbf{x}}^{(1)\prime}$, so $\bar{\mathbf{x}}^{(1)\prime}\mathbf{S}\mathbf{x}^{(j)} = \lambda\bar{\bar{\mathbf{x}}}^{(1)\prime}\mathbf{x}^{(j)} = 0$ if $j > 1$. Thus the top row equals $(\lambda, 0, \ldots, 0)$ as required.

But $(\overline{\mathbf{X}'\mathbf{S}\mathbf{X}})' = \bar{\mathbf{X}}'\mathbf{S}\mathbf{X}$ so the first column is $\begin{pmatrix} \lambda \\ 0 \\ \vdots \\ 0 \end{pmatrix}$, and this same relation shows that $\mathbf{S}^*$ satisfies $\bar{\mathbf{S}}^{*\prime} = \mathbf{S}^*$.

Now complete the proof by induction. □

Corollary. *If in addition* $\mathbf{S}^2 = \mathbf{I}$, *then* $\Lambda = \begin{pmatrix} \mathbf{I}_\rho & 0 \\ 0 & -\mathbf{I}_{n-\rho} \end{pmatrix}$.

Proof. $\mathbf{S}^2 = \mathbf{I} \Rightarrow \Lambda^2 = \mathbf{I} \Rightarrow \lambda_j^2 = 1$ for all j. Now let $\lambda_j = a_0 + \epsilon_1 a_1 + \epsilon_2 a_2 + \epsilon_3 a_3$ (even supposing it is a quaternion). Then

$$1 = \lambda_j^2 = (a_0^2 - a_1^2 - a_2^2 - a_3^2) + \epsilon_1 2.a_0 a_1 + \epsilon_2 2a_0 a_2 + \epsilon_3.2a_0 a_3$$

implies $a_0^2 = a_1^2 + a_2^2 + a_3^2 + 1$ so $a_0 \neq 0$, since a_1, a_2, a_3 are real, and $a_0 a_1 = 0$, $a_0 a_2 = 0$, $a_0 a_3 = 0$. Since $a_0 \neq 0$ so a_1, a_2, a_3 are all 0 so $a_0 = \pm 1$. Now shuffle up rows and $\Lambda = \begin{pmatrix} \mathbf{I}_\rho & 0 \\ 0 & \mathbf{I}_{n-\rho} \end{pmatrix}$ as required. □

Proposition 6. *Let* $\mathbf{T}$ *be a quaternion matrix satisfying* $\mathbf{T} + \bar{\mathbf{T}}' = 0$. *Then there exists a quaternion matrix* $\mathbf{O}$ *such that* $\mathbf{O}\bar{\mathbf{O}}' = \mathbf{I}_n$, $\mathbf{T} = \mathbf{O}\Lambda\bar{\mathbf{O}}'$, *where* $\Lambda = \begin{pmatrix} \lambda_1 & & & 0 \\ & \lambda_2 & & \\ & & \ddots & \\ 0 & & & \lambda_n \end{pmatrix}$ *and the* λ_j *are quaternions of the form* $\lambda_j = \mu_j \epsilon_1$ $(\mu_j \in \mathsf{R})$.

Proof. Consider the equation $\mathbf{T}\mathbf{x} = \lambda \mathbf{x}\epsilon_1$ and try to solve it for λ real. Write $\mathbf{T} = \mathbf{S}_0 + \epsilon_1\mathbf{S}_1 + \epsilon_2\mathbf{S}_2 + \epsilon_3\mathbf{S}_3$ so that $\mathbf{T} + \bar{\mathbf{T}}' = 0$ implies $\mathbf{S}_0$ is skew-symmetric and $\mathbf{S}_1$, $\mathbf{S}_2$, $\mathbf{S}_3$ are all (real) symmetric. Let $\mathbf{x} = \mathbf{u}_0 + \epsilon_1\mathbf{u}_1 + \epsilon_2\mathbf{u}_2 + \epsilon_3\mathbf{u}_3$. Then $\mathbf{T}\mathbf{x} = \lambda\mathbf{x}\epsilon_1$ becomes

$$\mathbf{S}_0\mathbf{u}_0 - \mathbf{S}_1\mathbf{u}_1 + \lambda\mathbf{u}_1 - \mathbf{S}_2\mathbf{u}_2 - \mathbf{S}_3\mathbf{u}_3 = 0 \qquad \text{(ii)}$$

$$\mathbf{S}_1\mathbf{u}_0 - \lambda\mathbf{u}_0 + \mathbf{S}_0\mathbf{u}_1 - \mathbf{S}_3\mathbf{u}_2 + \mathbf{S}_2\mathbf{u}_3 = 0 \qquad \text{(i)}$$

$$\mathbf{S}_2\mathbf{u}_0 + \mathbf{S}_3\mathbf{u}_1 + \mathbf{S}_0\mathbf{u}_2 - \mathbf{S}_1\mathbf{u}_3 - \lambda\mathbf{u}_3 = 0 \qquad \text{(iv)}$$

$$\mathbf{S}_3\mathbf{u}_0 - \mathbf{S}_2\mathbf{u}_1 + \mathbf{S}_1\mathbf{u}_2 + \lambda\mathbf{u}_2 + \mathbf{S}_0\mathbf{u}_3 = 0 \qquad \text{(iii)}$$

Number them as shown, then write them as numbered with signs of (ii), (iii) changed throughout. Then we get the equivalent matrix equation

$$\begin{pmatrix} \mathbf{S}_0 & -\mathbf{S}_1 + \lambda\mathbf{I} & -\mathbf{S}_2 & -\mathbf{S}_3 \\ \mathbf{S}_1 - \lambda\mathbf{I} & \mathbf{S}_0 & -\mathbf{S}_3 & \mathbf{S}_2 \\ \mathbf{S}_2 & \mathbf{S}_3 & \mathbf{S}_0 & -\mathbf{S}_1 - \lambda\mathbf{I} \\ \mathbf{S}_3 & -\mathbf{S}_2 & \mathbf{S}_1 + \lambda\mathbf{I} & \mathbf{S}_0 \end{pmatrix} \begin{pmatrix} \mathbf{u}_0 \\ \mathbf{u}_1 \\ \mathbf{u}_2 \\ \mathbf{u}_3 \end{pmatrix} = 0$$

i.e.

$$\begin{pmatrix} \mathbf{S}_1 - \lambda\mathbf{I} & \mathbf{S}_0 & -\mathbf{S}_3 & \mathbf{S}_2 \\ -\mathbf{S}_0 & \mathbf{S}_1 - \lambda\mathbf{I} & \mathbf{S}_2 & \mathbf{S}_3 \\ -\mathbf{S}_3 & \mathbf{S}_2 & -\mathbf{S}_1 - \lambda\mathbf{I} & -\mathbf{S}_0 \\ \mathbf{S}_2 & \mathbf{S}_3 & \mathbf{S}_0 & -\mathbf{S}_1 - \lambda\mathbf{I} \end{pmatrix} \begin{pmatrix} \mathbf{u}_0 \\ \mathbf{u}_1 \\ \mathbf{u}_2 \\ \mathbf{u}_3 \end{pmatrix} = 0$$

or say $(\mathbf{A} - \lambda\mathbf{I})\mathbf{u} = 0$. Here $\mathbf{A}$ is symmetric, so all its characteristic roots are real; let λ be one. Then for

$$\mathbf{x} = \mathbf{u}_0 + \epsilon_1\mathbf{u}_1 + \epsilon_2\mathbf{u}_2 + \epsilon_3\mathbf{u}_3$$

we have $\mathbf{T}\mathbf{x} = \lambda\mathbf{x}\epsilon_1$. Now normalize this $\mathbf{x}$ and write it as $\mathbf{x}^{(1)}$ so that $\mathbf{T}\mathbf{x}^{(1)} = \lambda\mathbf{x}^{(1)}\epsilon_1$ and complete to an orthonormal basis $\mathbf{x}^{(1)}, \ldots, \mathbf{x}^{(n)}$ of H^n/R.

Let $\mathbf{X} = (\mathbf{x}^{(1)}, \ldots, \mathbf{x}^{(n)})$ so that $\bar{\mathbf{X}}'\mathbf{X} = \mathbf{I}_n$. Then $\bar{\mathbf{X}}'\mathbf{T}\mathbf{X} = \begin{pmatrix} \lambda\epsilon_1 & 0 \\ 0 & \mathbf{T}^* \end{pmatrix}$ where again $\bar{\mathbf{T}}'^* + \mathbf{T}^* = 0$. Now proceed by induction as usual. □

Corollary. *If in addition,* $\mathbf{T}^* = -\mathbf{I}_n$, *then* $\Lambda = \epsilon_1\mathbf{I}$.

Proof. $\mathbf{T}^2 = -\mathbf{I}_n \Rightarrow \mathbf{O}\Lambda\bar{\mathbf{O}}'\mathbf{O}\Lambda\bar{\mathbf{O}}' = \mathbf{I}_n \Rightarrow \Lambda^2 = -\mathbf{I}_n \Rightarrow \lambda_j = \pm 1$ (for all j). So $\Lambda = \begin{pmatrix} \eta_1\epsilon_1 & & & 0 \\ & \eta_2\epsilon_1 & & \\ & & \ddots & \\ 0 & & & \eta_n\epsilon_1 \end{pmatrix}$ where $\eta_j = \pm 1$. But the -1's can all be made $+1$ by pre- and post- multiplication by $\begin{pmatrix} \ddots & 0 \\ 0 & \ddots \end{pmatrix}$ and $\begin{pmatrix} \ddots & 0 \\ 0 & \ddots \end{pmatrix}^{-1}$ where the diagonal entries are ϵ_1 if the corresponding sign is $+1$ and ϵ_2 if -1.

Appendix 2 (for Chapter 7): The Krein-Milman Theorem for convex cones

We assume the reader is familiar with convex subsets of $\mathbf{R}^n$ and their extreme points.

The set $S \subseteq \mathbf{R}^n$ is called a cone if for each $\underline{u} \in S$, $\lambda\underline{u} \in S$ for all $\lambda \geq 0$.

The result we require is the analogue of the Krein-Milman theorem for convex cones. We first state and prove this theorem for compact convex subsets of $\mathbf{R}^n$.

Theorem (Krein-Milman Theorem). *A compact convex subset k of $\mathbf{R}^n$ is the closed convex hull of its extreme points.*

Proof. Let $\mathfrak{p}$ denote the collection of all subsets X of k which are compact and which are such that

$$ty + (1-t)z \in X, \quad t > 0, \quad , y, z \in k \Rightarrow y, z \in X \qquad (*)$$

It is easy to check that if $X \in \mathfrak{p}$ and $\Lambda : \mathbf{R}^n \to \mathbf{R}$ is any linear map with maximum value λ on X, then

$$X_\Lambda = \{x \in X \mid \Lambda(x) = \lambda\}$$

is also in $\mathfrak{p}$; X_Λ is a hyperplane section of X. We now assert that each $X \in \mathfrak{p}$ contains an extreme point x. For consider $\mathfrak{p}_X$, the family of all sets in $\mathfrak{p}$ that are subsets of X. This family of closed subsets of the space X has the finite intersection property. Using the compactness of X it follows that $\mathfrak{p}_X$ must contain a minimal nonempty member Y. Such a Y must have only one element x, otherwise an appropriate Y_Λ would be still smaller. This x is clearly an extreme point by $(*)$.

Having proved, in particular, the existence of extreme points of k (note that $k \in \mathfrak{p}$) we now consider their closed convex hull H. We have shown in fact that H has a non-empty intersection with every member of $\mathfrak{p}$. This stronger fact implies $H = k$ for otherwise some $\Lambda : \mathbf{R}^n \to \mathbf{R}$, having maximum value λ on k, would have lesser values on all of H, and so H would not intersect k_Λ. □

Note. This proof can be found in W. Rudin's *Functional Analysis*, 1973, pages 70-71.

Now let C be a cone in $\mathbf{R}^n$. For each point $\underline{u} \in C$, we define the ray $[\underline{u}] \subset C$ as the set of all $\lambda\underline{u}$ $(\lambda > 0)$. Alternatively define an equivalence relation on points of C by saying $\underline{u} \sim \underline{v}$ if and only if $\underline{0}$, $\underline{u}$, $\underline{v}$ are collinear. Let $\tilde{C}$ be the quotient space (under this equivalence) and say C is compact if $\tilde{C}$ is compact. Further call the ray $[\underline{u}]$ extreme if $[\underline{u}]$ can not be written in the form $t[\underline{v}]+(1-t)[\underline{w}]$, $t > 0$ and $[\underline{v}]$, $[\underline{w}]$ distinct rays of C, both different from $[\underline{u}]$.

Then we have

Theorem 1′. *A "compact" convex cone is the closed convex hull of its extreme rays.*

The proof is exactly like that of Theorem 1. □

Finally we know that the set $\mathcal{P}_{n,m}$ of all positive semi-definite forms of degree m in n variables form a closed cone in $\mathbf{R}^k$ where $k = \binom{m+n-1}{n-1}$. It follows that $\mathcal{P}_{n,m}$ is the closed hull of its extreme rays as required.

Remarks. The case $\mathbf{R}^n$ considered above is older than Krein-Milman; indeed for $\mathbf{R}^n$ the theorem is already in H. Minkowski, *Gesammelte Abhandlungen* 1911 part II, page 160, lines 11-14.

The Krein-Milman paper is "On the extreme points of regularly convex sets", *Studia Math*, **9** (1940), 133-138.

A version that holds for k^n, k not complete, is in V.L. Klee, "Extreme points of convex sets without completeness of the scalar field", *Mathematika*, **11** (1964), 59-63.

References

[A1] J.F. Adams, Vector fields on spheres, *Annals of Maths*, **75** (1962), 603-632.

[A2] J. Adem, Construction of some normed maps, *Bol. Soc. Mat. Mexicana*, **20** (1975), 59-75.

[A3] J. Adem, On the Hurwitz problem over an arbitrary field I, *Bol. Soc. Mat. Mexicana*, **25** (1980), 29-51.

[A4] J. Adem, On the Hurwitz problem over an arbitrary field II, *Bol. Soc. Mat. Mexicana*, **26** (1981), 29-41.

[A5] A.A. Albert (editor), *Studies in Modern Algebra*, Vol. 2, MAA Studies in Maths (1963).

[A6] E. Artin, Über die Zerlegung definiter Funktionen in Quadrate, *Hamb. Abh.*, **5** (1927), 100-115.

[A7] L. Aubry, Solution de quelques questions d'analyse indéterminée, *Sphinxe Œdipe*, **7** (1912), 81-84.

[B1] F.W. Barnes, On the Stufe of an algebraic number field, *J. No. Th.*, **4** (1972), 474-478.

[B2] Z.I. Borevich and I.R. Shafarevich, *Number theory*, Academic Press (1973).

[B3] N. Bourbaki, *Algèbre commutative*, Hermann (1964).

[C1] A.P. Calderon, A note on biquadratic forms, *Lin. Alg. Appl.*, **7** (1973), 175-177.

[C2] J.W.S. Cassels, On the representation of rational functions as sums of squares, *Acta. Arith.*, **9** (1964), 79-82.

[C2]′ J.W.S. Cassels, *Local fields*, Cambridge University Press (1986).

[C3] J.W.S. Cassels, W.J. Ellison and A. Pfister, On sums of squares and elliptic curves over function fields, *J. No. Th.*, **3** (1971), 125-144.

[C4] A. Cayley, On Jacobi's elliptic functions, in reply to the Rev. Brice Bronwin, and on quaternions, *Philosophical Magazine and J. of Sc.*, **26** (1845), 208-211; Collected Papers **I**, 127.

[C5] P. Chawla, On the representation of -1 as a sum of squares in a cyclotomic field, *J. No. Th.*, **1** (1969), 208-210.

[C6] P. Chawla and S. Chawla, Determination of the Stufe of Cyclotomic fields, *J. No. Th.*, **2** (1970), 271-272.

[C7] M.D. Choi, Positive semi-definite biquadratic forms, *Lin. Alg. Appl.*, **12** (1975), 95-100.

[C8] M.D. Choi and T.Y. Lam, An old question of Hilbert, *Proc. of Conference on quadratic forms, Queen's papers on Pure and Applied Math.*, **46** (1976), 385-405, Kingston, Ontario, Queen's Univ.

[C9] M.D. Choi and T.Y. Lam, Extremal positive semidefinite forms, *Math. Ann.*, **231** (1977), 1-18.

[C10] M.D. Choi and T.Y. Lam, Symmetric positive semi-definite forms and sums of squares (in preparation).

[C11] M.D. Choi, T.Y. Lam, and B. Reznick, Even symmetric sextics, *Math. Zeit.*, **195** (1987), 559-580.

[C12] M.D. Choi, T.Y. Lam, and B. Reznick, Real zeros of positive semidefinite forms I, *Math. Zeit.*, **171** (1980), 1-26.

[C13] M.D. Choi, T.Y. Lam, and B. Reznick, Classification of symmetric positive semi-definite quartics (in preparation).

[C14] M.D. Choi, T.Y. Lam, and B. Reznick, Positive sextics and Schur's inequalities, *J. of Alg.* (to appear).

[C15] M.D. Choi, T.Y. Lam, B. Reznick and A. Rosenburg, Sums of squares in some integral domains, *J. of Alg.*, **65** (1980), 234-256.

[C16] M.D. Choi, M. Knebusch, T.Y. Lam and B. Reznick, Transversal zeros and positive semi-definite forms, *Géometrie Algébrique Réelle et Formes Quadratiques*, Proceedings Rennes (1981), 273-298, Springer Lecture Notes in Maths, Vol. 959 (1982).

[C17] I. Connell, The Stufe of number fields, *Math. Zeit.*, **124** (1972), 20-22.

[C18] M.R. Christie, Positive definite rational functions of two variables which are not the sum of three squares, *J. No. Th.*, **8** (1976), 224-232.

[D1] C.F. Degan, Adumbratio demonstrationis theorematis arithmetici maxime universalis. *Mémoires de l'Académie Impériale des Sciences de St. Pétersbourg*, **8** Années 1817 et 1818, (1822), 207-219.

[D2] L.E. Dickson, On quaternions and their generalizations and the history of the 8-square theorem, *Annals of Maths*, **20** (1919), 155-171.

[D3] D.W. Dubois, Note on Artin's solution of Hilbert's 17th problem, *Bull. Amer. Math. Soc.*, **73** (1967), 540-541.

[E1] B. Eckmann, Gruppentheoretischer Beweis des Satzes von Hurwitz-Radon über die Komposition der quadratischen Formen, *Comment. Math. Helv.*, **15** (1942/3), 358-366.

[E2] Leonard Euler, Letter to Goldbach, May 4, 1748, *Novi Comm. Acad. Petrop.*, (5), 1754-5, 3; 15, 1770, 75, *Comm. Arith. Coll.*, **I,** 230, 427.

[F1] B. Fein, G. Gordon and J.H. Smith, On the representation of -1 as a sum of 2 squares in an algebraic number field, *J. No. Th.*, **3** (1971), 310-315.

[G1] M.R. Gabel, Generic orthogonal stably free projectives, *J. of Alg.*, **29** (1974), 477-488.

[G2] M. Greenberg, *Lectures on forms in many variables*, W.A. Benjamin Inc., (1969), 15-23.

[G3] Emil Grosswald, *Representations of integers as sums of squares*, Springer-Verlag (1985).

[G4] A. Geramita and N.J. Pullman, A theorem of Hurwitz and Radon and orthogonal projective modules, *Proc. Amer. Math. Soc.*, **42** (1974), 51-56.

[G5] A. Geramita and J. Seberry, *Orthogonal designs*, Marcel Dekker, (1979).

[H1] W. Habicht, Über die Zerlegung strikte definiter Formen in Quadrate, *Comment. Math. Helv.*, **12** (1940), 317-322.

[H2] W.R. Hamilton, *Lectures on quaternions*, Dublin, (1853).

[H3] David Hilbert, Über die Darstellung definiter Formen als Summe von Formenquadraten, *Math. Ann.*, **32** (1888), 342-350; = *Ges. Abh.*, **2** 154-161.

[H4] David Hilbert, Über ternäre definite Formen, *Acta. Math.*, **17** (1843), 169-198; = *Ges. Abh.*, **2** 345-366.

[H4]′ D. Hilbert, Mathematische Probleme, *Göttinger Nachrichten* (1906), 253-297; = *Ges. Abh.*, **3** 290-329.

[H5] J.S. Hsia and R.P. Johnson, On the representation in sums of squares for definite functions in one variable over an algebraic number field, *Amer. J. Math.*, **96** (1974), 448-453.

[H6] Adolf Hurwitz, Über die Komposition der quadratischen Formen, *Math. Ann.*, **88** (1923), 1-25; = *Math. Werke*, **II** 641-666.

[H7] Adolf Hurwitz, Über der Komposition der quadratischen Formen von beliebig vielen Variabeln, *Nachrichten von der Königlichen*

Gesellschaft der Wissenschaften in Göttingen (1898), 309-316; = *Math. Werke*, **II** 565-571.

[J1] N. Jacobson, *Basic algebra*, Vol. 2., Freeman.

[K1] Kazuya Kato, A Hasse principle for two dimensional global fields, *J. reine angew. Math.*, **366** (1986), 142-183.

[K2] M. Knebusch, On the uniqueness of real closures and the existence of real places, *Comment. Math. Helv.*, **47** (1972), 260-269.

[K3] T. Koya, Synthesis of finite passive n-ports with prescribed positive real matrices of several variables, *IEEE Trans. Circ. Theory*, **CT-15** (1968), 2-23.

[K4] G. Kreisel, Hilbert's 17th problem, Summaries of talks presented at the Summer Institute of Symbolic Logic (1957) at Cornell University, 313-320.

[K5] T. Kirkman, On pluquaternions and homoid products of sums of n squares; *Philos. Mag.* (Ser. 3) **33** (1848), 447-459; 494-509.

[L1] T.Y. Lam, Construction of non-singular bilinear maps, *Topology*, **6** (1967), 423-426.

[L2] T.Y. Lam, *The algebraic theory of quadratic forms*, Benjamin, (1973).

[L3] T.Y. Lam, An introduction to real algebra, *Rock. Mt. J. Maths.*, **14** (1984), 767-814.

[L4] E. Landau, Über die Darstellung definiter Funktionen durch Quadrate, *Math. Ann.*, **62** (1906), 272-285.

[L5] E. Landau, *Elementary number theory*, Chelsea, (1958).

[L6] S. Lang, On quasi-algebraic closure, *Annals of Maths.*, **55** (1952), 373-390.

[L7] E. Landau, Über die Zerlegung total positiver Zahlen in Quadrate, *Göttinger Nachrichten*, (1919), 392-396.

[M1] C. Moser, Representation de -1 comme somme de carrés dans un corps cyclotomique quelconque, *J. No. Th.*, **5** (1973), 139-141.

[M2] T.S. Motzkin, The arithmetic-geometric inequality; in *Inequalities*, Oved Shisha (ed) Academic Press, (1967), 205-224.

[O1] O.T. O'Meara, *An introduction to quadratic forms*, Springer-Verlag, (1963).

[P1] A. Pfister, Zur Darstellung von -1 also Summe von Quadraten in einem Körper, *Journal L.M.S.*, **40** (1965), 159-165.

[P2] A. Pfister, Zur Darstellung definiter Funktionen als Summe von Quadraten, *Inventiones Math.*, **4** (1967), 229-236.

[P3] A. Pfister, Quadratic forms over fields, *Proc. of Symposia in Pure Maths.*, **20** (1969), 150-160.

[P4] A. Pfister, Hilbert's 17th problem and related problems on definite forms, *Proc. of Symposia in Pure Maths.*, **28** (1976), 483-489.

[P5] A. Pfister, Multiplicative quadratische Formen, *Arch. Math.*, **16** (1965), 363-370.

[P5]′ A. Pfister, Quadratische Formen in beliebigen Körpern, *Inventiones Math.*, **1** (1966), 116-132.

[P6] Y. Pourchet, Sur la répresentation en somme de carrés des polynômes a une indeterminée sur un corps de nombres algébriques, *Acta. Arith.*, **19** (1971), 89-109.

[P7] J.C. Parnami, M.K. Agarwal and A.R. Rajwade, On the Stufe of quartic fields, *J. No. Th.*, **38** (1991), 106-109.

[R1] H. Rademacher and O. Toeplitz, *The enjoyment of mathematics*, Princeton University Press (1956).

[R2] J. Radon, Lineare Scharen orthogonaler Matrizen, *Abh. Math. Sem. Univ. Hamburg*, **1** (1922), 1-14.

[R3] A.R. Rajwade, A note on the Stufe of quadratic fields, *Indian J. Pure and App. Maths*, **6** (1975), 725-726.

[R4] B. Reznick, Extremal positive semi-definite forms with few terms, *Duke Math. J.*, **45** (1978), 363-374.

[R5] P. Ribenboim, *L'arithmétique des corps*, Hermann, (1972).

[R6] L.J. Risman, A new proof of the 3-square theorem, *J. No. Th.*, **6** (1974), 282-283.

[R7] A. Robinson, On ordered fields and definite functions, *Math. Ann.*, **130** (1955), 257-271.

[R8] R.M. Robinson, Some definite polynomials which are not sums of squares of real polynomials; in *Selected questions of algebra and logic*, Acad. Sci. USSR (1973), 264-282.

[S1] W. Scharlau, *Quadratic forms and Hermitian forms*, Springer-Verlag (1988).

[S2] J.P. Serre, *A course of arithmetic*, Springer-Verlag (1973).

[S3] D.B. Shapiro, Spaces of similarities I, II, *J. of Alg.*, **46** (1977), 148-181.

[S4] D.B. Shapiro, Spaces of similarities IV, *Pacific J. Maths.*, **69** (1977), 233-244.

[S5] D.B. Shapiro, On the Hurwitz problem over an arbitrary field, *Bol. Soc. Mat. Mexicana*, **29** (1984), 1-4.

[S6] D.B. Shapiro, Products of sums of squares, *Exp. Math.*, **2** (1989), 235-261.

[S7] C.L. Siegel, Darstellung total positiver Zahlen durch Quadrate, *Math. Zeit.*, **11** (1921), 246-275.

[S8] Charles Small, Sums of 3 squares and levels of quadratic number fields, *Am. Math. Month.*, **93** (1986), 276-279.

[T1] Olga Taussky, History of sums of squares in algebra, *Amer. Math.*

Heritage, Alg. and Applied Maths. Texas Tech. Univ. Math. Series, **13** (1981), 73-90.

[T2] C.C. Tsen, Zur Stufentheorie der Quasi-algebraisch-abegeschlossenheit kommutativer Körper, *J. Chinese Math. Soc.*, **1** (1936), 81-92.

[T3] J.A. Todd, *Projective and analytical geometry*, Pitman, 1947.

[W1] R. Walker, *Algebraic curves*, Dover (1959).

[W2] E. Witt, Theorie der quadratischen Formen in beliebigen Körpern. *J. reine angew. Math.*, **176** (1937), 31-44.

[W3] J.A. Wolf, Geodesic spheres in Grassmann manifolds, *Illinois J. Math.*, **7** (1963), 425-446.

[W4] B.L. van der Waerden, *Modern Algebra*, Vol 1 Ungar, (1949).

[Y1] Paul Y.H. Yiu, Sums of squares formulae with integer coefficients, *Canadian Math. Bull*, **30** (1987), 318-324.

[Y2] Paul Y.H. Yiu, On the product of 2 sums of 16 squares as a sum of squares on integral bilinear forms, *Quart. J. Math.* (2), **41** (1990), 463-500.

[Y3] S. Yuzvinsky, A series of monomial pairings, preprint (1982).

[Y4] S. Yuzvinsky, On the Hopf condition over an arbitrary field, *Bol. Soc. Mat. Mexicana*, **28** (1993), 1-8.

Index